DYNAMICS AND BIFURCATION OF PATTERNS IN DISSIPATIVE SYSTEMS

WORLD SCIENTIFIC SERIES ON NONLINEAR SCIENCE

Editor: Leon O. Chua
University of California, Berkeley

Series B. SPECIAL THEME ISSUES AND PROCEEDINGS

Volume 1: Chua's Circuit: A Paradigm for Chaos
Edited by *R. N. Madan*

Volume 2: Complexity and Chaos
Edited by *N. B. Abraham, A. M. Albano, A. Passamante, P. E. Rapp, and R. Gilmore*

Volume 3: New Trends in Pattern Formation in Active Nonlinear Media
Edited by *V. Perez-Villar, V. Perez-Munuzuri, C. Perez Garcia, and V. I. Krinsky*

Volume 4: Chaos and Nonlinear Mechanics
Edited by *T. Kapitaniak and J. Brindley*

Volume 5: Fluid Physics — Lecture Notes of Summer Schools
Edited by *M. G. Velarde and C. I. Christov*

Volume 6: Dynamics of Nonlinear and Disordered Systems
Edited by *G. Martínez-Mekler and T. H. Seligman*

Volume 7: Chaos in Mesoscopic Systems
Edited by *H. A. Cerdeira and G. Casati*

Volume 8: Thirty Years After Sharkovskiĭ's Theorem: New Perspectives
Edited by *L. Alsedà, F. Balibrea, J. Llibre and M. Misiurewicz*

Volume 9: Discretely-Coupled Dynamical Systems
Edited by *V. Pérez-Muñuzuri, V. Pérez-Villar, L. O. Chua and M. Markus*

Volume 10: Nonlinear Dynamics & Chaos
Edited by *S. Kim, R. P. Behringer, H.-T. Moon & Y. Kuramoto*

Volume 11: Chaos in Circuits and Systems
Edited by G. Chen & T. Ueta

WORLD SCIENTIFIC SERIES ON NONLINEAR SCIENCE Series B Vol. 12
Series Editor: Leon O. Chua

DYNAMICS AND BIFURCATION OF PATTERNS IN DISSIPATIVE SYSTEMS

edited by

Gerhard Dangelmayr
Colorado State University, USA

Iuliana Oprea
Colorado State University, USA
and
University of Bucharest, Romania

World Scientific

NEW JERSEY • LONDON • SINGAPORE • BEIJING • SHANGHAI • HONG KONG • TAIPEI • CHENNAI

Published by

World Scientific Publishing Co. Pte. Ltd.
5 Toh Tuck Link, Singapore 596224
USA office: 27 Warren Street, Suite 401-402, Hackensack, NJ 07601
UK office: 57 Shelton Street, Covent Garden, London WC2H 9HE

British Library Cataloguing-in-Publication Data
A catalogue record for this book is available from the British Library.

DYNAMICS AND BIFURCATION OF PATTERNS IN DISSIPATIVE SYSTEMS

ISBN 981-238-946-6

Printed in Singapore by World Scientific Printers (S) Pte Ltd

PREFACE

The subject of the spontaneous formation of spatio-temporal structures in systems far from equilibrium, generally referred to as patterns, is an exciting and fast-growing branch of applied mathematics and physics, with strong impact on fields as diverse as ecology, chemistry, engineering, social sciences, as well as new technologies and processes. Systems with different microscopic descriptions frequently exhibit similar patterns on a macroscopic level. The emergence of macroscopic order and the spontaneous breaking of spatiotemporal symmetries are features common to most natural phenomena, and many structures seen in our world can be considered the result of a sequence of successive symmetry-breaking instabilities caused by nonlinear processes under non-equilibrium conditions.

Although studied intensively for most of the last century, it has only been during the past thirty years that pattern formation has emerged as an own branch of science. Understanding the spontaneous formation and dynamics of spatio-temporal patterns in dissipative non-equilibrium systems is one of the major challenges in nonlinear science, and it is to be expected that new concepts and methods resulting from this field will influence future developments in many disciplines. Recent experimental results have demonstrated a variety of new patterns that can be observed in macroscopic as well as microscopic systems far from equilibrium, which are mostly unexplored and demand a theoretical description.

Studies of pattern formation and pattern dynamics use a common set of fundamental concepts to describe how non-equilibrium processes cause structures to appear in a wide range of systems in nature and technology. A good step towards the description of patterns and the fundamental understanding of nonlinear processes in the large is provided by studying dissipative systems, in which energy is supplied through a gradient such as in temperature, velocity, concentration etc. that maintains the system away from equilibrium.

Mathematical Approaches. Experimental and theoretical investigations of patterns are aimed at understanding the mechanisms involved in the formation and the selection of patterns, as well as their stability, temporal evolution and control. Addressing these issues mathematically requires a wide range of techniques involving ideas from dynamical systems, functional analysis, group theory, perturbation methods, ordinary and partial differential equations, geometry and singularity theory, as well as the ability to find the appropriate mathematical framework that adequately describes an experimentally observed phenomenon.

Equivariant Bifurcation Theory. Since the seminal paper of Turing appeared in 1952, it is well known that pattern formation is intimately connected with instabilities and spontaneously broken symmetries. This observation has led to the development of equivariant bifurcation theory, initiated by Sattinger in the 1970's and established as a branch of mathematics by Golubitsky and Stewart in the 1980's. In equivariant bifurcation theory the patterns studied are usually spatially or spatio-temporally periodic. This restriction allows to reduce an extended system to a finite dimensional system of "normal forms" near an instability, and the application of equivariant bifurcation theorems allows to predict the patterns that can be expected above threshold and to study their stability against periodic perturbations. Since the discretization of spatial systems usually leads to finite but high dimensional systems of equations, the numerical continuation of these patterns and the detection of secondary bifurcations require efficient algorithms which are a challenge for numerical analysts.

Modulation Approach. On the other hand, the modulation or envelope approach, also referred to as "Ginzburg Landau formalism", takes spatial modulations of periodic patterns into account, which are governed by space and time dependent amplitude or envelope equations. This approach was initiated by Newell, Whitehead and Segal in the late 1960's in the context of fluid mechanics. Many of the qualitative and quantitative theoretical predictions based on the amplitude equation approach have been successfully confronted with experiments, but a rigorous mathematical justification was lacking until the 1990's. In the meantime a number of theorems that justify this approach rigorously are available, however all of these results are for autonomous systems and so exclude time periodic systems such as the equations describing the Farady experiment or ac–driven electroconvection in liquid crystals.

Cellular Automata and Lattice Gases. The above approaches to study pattern formation and pattern dynamics rely on a macro- or mesoscopic, i.e.

deterministic level of description of the physical system through a governing set of nonlinear partial differential equations. For most of the systems studied in pattern formation, a fully microscopic description on the basis of statistical physics or molecular dynamics is untractable. Simplified models designed to understand the spatio-temporal evolution of macroscopic systems on a microscopic level are provided by cellular automata and lattice gases, which date back to von Neumann in the 1930's. Triggered by the availability of high power computing systems, there has been growing interest in simulations of these models to study collective phenomena. Cellular automata and lattice gases show a rich variety of patterns already for very simple rules of evolution.

Complex Patterns. In the past fifteen years we have witnessed significant progress in the field, and the focus shifted to more and more complex patterns. Progress in equivariant bifurcation theory encompasses superlattice patterns and quasi-patterns, and the extension to non-compact Euclidean symmetry resulted in a solid mathematical theory for the bifurcation of spiral and target waves. Studies of amplitude and phase diffusion equations, in particular the complex Ginzburg Landau equation, have led to a good qualitative understanding of different types of weak turbulence and transitions between them, see Chapter 8 for a survey. These phenomena are intrinsically spatio-temporal, and so cannot be described by finite dimensional dynamical systems. On the other hand, the observation of heteroclinic cycles as robust phenomena in finite dimensional systems with symmetry triggered much research in equivariant dynamics. Various new types of intermittency have been described and put on a solid mathematical basis.

While much progress towards understanding pattern formation and dynamics has been made in recent years, fundamental challenges remain. The basic question of whether universality classes exist for patterning behavior is still unanswered, and the characterization of patterns that are complex in both space and time (spatiotemporal chaos) is far from being complete. Observation of localized structures confined to a small spatial region of the system (e.g. "oscillons" and "worms") have also led to additional questions, for example why such states do not expand to fill the entire domain. Progress towards resolving fundamental questions of pattern formation also has significant practical implications for control since many technological processes involve pattern formation at some stage.

Contents of the Book. The present book emerged from a workshop on *Dynamics and Bifurcation of Patterns in Dissipative Systems* organized by the editors in May 2003 at Colorado State University. The book focuses on key ideas, new advances and open questions in the description and analysis of spatiotemporal patterns in dissipative extended systems. In a collection of expository papers and advanced research articles, written by leading experts ranging from applied mathematicians to theoretical and experimental physicists, the book provides an overview of the current state of the art in dynamics and bifurcation of patterns. The topics include mathematical issues related to bifurcations and instabilities in spatio-temporally continuous systems and the role of symmetry, advances in the study of localized patterns, dissipative waves, and weak turbulence, and new approaches to the modeling and characterization of spatio-temporal complexity. The applications encompass a remarkable variety of applied fields such as magneto- and binary fluid convection, liquid crystals, granular media, Faraday waves, multiscale biological patterns, visual hallucinations, and biological pacemakers.

The book is divided into three major parts, in which contributions using common mathematical methods or addressing related mathematical questions are grouped together. Each part begins with a relatively broad survey on recent research results, followed by contributions addressing more specific questions.

Part I is dedicated to **instabilities, bifurcation and the role of symmetry**. Chapter 1 gives a survey on recent and previous results on pattern formation in the visual system, and shows how these results lead to a classification of visual hallucinations. In Chapter 2 the authors discuss new efficient numerical algorithms for the continuation of periodic solutions of high dimensional systems, the detection of bifurcations, and the analysis of instabilities, including an application to electroconvection in nematic liquid crystals. Chapter 3 makes a first step towards the rigorous justification of the Ginzburg Landau formalism for time periodically driven systems. Chapter 4 discusses the stability and bifurcation of families of equilibria and periodic orbits due to continuous symmetries (referred to as relative equilibria and relative periodic orbits), which are important for analyzing bifurcations of spiral and target waves. In Chapter 5 the problem of rotating magnetoconvection with magnetostrophic balance, originally formulated by Chandrasekhar, is studied. The authors notice that Chandrasekhar's discussion is incomplete and find a complex sequence of transitions involving three oscillatory modes as parameters are varied in certain regimes of astrophysical interest. Chapter 6 reports on new results on pattern formation on a sphere for the even l mode representations of the spherical symmetry group $O(3)$. Chapter 6 discusses the convergence properties of Fourier mode representations of quasipatterns and the problems encountered by virtue of the presence of small divisors in a perturbation analysis.

The subject of **Part II** are **localized patterns, waves, and weak turbulence**. The part begins in Chapter 8 with a survey on phase diffusion, phase instabilities, and weak turbulence as described by classical envelope equations, in particular the complex Ginzburg Landau equation, that also includes a brief review of statistical approaches in the description of patterns. Chapter 9 discusses the merger of pattern formation and parametric resonance in terms of a new prototype system, the Mathieu partial differential equation, aiming at a description of localized patterns observed in the Faraday experiment and in shaked granular media. Averaging leads here to a dissipatively perturbed nonlinear Schrödinger equation. Related to Chapter 9 is Chapter 10, in which mean flow effects in model equations for the Faraday waves are analyzed that lead to new types of instabilities. In Chapter 11 rogue waves and the Benjamin-Feir instability are studied in the framework of a dispersively perturbed nonlinear Schrödinger equation. Here a Melnikov analysis shows that rogue waves are well approximated by homoclinic solutions of the unperturbed equation. The subject of Chapter 12 are target patterns produced by heterogeneous pacemakers in oscillatory media described by reaction diffusion equations with space dependent oscillation frequency.

Part III deals with the **modelling and characterization of spatio-temporal complexity**. The first chapter, Chapter 13, gives a survey on recent results in which bursting behaviour observed in different fluid systems has been successfully described by a common finite dimensional mechanism. Related to Chapter 13 are Chapters 18 and 19. In Chapter 18 the destruction of symmetry-enforced, robust heteroclinic cycles through symmetry breaking imperfections and the resulting complex dynamics are studied, whereas in Chapter 19 the new concept of internal dynamics of intermittency is introduced and illustrated by examples. Chapters 15–17 are devoted to the characterization of complex spatio-temporal behavior by extracting certain characteristics from computed or measured data. The subject of Chapter 15 is the computation of coherent structures and their temporal evolution from simulated hurricane data. The authors apply the classical Karhunen Loeve projection and the newly introduced signal fraction analysis projection and compare the results obtained with these two methods. In Chapter 16 the characteristics are based on an extension of the Nusselt number used in Rayleigh Benard convection to ac-driven electroconvection in nematic liquid crystals, which leads here to a distinction in linear and higher order nonlinear Nusselt numbers. In Chapter 17 the extraction of characteristics of far from equilibrium structures using their contours is described and two measures are discussed that can be used to describe labyrinthine patterns and growth interfaces. Chapters 14 and 20 discuss pattern formation and morphogenesis in developmental biology from a modeling and experimental perspective, respectively. In Chapter 14 several stochastic lattice gas models capable of describing different biological systems are reviewed, and in Chapter 20 the formation of aggregation mounds of a certain amoebae species under different geometrical constraints is studied.

The unitary presentations in the book, guiding the reader from basic fundamental concepts to most recent research results, makes it suitable for a wide group of graduate students and postgraduates of applied mathematics and theoretical physics, as well as any researcher interested in pattern formation and nonlinear instabilities.

Acknowledgements: We are grateful to the authors for their efforts to produce their valuable contributions in a timely fashion. This book is based upon work supported partly by the National Science Foundation under Grant No. 0228181, and partly by Colorado State University. Any opinions, findings, and conclusions or recommendations expressed in this book are those of the authors and do not reflect the views of the National Science Foundation.

Fort Collins, August 2004

Gerhard Dangelmayr
Iuliana Oprea

CONTENTS

PART I

Instabilities, Bifurcation, and the Role of Symmetry

CHAPTER 1

SYMMETRY AND PATTERN FORMATION ON THE VISUAL CORTEX

Martin Golubitsky*, LieJune Shiau†, and Andrei Török*

* *Department of Mathematics, University of Houston, Houston, TX 77204-3008, USA*

† *Department of Mathematics, University of Houston – Clear Lake, Houston, TX 77058, USA*

Mathematical studies of drug induced geometric visual hallucinations include three components: a model that abstracts the structure of the primary visual cortex V1; a mathematical procedure for finding geometric patterns as solutions to the cortical models; and a method for interpreting these patterns as visual hallucinations. In this note we survey the symmetry based ways in which geometric visual hallucinations have been modelled. Ermentrout and Cowan model the activity of neurons in the primary visual cortex. Bressloff, Cowan, Golubitsky, Thomas, and Wiener include the orientation tuning of neurons in V1 and assume that lateral connections in V1 are *anisotropic*. Golubitsky, Shiau, and Török assume that lateral connections are *isotropic* and then consider the effect of perturbing the lateral couplings to be weakly anisotropic.

These models all have planar Euclidean $\mathbf{E}(2)$ symmetry. Solutions are assumed to be spatially periodic and patterns are formed by symmetry-breaking bifurcations from a spatially uniform state. In the Ermentrout-Cowan model $\mathbf{E}(2)$ acts in its standard representation on $\mathbf{R}^2$, whereas in the Bressloff *et al.* model $\mathbf{E}(2)$ acts on $\mathbf{R}^2 \times \mathbf{S}^1$ via the *shift-twist* action. Isotropic coupling introduces an additional $\mathbf{S}^1$ symmetry, and weak anisotropy is then thought of as forced symmetry-breaking from $\mathbf{E}(2)\dot{+}\mathbf{S}^1$ to $\mathbf{E}(2)$ in its shift-twist action. We outline the way symmetry appears in bifurcations in these different models.

1. Introduction to Geometric Visual Hallucinations

When describing drug induced geometric visual hallucinations Klüver [17] states on p. 71: "We wish to stress merely one point, namely, that under diverse conditions the visual system responds in terms of a limited number of form constants." Klüver then classified geometric visual hallucinations into four groups or *form constants*: honeycombs, cobwebs, funnels and tunnels, and spirals. See Figure 1.

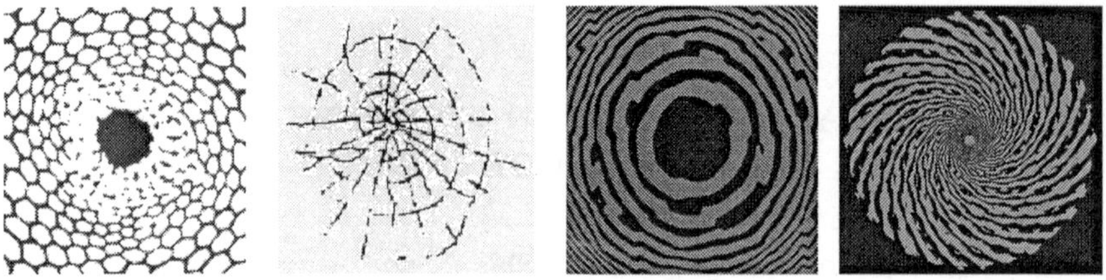

(a) (b) (c) (d)

Fig. 1. (a) Honeycomb by marihuana; [8] (b) cobweb petroglyph; [20] (c) tunnel [21], (d) spiral by LSD [21].

Ermentrout and Cowan [10] pioneered an approach to the mathematical study of geometric patterns produced in drug induced hallucinations. They assumed that the drug uniformly stimulates an inactive cortex and produces, by spontaneous symmetry-breaking, a patterned activity state. The mind then interprets the pattern as a visual image — namely the visual image that would produce the same pattern of activity on the primary visual cortex V1.[a] The Ermentrout-Cowan analysis assumes that a differential equation governs the symmetry-breaking transition from an inactive to an active cortex and then studies abstractly the transition using standard pattern formation arguments developed for reaction-diffusion equations. Their cortical patterns are obtained by thresholding (points where the solution is greater than some threshold are colored black, whereas all other points are colored white). These cortical patterns are then transformed to retinal patterns using the inverse of the retino-cortical map described in (4), and these retinal patterns are similar to some of the geometric patterns of visual hallucinations, namely, funnels and spirals.

In this note we survey recent work of Bressloff, Cowan, Golubitsky, Thomas, and Wiener [4–6] and Golubitsky, Shiau, and Török [13] who refine the Ermentrout-Cowan model to include more of the structure of V1. Neurons in V1 are known to be sensitive to orientations in the visual field[b] and it is mathematically reasonable to assign an orientation preference to

[a]The *primary visual cortex* is the area of the visual cortex that receives electrical signals directly from the retina.

[b]Experiments show that most V1 cells signal the local orientation of a contrast edge or bar; these neurons are tuned to a particular local orientation. See [1, 3, 12, 16] and [5] for further discussion.

each neuron in V1. Hubel and Wiesel [16] introduced the notion of a *hypercolumn* — a region in V1 containing for each orientation at a single point in the visual field (a mathematical idealization) a neuron sensitive to that orientation.

Bressloff *et al.* [5] studied the geometric patterns of drug induced hallucinations by including orientation sensitivity. As before, the drug stimulation is assumed to induce spontaneous symmetry-breaking, and the analysis is local in the sense of bifurcation theory. There is one major difference between the approaches in [5] and [10]. Ignoring lateral boundaries Ermentrout and Cowan [10] idealize the cortex as a plane, whereas Bressloff *et al.* [5] take into account the orientation tuning of cortical neurons and idealize the cortex as $\mathbf{R}^2 \times \mathbf{S}^1$. This approach recovers thin line hallucinations such as cobwebs and honeycombs, in addition to the threshold patterns found in the Ermentrout-Cowan theory.

There are two types of connections between neurons in V1: local and lateral. Experimental evidence suggests that neurons within a hypercolumn are all-to-all connected, whereas neurons in different hypercolumns are connected in a very structured way. This structured lateral coupling is called *anisotropic*, and it is the bifurcation theory associated with anisotropic coupling that is studied in Bressloff *et al.* [4, 5].

Golubitsky, Shiau, and Török [13] study generic bifurcations when lateral coupling is weakly anisotropic. First, they study bifurcations in models that are isotropic showing that these transitions lead naturally to a richer set of planforms than is found in [4, 5] and, in particular, to time-periodic states. (Isotropic models have an extra $\mathbf{S}^1$ symmetry and have been studied by Wolf and Geisel [24] as a model for the development of anisotropic lateral coupling.) There are three types of time dependent solutions: slowly *rotating* spiral and funnel shaped retinal images; *tunneling* images where the retinal image appears to rush into or spiral into the center of the visual field; and *pulsating* images where the spatial pattern of the solution changes periodically in time. Movies of these states may be found in [13]. Such images have been reported in the psychophysics literature, see Klüver [17], p. 24. (Note that near death experiences are sometimes described as traveling down a tunnel toward a central area.) Second, they consider weak anisotropy as forced symmetry breaking from isotropy.

The remainder of this note is divided into three sections. Section 2 discusses the basic structure of the continuum models of the visual cortex, the symmetries of these models, and some of the resulting cortical patterns. Section 3 outlines how Euclidean symmetry gives structure to the pattern forming bifurcations by constraining the form of the possible eigenfunctions. Finally, in Section 4, we discuss the specific group actions and bifurcating branches of solutions that occur in symmetry-breaking bifurcations in the three different cortical models. We emphasize that the lists of solutions are model-independent; they depend on the way that Euclidean symmetry is present in the models and not on a specific set of differential equations.

2. Models, Symmetry, and Planforms

The Ermentrout and Cowan [10] model of V1 consists of neurons located at each point $\mathbf{x}$ in $\mathbf{R}^2$. Their model equations, variants of the Wilson-Cowan equations [23], are written in terms of a real-valued *activity variable* $a(\mathbf{x})$, where a represents, say, the voltage potential of the neuron at location $\mathbf{x}$.

Bressloff *et al.* [5] incorporate the Hubel-Weisel hypercolumns [16] into their model of V1 by assuming that there is a hypercolumn centered at each location $\mathbf{x}$. Here a *hypercolumn* denotes a region of cortex that contains neurons sensitive to orientation φ for each direction φ. Their models, also adaptations of the Wilson-Cowan equations [23], are written in terms of a real-valued *activity variable* $a(\mathbf{x}, \varphi)$ where a represents, say, the voltage potential of the neuron tuned to orientation φ in the hypercolumn centered at location $\mathbf{x}$. Note that angles φ and $\varphi + \pi$ give the same orientation; so

$$a(\mathbf{x}, \varphi + \pi) = a(\mathbf{x}, \varphi).$$

The cortical planform associated to $a(\mathbf{x}, \varphi)$ is obtained in a way different from the Ermentrout-Cowan approach. For each fixed $\mathbf{x} \in \mathbf{R}^2$, $a(\mathbf{x}, \cdot)$ is a function on the circle. The planform associated to a is obtained through a *winner-take-all* strategy. The neuron that is most active in its hypercolumn is presumed to suppress the activity of other neurons within that hypercolumn. The winner-take-all strategy chooses, for each $\mathbf{x}$, the directions φ that maximize $a(\mathbf{x}, \cdot)$, and results in a field of directions. The two approaches to creating planforms can be combined by assigning directions only to those locations $\mathbf{x}$ where the associated maximum of $a(\mathbf{x}, \cdot)$ is larger than a given threshold. We will call these models the Ermentrout-Cowan model and the Wiener-Cowan model.

Euclidean Symmetry

The Euclidean group $\mathbf{E}(2)$ is crucial to the analyses in both [10] and [5] — but the way that group acts is different. In Ermentrout-Cowan the Euclidean group acts on the plane by its standard action, whereas in Wiener-

Cowan the Euclidean group acts on $\mathbf{R}^2 \times \mathbf{S}^1$ by the so-called shift-twist representation, as we now explain.

Bressloff *et al.* [5] argue that the lateral connections between neurons in neighboring hypercolumns are anisotropic. *Anisotropy* means that the *strength* of the connections between neurons in two neighboring hypercolumns depends on the orientation tuning of both neurons and on the relative locations of the two hypercolumns. Anisotropy is idealized to the one illustrated in Figure 2 where only neurons with the same orientation selectivity are connected and then only neurons that are oriented along the direction of their cells preference are connected. In particular, the symmetries of V1 model equations are those that are consistent with the idealized structure shown in Figure 2.

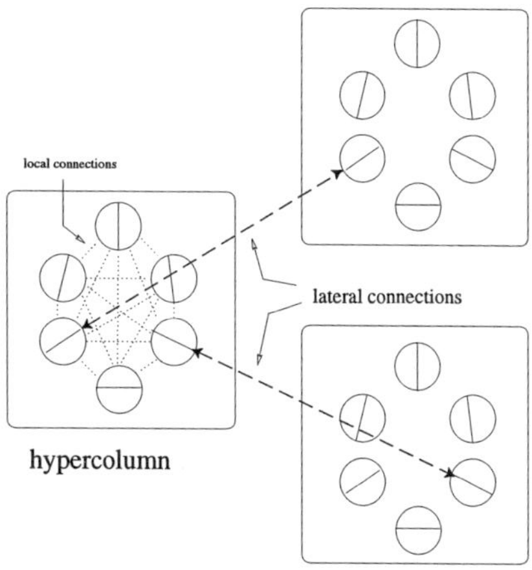

Fig. 2. Illustration of isotropic local and anisotropic lateral connection patterns.

The Euclidean group $\mathbf{E}(2)$ is generated by translations, rotations, and a reflection. The action of $\mathbf{E}(2)$ on $\mathbf{R}^2 \times \mathbf{S}^1$ that preserves the structure of lateral connections illustrated in Figure 2 is the *shift-twist* action. This action is given by:

$$\begin{aligned} \mathcal{T}_{\mathbf{y}}(\mathbf{x}, \varphi) &\equiv (\mathbf{x} + \mathbf{y}, \varphi) \\ \mathcal{R}_{\theta}(\mathbf{x}, \varphi) &\equiv (R_{\theta}\mathbf{x}, \varphi + \theta) \\ \mathcal{M}_{\kappa}(\mathbf{x}, \varphi) &\equiv (\kappa\mathbf{x}, -\varphi), \end{aligned} \tag{1}$$

where $(\mathbf{x}, \varphi) \in \mathbf{R}^2 \times \mathbf{S}^1$, $\mathbf{y} \in \mathbf{R}^2$, κ is the reflection $(x_1, x_2) \mapsto (x_1, -x_2)$, and $R_\theta \in \mathbf{SO}(2)$ is rotation of the plane counterclockwise through angle θ.

Isotropy of Lateral Connections

The anisotropy in lateral connections pictured in Figure 2 can be small in the sense that it is close to isotropic. We call the lateral connections between hypercolumns *isotropic*, as is done in Wolf and Geisel [24], if the strength of lateral connections between neurons in two neighboring hypercolumns depends only on the difference between the angles of the neurons' orientation sensitivity. Lateral connections in the isotropic model are illustrated in Figure 3. In this model, equations admit, in addition to Euclidean symmetry, the following $\mathbf{S}^1$ symmetry:

$$\mathcal{I}_{\widehat{\varphi}}(\mathbf{x}, \varphi) = (\mathbf{x}, \varphi + \widehat{\varphi}). \tag{2}$$

Note that $\widehat{\varphi} \in \mathbf{S}^1$ commutes with $\mathbf{y} \in \mathbf{R}^2$ and $R_\theta \in \mathbf{SO}(2)$, but $\kappa\widehat{\varphi} = (-\widehat{\varphi})\kappa$.

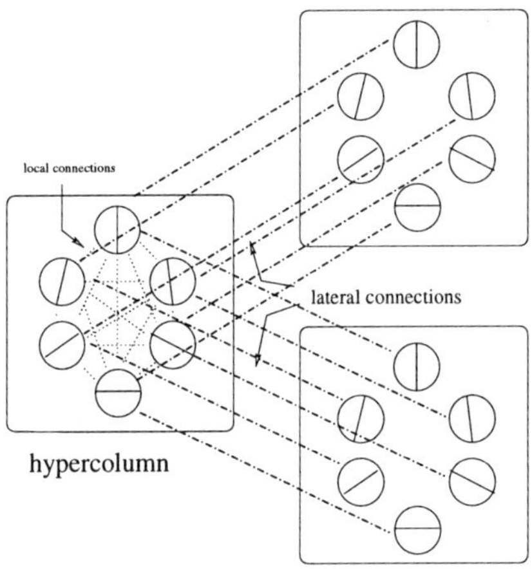

Fig. 3. Illustration of isotropic local and isotropic lateral connection patterns.

The action of $\gamma \in \mathbf{E}(2)\dot{+}\mathbf{S}^1$ on the activity function a is given by

$$\gamma a(\mathbf{x}, \varphi) = a(\gamma^{-1}(\mathbf{x}, \varphi)).$$

For example, $R_\theta \in \mathbf{SO}(2)$ acts by

$$(\mathcal{R}_\theta a)(\mathbf{x}, \varphi) = a(R_{-\theta}\mathbf{x}, \varphi - \theta).$$

Symmetry-Breaking Bifurcations on Lattices

Spontaneous symmetry-breaking in the presence of a noncompact group such as the Euclidean group is far from understood. The standard approach is to reduce the technical difficulties by looking only for solutions that are spatially doubly periodic with respect to some planar lattice (see Golubitsky and Stewart [14]); this is the approach taken in [5, 10, 13].

The first step in such an analysis is to choose a lattice type; in this paper we only describe transitions on square lattices. The second step is to decide on the size of the lattice. Euclidean symmetry guarantees that at bifurcation, critical eigenfunctions will have *plane wave* factors $e^{2\pi i \mathbf{k}\cdot\mathbf{x}}$ for some critical dual wave vector $\mathbf{k}$. See [4], Chapter 5 of [14], or Section 3. Typically, the lattice size is chosen so that the critical wave vectors will be vectors of shortest length in the dual lattice; that is, the lattice has the smallest possible size that can support doubly periodic solutions.

By restricting the bifurcation problem to a lattice, the group of symmetries is transformed to a compact group. First, translations in $\mathbf{E}(2)$ act modulo the spatial period as a 2-torus $\mathbf{T}^2$. Second, only those rotations and reflections in $\mathbf{E}(2)$ that preserve the lattice (namely, the holohedry $\mathbf{D}_4$ for the square lattice) are symmetries of the lattice restricted problem. Thus, the symmetry group of the square lattice problem is $\Gamma = \mathbf{D}_4 \dot{+} \mathbf{T}^2$. Recall that at bifurcation Γ acts on the kernel of the linearization, and a subgroup of Γ is *axial* if its fixed-point subspace in that kernel is one-dimensional. Solutions are guaranteed by the Equivariant Branching Lemma (see [14, 15]) which states: generically there are branches of equilibria to the nonlinear differential equation for every axial subgroup of Γ. The nonlinear analysis in [4, 10, 13] proceeds in this fashion.

Previous Results on the Square Lattice

In Ermentrout and Cowan [10] translation symmetry leads to eigenfunctions that are linear combinations of plane waves, and, on the square lattice, to two axial planforms: stripes and squares. See Figure 4.

In Bressloff *et al.* [4, 5] translation symmetry leads to critical eigenfunctions that are linear combinations of functions of the form $u(\varphi)e^{2\pi i \mathbf{k}\cdot\mathbf{x}}$. These eigenfunctions correspond to one of two types of representations of $\mathbf{E}(2)$ (restricted to the lattice): *scalar* (u even in φ) and *pseudoscalar* (u odd). The fact that two different representations of the Euclidean group can appear in bifurcations was first noted by Bosch Vivancos, Chossat, and Melbourne [2]. Bressloff *et al.* [5] also show that a trivial solution to the Wilson-Cowan equation will lose stability via a scalar or a pseudoscalar bifurcation depending on the exact form of the lateral coupling. Thus, each of these representations is, from a mathematical point of view, equally likely

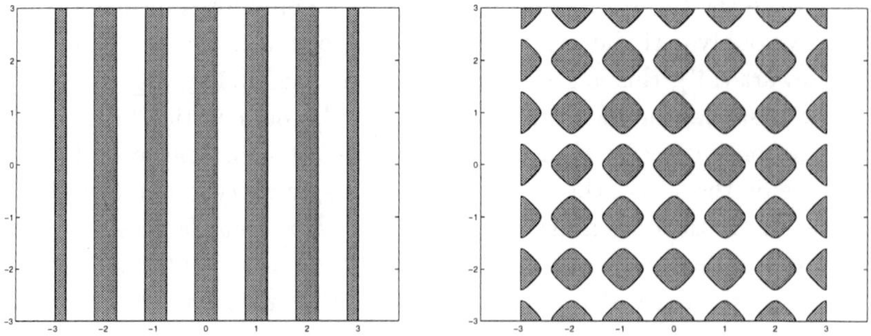

Fig. 4. Thresholding of eigenfunctions: (left) stripes, (right) squares.

to occur. On the square lattice, in [2,4] it is shown that there are two axial planforms each in the scalar and pseudoscalar cases: stripes and squares.

To picture the planforms in these cases, we must specify the function $u(\varphi)$, and this can be accomplished by assuming that anisotropy is small. When anisotropy is zero, the $\mathbf{S}^1$ symmetry in (2) forces $u(\varphi) = \cos(2m\varphi)$ in the scalar case and $u(\varphi) = \sin(2m\varphi)$ in the pseudoscalar case. (This point will be discussed in more detail when we review representation theory in Section 4.) The assumptions in Bressloff *et al.* [5] imply that u is a small perturbation of sine or cosine. Note that the Ermentrout-Cowan planforms are recovered in the scalar case when $m = 0$ — in this case u is constant and all directions are equally active. As often happens in single equation models, the first instability of a trivial (spatially constant) solution is to eigenfunctions with m small — and that is what occurs in certain models based on the Wilson-Cowan equation (see [5]). Planforms for the scalar and pseudoscalar planforms when $m = 1$ are shown in Figure 5.

stripes squares stripes squares

Fig. 5. Direction fields: scalar eigenfunctions (left) and pseudoscalar eigenfunctions (right).

New Planforms When Lateral Connections are Isotropic

In our analysis of the isotropic case ($\widetilde{\Gamma} = \Gamma\dot{+}\mathbf{S}^1$ symmetry) we find four axial subgroups (Σ_1-Σ_4) and one maximal isotropy subgroup Σ_5 with a two-dimensional fixed-point subspace. The axial subgroups lead to group orbits of equilibria. This fact must be properly interpreted to understand how the new planforms relate to the old. A phase shift of $\sin(2\varphi)$ yields $\cos(2\varphi)$. Thus, the extra $\mathbf{S}^1$ symmetry based on isotropic lateral connections identifies scalar and pseudoscalar planforms; up to this new symmetry the planforms are the same. Thus, the axial subgroup Σ_3 corresponds to stripes (both scalar and pseudoscalar) and the axial subgroup Σ_1 corresponds to squares (both scalar and pseudoscalar). The axial subgroups Σ_2 and Σ_4 correspond to new types of planforms. Finally, the maximal isotropy subgroup Σ_5 with its two-dimensional fixed-point subspace leads to a time-periodic rotating wave whose frequency is zero at bifurcation. The planforms associated with these new types of solutions are pictured in Figure 6.

It is unusual for a steady-state bifurcation (eigenvalues of a linearization moving through 0) to lead to time periodic states. It is well known that in systems without symmetry, time periodic states will appear in unfoldings of codimension two Takens-Bogdanov singularities (a double zero eigenvalue with a nilpotent normal form). It is less well known that codimension one steady-state bifurcations with symmetry can also lead to time periodic states. Field and Swift [11] were the first to find such a bifurcation (in a system with finite symmetry). Melbourne [19] was the first to find an example of a rotating wave in a steady-state bifurcation in a system with continuous symmetry. Nevertheless, the documented cases where time periodic states occur in codimension one steady-state bifurcations are relatively rare and our work provides the first example where this mathematical phenomenon appears in model equations.

Weak Anisotropy in Lateral Connections

Next, we discuss what happens to the bifurcating solutions to the isotropic nonlinear equation when anisotropy is added as a small symmetry-breaking parameter. As was noted in Bressloff *et al.* [5], the linear effect of anisotropy is to split the eigenfunctions into scalar and pseudoscalar representations. The effect on solutions to the nonlinear equation can also be established using the methods of Lauterbach and Roberts [18]. This method is applied independently to each branch of (group orbits of) solutions found in the

Fig. 6. Direction fields of new planforms in isotropic model: (left) axial planform Σ_2; (center) axial planform Σ_4; (right) rotating wave Σ_5 (direction of movement is up and to the left).

isotropic case. The results for square lattice solutions are easily described.

Generically, the dynamics on the $\widetilde{\Gamma}$ group orbit of equilibria corresponding to the axial subgroup Σ_3 has two (smaller) Γ group orbits of equilibria: scalar stripes and pseudoscalar stripes. There may be other equilibria coming from the $\widetilde{\Gamma}$ group orbit; but at the very least scalar and pseudoscalar stripes always remain as solutions.

Similarly, the dynamics on the group orbit of equilibria corresponding to the axial subgroup Σ_1 generically have two equilibria corresponding to scalar and pseudoscalar squares.

The dynamics on the group orbit of the axial subgroups Σ_2 and Σ_4 and the fifth maximal isotropy subgroup Σ_5 do not change substantially when anisotropy is added. These group orbits still remain as equilibria and rotating waves.

Retinal Images

Finally, we discuss the geometric form of the cortical planforms in the visual field, that is, we try to picture the corresponding visual hallucinations. It is known that the density of neurons in the visual cortex is uniform, whereas the density of neurons in the retina fall offs from the fovea[c] at a rate of $1/r^2$. Schwartz [22] observed that there is a unique conformal map taking a disk with $1/r^2$ density to a rectangle with uniform density, namely, the complex logarithm. This is also called the *retino-cortical* map. It is thought that using the inverse of the retino-cortical map, the complex exponential, to push forward the activity pattern from V1 to the retina is a reasonable way to form the hallucination image — and this is the approach

[c]The *fovea* is the small central area of the retina that gives the sharpest vision.

used in Ermentrout and Cowan [10] and in Bressloff *et al.* [5,6]. Specifically, the transformation from polar coordinates (r, θ) on the retina to cortical coordinates (x, y) is given in Cowan [9] to be:

$$\begin{aligned} x &= \tfrac{1}{\varepsilon} \ln \left(\tfrac{1}{\omega} r\right) \\ y &= \tfrac{1}{\varepsilon} \theta \end{aligned} \tag{3}$$

where ω and ε are constants. See Bressloff *et al.* [6] for a discussion of the values of these constants. The inverse of the retino-cortical map (3) is

$$\begin{aligned} r &= \omega \exp(\varepsilon x) \\ \theta &= \varepsilon y \end{aligned} \tag{4}$$

In our retinal images we take

$$\omega = \frac{30}{e^{2\pi}} \quad \text{and} \quad \varepsilon = \frac{2\pi}{n_h}$$

where n_h is the number of hypercolumn widths in the cortex, which we take to be 36.

The images of these cortical patterns in retinal coordinates, as well as the discussion of additional issues concerning the construction of these images, are given in [13].

3. A Brief Outline of Local Equivariant Steady-State Bifurcation Theory

In finite dimensions steady-state bifurcation reduces to finding all zeros of a parameterized map $f : \mathbf{R}^n \times \mathbf{R} \to \mathbf{R}^n$ near a known zero, which we can take to be $f(0,0) = 0$. In equivariant bifurcation theory we assume that f commutes with the action of a compact Lie group Γ, that is, $\Gamma \subset \mathbf{O}(n)$ and

$$f(\gamma x, \lambda) = \gamma f(x, \lambda) \tag{5}$$

for all $\gamma \in \Gamma$ and that f has a *trivial* Γ-invariant solution for all λ, that is, $f(0, \lambda) = 0$. It follows from (5) and the chain rule that the Jacobian at a trivial solution also commutes with Γ, that is,

$$(df)_{0,\lambda}\gamma = \gamma (df)_{0,\lambda} \tag{6}$$

for all $\gamma \in \Gamma$.

A steady-state bifurcation occurs at $\lambda = 0$ when $\mathcal{K} \equiv \ker L \neq 0$, where $L = (df)_{0,0}$. It follows from (6) that $\mathcal{K}$ is Γ-invariant. Indeed, generically $\mathcal{K}$ is an (absolutely) irreducible representation of Γ. Liapunov-Schmidt and center manifold reductions can be performed in a way that preserves symmetry. In either case, finding the zeros of f reduces to finding the zeros of a map $g : \mathcal{K} \times \mathbf{R} \to \mathcal{K}$ near the trivial solution $g(0, \lambda) = 0$. Moreover, g commutes with the action of Γ on the kernel $\mathcal{K}$ and $(dg)_{0,0} = 0$.

In general $\mathcal{K}$ can still be a large dimensional space and finding the zeros of g can be a daunting task. However, the Equivariant Branching Lemma [14, 15] enables one to find many (though not all) of the zeros of g. This lemma states that generically for each axial subgroup Σ of the action of Γ on $\mathcal{K}$, there exists a unique (local) branch of zeros of g whose solutions have Σ symmetry. An *axial* subgroup is an isotropy subgroup whose fixed-point subspace

$$\mathrm{Fix}(\Sigma) = \{x \in \mathcal{K} : \sigma x = x \quad \forall \sigma \in \Sigma\}$$

is one-dimensional.

It is important to note that the existence of axial solutions is *model independent.* The existence of axial solutions does not depend (in any essential way) on f — just on the form of the action of Γ on $\mathcal{K}$. Detailed statements and proofs can be found in Chapter 1 of [14].

Euclidean Symmetry and Eigenfunctions

Planar pattern formation arguments begin by assuming that a uniform (Euclidean invariant) equilibrium for a Euclidean equivariant system of differential equations loses stability as a parameter is varied. For hallucination models we assume that the system of differential equations is defined on the space $\mathbf{R}^2 \times \Omega$ where Ω is a point in the Ermentrout-Cowan model and $\Omega = \mathbf{S}^1$ in the orientation tuning models. More precisely, the system of (partial) differential equations should be viewed as an operator f on the space of real-valued functions $\mathcal{F}$ defined on $\mathbf{R}^2 \times \Omega$.

There is a fundamental complication that appears when trying to apply the outline of equivariant steady-state bifurcation theory to planar pattern formation: $\mathcal{K}$ need not be finite-dimensional. To understand this difficulty, and one way around it, we next discuss how Euclidean symmetry determines the eigenfunctions of L.

Let $\mathbf{k} \subset \mathbf{R}^2$ and let

$$W_{\mathbf{k}} = \{u(\varphi)e^{i\mathbf{k}\cdot\mathbf{x}} + \text{c.c.} : u : \Omega \to \mathbf{C}\}$$

Observe that translations act on $W_{\mathbf{k}}$ by

$$T_{\mathbf{y}}(u(\varphi)e^{i\mathbf{k}\cdot\mathbf{x}}) = u(\varphi)e^{i\mathbf{k}\cdot(\mathbf{x}+\mathbf{y})} = \left[e^{i\mathbf{k}\cdot\mathbf{y}}u(\varphi)\right] e^{i\mathbf{k}\cdot\mathbf{x}}$$

It follows that $L : W_{\mathbf{k}} \to W_{\mathbf{k}}$ and that eigenfunctions of L have *plane wave* factors. The vectors $\mathbf{k}$ are called *dual wave vectors.*

Let ρ be the reflection such that $\rho\mathbf{k} = \mathbf{k}$. Euclidean equivariance implies

$$\rho\left(u(\varphi)e^{i\mathbf{k}\cdot\mathbf{x}}\right) = \rho(u(\varphi))e^{i\mathbf{k}\cdot\mathbf{x}}$$

So $\rho : W_\mathbf{k} \to W_\mathbf{k}$. It follows from $\rho^2 = 1$ that

$$W_\mathbf{k} = W_\mathbf{k}^+ \oplus W_\mathbf{k}^-$$

where ρ acts as $+1$ on $W_\mathbf{k}^+$ and -1 on $W_\mathbf{k}^-$. Hence, eigenfunctions of L are either even $(W_\mathbf{k}^+)$ or odd $(W_\mathbf{k}^-)$. When $\mathbf{k} = (1, 0)$

$$u(-\varphi) = \begin{cases} u(\varphi) & u \in W_\mathbf{k}^+ \\ -u(\varphi) & u \in W_\mathbf{k}^- \end{cases}$$

Finally, rotations act on the subspaces $W_\mathbf{k}$ by

$$R_\theta\left(u(\varphi)e^{i\mathbf{k}\cdot\mathbf{x}}\right) = R_\theta(u(\varphi))e^{iR_\theta(\mathbf{k})\cdot\mathbf{x}}$$

Therefore

$$R_\theta(W_\mathbf{k}) = W_{R_\theta(\mathbf{k})}$$

Hence, for each eigenfunction in $W_\mathbf{k}$, there is an independent eigenfunction in $W_{R_\theta(\mathbf{k})}$ and $\ker L$ is ∞-dimensional. Standard reductions theorems then fail.

Euclidean Equivariant Bifurcation Theory

A standard way around this difficulty is to look for solutions in the space of double-periodic functions. Let $\mathcal{L}$ be a planar lattice and let

$$\mathcal{F}_\mathcal{L} = \{a : \mathbf{R}^2 \times \Omega \to \mathbf{R} : a(\mathbf{x} + \boldsymbol{\ell}, \varphi) = h(\mathbf{x}, \varphi) \quad \forall \boldsymbol{\ell} \in \mathcal{L}\}$$

There are only a finite number of rotations that leave $\ker(L|\mathcal{F}_\mathcal{L})$ invariant, namely, the rotations that preserve the lattice $\mathcal{L}$. Thus, $\ker(L|\mathcal{F}_\mathcal{L})$ is generically finite-dimensional and we can choose the size of the lattice so that the shortest dual wave vectors $\mathbf{k}$ are the critical eigenvectors.

Note that the symmetries that act on the function space $\mathcal{F}_\mathcal{L}$ are different from those that act on the base space $\mathbf{R}^2 \times \Omega$ in two ways: translations act on $\mathcal{F}_\mathcal{L}$ as a 2-torus $\mathbf{T}^2$ and only those rotations and reflections that preserve the lattice $\mathcal{L}$ (often called the *holohedry* subgroup H) act on $\mathcal{F}_\mathcal{L}$. Thus the group of symmetries acting on the reduction to doubly-periodic states is the compact group $H\dot{+}\mathbf{T}^2$. For square lattices $H = \mathbf{D}_4$ and for hexagonal lattices $H = \mathbf{D}_6$. We report here on the simpler square lattice case. The symmetry group for the Ermentrout-Cowan model and the anisotropic model is $\Gamma = \mathbf{D}_4\dot{+}\mathbf{T}^2$; the symmetry group for the isotropic model is $\widetilde{\Gamma} = \Gamma\dot{+}\mathbf{S}^1$.

4. Square Lattice Planforms

In this section we discuss the spatially doubly periodic solutions that must emanate from the simplest bifurcations of Euclidean invariant differential equations restricted to a square lattice. To do this we describe the simplest kernel $\mathcal{K}$, the representations of Γ and $\widetilde{\Gamma}$ on $\mathcal{K}$, and the axial subgroups of these actions.

4.1. *Representation Theory of* $\mathbf{\Gamma}$ *and* $\widetilde{\mathbf{\Gamma}}$

Without loss of generality, we assume that the square lattice $\mathcal{L}$ consists of squares of unit length. The action of $\widetilde{\Gamma}$ on $\mathcal{F}_{\mathcal{L}}$ is the one induced from the action of $\mathbf{E}(2)\dot{+}\mathbf{S}^1$ on $\mathbf{R}^2 \times \mathbf{S}^1$ given in (1) and (2).

We expect the simplest square lattice bifurcations to be from equilibria whose linearizations have kernels that are irreducible subspaces of $\mathcal{F}_{\mathcal{L}}$ and we only consider bifurcations based on dual wave vectors of shortest (unit) length. We write $\mathbf{x} = (x_1, x_2)$.

The eigenfunctions corresponding to Γ in the Ermentrout-Cowan models have the form

$$a(\mathbf{x}) = z_1 e^{2\pi i x_1} + z_2 e^{2\pi i x_2} + c.c.$$

where $(z_1, z_2) \in \mathbf{C}^2$.

The eigenfunctions corresponding to Γ in the Wiener-Cowan models have the form

$$a(\mathbf{x}, \varphi) = z_1 u(\varphi) e^{2\pi i x_1} + z_2 u(\varphi - \frac{\pi}{2}) e^{2\pi i x_2} + c.c.$$

where $(z_1, z_2) \in \mathbf{C}^2$ and $u(\varphi)$ is π-periodic, real-valued, and either odd or even. See [4].

The eigenfunctions corresponding to $\widetilde{\Gamma}$ have the form

$$\begin{aligned} a(\mathbf{x}, \varphi) = &\left(z_1 e^{2i\varphi} + w_1 e^{-2i\varphi}\right) e^{2\pi i \mathbf{k}_1 \cdot \mathbf{x}} + \\ &\left(z_2 e^{2i(\varphi - \pi/2)} + w_2 e^{-2i(\varphi - \pi/2)}\right) e^{2\pi i \mathbf{k}_2 \cdot \mathbf{x}} + c.c. \end{aligned} \tag{7}$$

where $(z_1, w_1, z_2, w_2) \in \mathbf{C}^4$. See [13].

4.2. *Group Actions on* $\mathcal{K}$ *and their Axial Subgroups*

A calculation shows that the actions of Γ on $\mathbf{C}^2$ (in the scalar and pseudoscalar representations) and of $\widetilde{\Gamma}$ on $\mathbf{C}^4$ (when $m = 1$) are as given in Table 1.

Table 1. For Ermentrout-Cowan and for $u(\varphi)$ even, let $\varepsilon = +1$; for $u(\varphi)$ odd, let $\varepsilon = -1$.

$\mathbf{D}_4$	Action on $\mathbf{C}^2$	Isotropic Case Action on $\mathbf{C}^4$
$\mathbf{1}$	(z_1, z_2)	(z_1, w_1, z_2, w_2)
ξ	$(\overline{z_2}, z_1)$	$(\overline{w}_2, \overline{z}_2, z_1, w_1)$
ξ^2	$(\overline{z_1}, \overline{z_2})$	$(\overline{w}_1, \overline{z}_1, \overline{w}_2, \overline{z}_2)$
ξ^3	$(z_2, \overline{z_1})$	$(z_2, w_2, \overline{w}_1, \overline{z}_1)$
κ	$\varepsilon(z_1, \overline{z_2})$	$(w_1, z_1, \overline{z}_2, \overline{w}_2)$
$\kappa\xi$	$\varepsilon(\overline{z_2}, \overline{z_1})$	$(\overline{z}_2, \overline{w}_2, \overline{z}_1, \overline{w}_1)$
$\kappa\xi^2$	$\varepsilon(\overline{z_1}, z_2)$	$(\overline{z}_1, \overline{w}_1, w_2, z_2)$
$\kappa\xi^3$	$\varepsilon(z_2, z_1)$	(w_2, z_2, w_1, z_1)
$[\theta_1, \theta_2]$	$(e^{-2\pi i\theta_1} z_1, e^{-2\pi i\theta_2} z_2)$	$(e^{2\pi i\theta_1} z_1, e^{2\pi i\theta_1} w_1, e^{2\pi i\theta_2} z_2, e^{2\pi i\theta_2} w_2)$
$[0, 0, \hat{\varphi}]$		$(e^{-2i\hat{\varphi}} z_1, e^{2i\hat{\varphi}} w_1, e^{-2i\hat{\varphi}} z_2, e^{2i\hat{\varphi}} w_2)$

The axial subgroups of Γ acting on $\mathbf{C}^2$ are given in Table 2 and the maximal isotropy subgroups of $\widetilde{\Gamma}$ acting on $\mathbf{C}^4$ are given in Table 3. Observe that the action of $\widetilde{\Gamma}$ on W_1 has four axial subgroups Σ_1-Σ_4 and one maximal isotropy subgroup with a two-dimensional fixed-point subspace Σ_5.

Table 2. Square lattice axial subgroups of Γ acting on $\mathbf{C}^2$.

$\varepsilon = +1$	$\varepsilon = -1$	Fixed Subspace
$\Sigma_1 = \langle \xi, \kappa \rangle$	$\langle \xi, \kappa[\frac{1}{2}, \frac{1}{2}] \rangle$	$\mathbf{R}\{(1,1)\}$
$\Sigma_2 = \langle \xi^2, \kappa, [0, \theta_2] \rangle$	$\langle \xi^2, \kappa[\frac{1}{2}, 0], [0, \theta_2] \rangle$	$\mathbf{R}\{(1,0)\}$

Table 3. Square lattice maximal isotropy subgroups of $\widetilde{\Gamma}$ acting on $\mathbf{C}^4$; $u \in \mathbf{C}$.

Isotropic case	Fixed Subspace
$\Sigma_1 = \langle \kappa, \xi \rangle$	$\mathbf{R}\{(1,1,1,1)\}$
$\Sigma_2 = \langle \kappa, \left[\frac{3}{4}, \frac{1}{4}, \frac{\pi}{4}\right] \xi \rangle$	$\mathbf{R}\{(1,1,1,-1)\}$
$\Sigma_3 = \langle \kappa, \xi^2, [0, \theta_2, 0] \rangle$	$\mathbf{R}\{(1,1,0,0)\}$
$\Sigma_4 = \langle \kappa\xi^2, [0, \theta_2, 0], [\theta_1, 0, \pi\theta_1] \rangle$	$\mathbf{R}\{(1,0,0,0)\}$
$\Sigma_5 = \langle \kappa\xi, [\theta_1, \theta_1, \pi\theta_1] \rangle$	$\{(u, 0, \overline{u}, 0)\}$

The Effect of Weak Anisotropy We discuss how solutions corresponding to $\widetilde{\Gamma}$-bifurcations behave generically when the isotropy of the lateral connections is broken, that is, when the $\widetilde{\Gamma}$-equivariant vector field is perturbed to a Γ-equivariant field. Detailed arguments in [13] show that equilibria corresponding to the axial subgroups $\Sigma_1 - \Sigma_4$ persist on symmetry-breaking perturbations. More interesting, the maximal isotropy subgroup Σ_5 with a two-dimensional fixed-point subspace in the isotropic case leads to circles of equilibria and to periodic solutions on the breaking of $\widetilde{\Gamma}$ symmetry to Γ symmetry. Moreover the period of this periodic solution tends to ∞ at the bifurcation point ($\lambda = 0$).

Concluding Remarks We note that pattern formation on the hexagonal lattice is also treated in [5,6,13]. The calculations are more difficult and the results more complicated but the basic ideas are the same. Note that tunnelling and pulsating periodic solutions occur only on a hexagonal lattice. We expect pseudoscalar representations, as well as the usual scalar representations to occur in a variety of pattern formation problems, particularly those where the pattern produces a line-field rather than just a threshold or level contour. Another example occurs in liquid crystals; see [7].

Acknowledgements

We are grateful to Jack Cowan, Paul Bressloff, and Peter Thomas for many helpful conversations about the delightful structure of and pattern formation on the visual cortex. This work was supported in part by NSF Grant DMS-0244529.

References

[1] G.G. Blasdel. *J. Neurosci.* **12** (1992) 3139–3161.
[2] I. Bosch Vivancos, P. Chossat, and I. Melbourne, *Arch. Rational Mech. Anal.* **131** (1995) 199–224.
[3] W.H. Bosking, Y. Zhang, B. Schofield, and D. Fitzpatrick, *J. Neurosci.* **17** 6 (1997) 2112–2127.
[4] P.C. Bressloff, J.D. Cowan, M. Golubitsky, and P.J. Thomas, *Nonlinearity* **14** (2001) 739–775.
[5] P.C. Bressloff, J.D. Cowan, M. Golubitsky, P.J. Thomas, and M.C. Wiener, *Phil. Trans. Royal Soc. London* ***B*** **356** (2001) 299–330.
[6] P.C. Bressloff, J.D. Cowan, M. Golubitsky, P.J. Thomas and M.C. Wiener, *Neural Computation* **14** (2002) 473–491.
[7] D. Chillingworth and M. Golubitsky, *J. Math. Phys.* **44** (9) (2003) 4201–4219.
[8] J. Clottes and D. Lewis-Williams. *The Shamans of Prehistory: Trance and Magic in the Painted Caves.* Abrams, New York, 1998.
[9] J.D. Cowan, *Neurosciences Research Program Bull.*, **15** (1977) 492–517.
[10] G.B. Ermentrout and J.D. Cowan, *Biol. Cybernetics* **34** (1979) 137–150.
[11] M. Field and J.W. Swift, *Nonlinearity* **4** (1991) 1001–1043.
[12] C.D. Gilbert, *Neuron* **9** (1992) 1–13.
[13] M. Golubitsky, L-J. Shiau and A. Török, *SIAM J. Appl. Dynam. Sys.* **2** (2003) 97–143.

[14] M. Golubitsky and I. Stewart. *The Symmetry Perspective: From Equilibrium to Chaos in Phase Space and Physical Space*, Progress in Mathematics **200**, Birkhäuser, Basel, 2002.

[15] M. Golubitsky, I.N. Stewart and D.G. Schaeffer. *Singularities and Groups in Bifurcation Theory: Vol. II*, Applied Mathematical Sciences **69**, Springer-Verlag, New York, 1988.

[16] D.H. Hubel and T.N. Wiesel, *J. Comp. Neurol.* **158** No. 3 (1974) 267–294; *J. Comp. Neurol.* **158** No. 3 (1974) 295–306; *J. Comp. Neurol.* **158** No. 3 (1974) 307–318.

[17] H. Klüver. *Mescal and Mechanisms of Hallucinations.* University of Chicago Press, Chicago, 1966.

[18] R. Lauterbach and M. Roberts, *J. Diff. Eq.* **100** (1992) 22–48.

[19] I. Melbourne, *Nonlinearity* **7** (1994) 1385–1393.

[20] A. Patterson. *Rock Art Symbols of the Greater Southwest.* Johnson Books, Boulder, 1992.

[21] G. Oster, *Scientific American* **222** No. 2 (1970) 83–87.

[22] E. Schwartz, *Biol. Cybernetics* **25** (1977) 181–194.

[23] H.R. Wilson and J.D. Cowan, *Biophys. J.* (1972) **12** 1–24.

[24] F. Wolf and T. Geisel,*Nature* **395** (1998) 73–78.

CHAPTER 2

MATRIX FREE APPROACH IN THE NUMERICAL ANALYSIS OF BIFURCATIONS AND INSTABILITIES

Eugene L. Allgower, Gerhard Dangelmayr, Kurt Georg, and Iuliana Oprea

Department of Mathematics, Colorado State University, Fort Collins, CO 80523-1874, USA

In this paper we review some recent techniques which have been studied in the numerical treatment of bifurcation problems. We discuss matrix-free continuation methods and the incorporation of test functions for various types of bifurcation points. In the approach stemming from Lyapunov-Schmidt reductions, defining equations for bifurcations and other singularities are obtained. In this approach, the dimension is low and now the task is to find all of the solutions to a bifurcation equation or a system of defining equations. For this we discuss the cellular exclusion algorithms for finding all of the real solutions of a system of nonlinear equations in a specified cell. Cellular exclusion algorithms can also be adapted to handle the task of global optimization over a specified cell. This technique has been applied to find the global minimum of the oscillatory neutral stability surface arising in the bifurcation study of electroconvection in nematic liquid crystals.

1. Introduction

One approach to the numerical treatment of bifurcation problems is to trace solution branches, and while tracing, to monitor for bifurcations via test functions. Since many numerical bifurcation problems arise from discretizations of operator equations, this leads to large finite dimensional systems. Hence one wishes to carry out the tracing and testing in a numerically efficient way. Georg [10] has given a description for matrix-free techniques, using only the action of a linear transformation upon a vector, for carrying out this approach. In [2] this approach was applied to handling torus bifurcations. Numerical tests on the Brusselator problem in one and two space dimensions were carried out. Another approach to numerical bifurcation is to derive defining equations for bifurcations and other singularities via, e.g., Lyapunov-Schmidt reductions. In this approach, the dimension is low. Of interest now is the task of finding all of the solutions to a bifurcation equation or a system of defining equations. Recently, cellular exclusion algorithms for finding all of the real solutions of a system of nonlinear equations in a specified cell have been studied and improved [11]. Analogously, cellular exclusion algorithms can be adapted to handle the task of global optimization over a specified cell [1]. This technique has been applied to find the global minimum of the oscillatory neutral stability surface arising in the study of electroconvection in nematic liquid crystals [4]. The global minimum is associated with a Hopf bifurcation. In this paper we describe cellular exclusion algorithms for finding all zeros in a cell and for finding the global minimum in a cell.

2. A Dynamical System Setting

To illustrate the matrix-free approach, let us consider systems of the form

$$\dot{\varphi} = f(\varphi, \lambda) \tag{1}$$

where $f : \mathbb{R}^n \times \mathbb{R} \to \mathbb{R}^n$. Such systems often arise from space discretizations of parabolic PDEs. We are interested in branches (with respect to λ) of periodic solutions of such systems.

We denote a solution of (1) which has initial value u for $t = 0$ with $\varphi(u, t, \lambda)$. We have that $t \mapsto \varphi(u, t, \lambda)$ is periodic with period T iff

$$\varphi(u, T, \lambda) - u = 0. \tag{2}$$

However, even fixing λ, all points u of the same periodic orbit would satisfy this equation, hence we need an additional *phase condition*, say

$$h(u, T, \lambda) = 0 \tag{3}$$

to single out, at least locally, one point per orbit (see, e.g., [19]). For our present purpose we use the Poincaré phase condition

$$(u - \hat{u})^T f(\hat{u}, \lambda) = 0 \tag{4}$$

where $\hat{u}$ is some current point close to some orbit for given λ. This point will, of course, need to be adapted regularly. Let us remark that the Poincaré phase condition has proven to be an effective choice, see [16, 17].

Let us form the equation

$$H(u, T, \lambda) := \begin{bmatrix} \varphi(u, T, \lambda) - u \\ h(u, T, \lambda) \end{bmatrix} = 0. \tag{5}$$

For almost all choices of λ there is a neighborhood of the orbit such that 0 is a regular value of H, if $\hat{u}$ is in that neigborhood. Hence, the periodic orbits of (1) can be traced by using numerical continuation methods (arising from varying λ) on equation (5). In particular, a matrix-free approach using only the action of the Jacobian is very suitable if we are interested in allowing large dimensions n, since an action of the (full) Jacobian H' can be readily obtained, as we will see in Section 4. On the other hand, an explicit evaluation of the full Jacobian for large n is prohibitively expensive (see, e.g., [16, 17]). Hence, in this case direct linear solving methods are generally out of the question.

3. Numerical Continuation

We consider the numerical tracing of a solution branch

$$s \mapsto (u(s), T(s), \lambda(s))$$

of Equation (5). For simplicity, we view s as an arc-length parameter. Numerically, one usually performs pseudo-arclength steps, see, e.g., [3, 14]. A numerical continuation (predictor-corrector) method repeats two steps:

(1) A predictor step generates an approximate point further along the solution curve, typically by linear extrapolation.
(2) A corrector step finds a point approximately on the solution curve and close to the predicted point, typically by Newton-like steps.

Algorithm 1: **(Matrix-Free Predictor-Corrector)**

(1) **Initialization**
  choose x such that $H(x) \approx 0$
  choose approximate tangent S such that $H'(x)S \approx 0$, $||S|| = 1$
  choose step size $h > 0$
  choose small reduction factor $1 \gg \eta > 0$
(2) repeat
  (a) **Predictor**
    $\tilde{x} \leftarrow x + hS$
  (b) **Corrector**
    find Δx such that

$$\left\| \begin{bmatrix} H(\tilde{x}) \\ 0 \end{bmatrix} + \begin{bmatrix} H'(\tilde{x}) \\ S^T \end{bmatrix} \Delta x \right\| \leq \eta \left\| \begin{bmatrix} H(\tilde{x}) \\ 0 \end{bmatrix} \right\|$$

    via a transpose-free iterative linear solver.
    $\tilde{x} \leftarrow \tilde{x} + \Delta x$

(c) determine new h
$S \leftarrow (\tilde{x} - x)/||\tilde{x} - x||$
$x \leftarrow \tilde{x}$

Remark 1: The corrector step is an inexact Newton step. In practice, one uses several such steps while reducing η.

4. Implementing the action of $H'(u, T, \lambda)$

In order to perform the predictor-corrector steps of numerical continuation and the evaluation of the test functions for bifurcations, it is necessary to implement the action of the Jacobian $H'(u, T, \lambda) = \begin{bmatrix} \partial_1\varphi - I & \partial_2\varphi & \partial_3\varphi \\ \partial_1 h & \partial_2 h & \partial_3 h \end{bmatrix} (u, T, \lambda)$ efficiently, see also [9, 16, 17].

(1) The action $v \mapsto \partial_1\varphi(u, T, \lambda)v$ is obtained in the following way: Defining $z(t) := \partial_1\varphi(u, t, \lambda)v$ and differentiating the equations

$$\partial_2\varphi(u, t, \lambda) = f(\varphi(u, t, \lambda), \lambda), \quad \varphi(u, 0, \lambda) = u \tag{6}$$

with respect to u leads to the following description: We solve

$$\dot{z} = \partial_1 f(\varphi(u, t, \lambda), \lambda)\, z, \quad z(0) = v \tag{7}$$

and obtain $\partial_1\varphi(u, T, \lambda)v = z(T)$.

(2) The vector $\partial_3\varphi(u, T, \lambda)$ is obtained in the following way: Defining $\xi(t) = \partial_3\varphi(u, t, \lambda)$ and differentiating the equations (6) with respect to λ leads to the following description: We solve

$$\dot{\xi} = \partial_1 f(\varphi(u, t, \lambda), \lambda)\, \xi + \partial_2 f(\varphi(u, t, \lambda), \lambda), \quad \xi(0) = 0 \tag{8}$$

and obtain $\partial_3\varphi(u, T, \lambda) = \xi(T)$.

(3) The vector

$$\partial_2\varphi(u, T, \lambda) = f(\varphi(u, T, \lambda), \lambda) \tag{9}$$

is immediately obtained from (6)

The action of the other derivatives contained in $H'(u, T, \lambda)$ are more obvious.

The previously indicated implementations all rely on an approximation of the orbit $t \mapsto \varphi(u, t, \lambda)$. Our present aim is only to demonstrate the usefulness and applicability of the *matrix-free approach.* We need to approximate the orbit on a grid $t_0 = 0, t_1, \ldots, t_{m-1}, t_m = T$. For simplicity and convenience, we consider an equidistant grid $t_i = ih$ and note that for greater efficiency an adaptive grid should be taken into consideration.

Since we are mainly interested in cases where $\dot{\varphi} = f(\varphi, \lambda)$ is obtained from parabolic PDEs via space discretizations, we concentrate on the case where $\dot{\varphi} = f(\varphi, \lambda)$ is stiff, hence we have to consider the use of an implicit solver for the time steps. The main point here is that we need to solve a nonlinear system for each time step.

For simplicity, let us consider the implicit midpoint rule, i.e., we approximate $\varphi(u, t_i, \lambda) \approx \varphi_i$ via

$$\frac{\varphi_i - \varphi_{i-1}}{h} = f\left(\frac{\varphi_i + \varphi_{i-1}}{2}, \lambda\right).$$

We solve this for φ_i using an inexact Newton's method where the linear equations (per Newton step) are solved with an iterative linear solver. For a preconditioner we use an (occasionally updated) sparse (possibly incomplete) LU factorization of the approximate Jacobian $\mathrm{I} - h\partial_1 f\,(v, \lambda)$ where $v \approx \varphi_i$. A few preconditioners are stored along an orbit, and are used again for small variations in u and λ.

The linear differential equation (7) is solved by the same implicit midpoint rule. This leads to the linear systems

$$\left(\mathrm{I} - \frac{h}{2}\partial_1 f(\varphi(u, \frac{t_{i+1} + t_i}{2}, \lambda), \lambda)\right) z_{i+1} = \left(\mathrm{I} + \frac{h}{2}\partial_1 f(\varphi(u, \frac{t_{i+1} + t_i}{2}, \lambda), \lambda)\right) z_i$$

which one needs to solve for z_{i+1}. Typically we would replace the unknown $\varphi(u, (t_{i+1} + t_i)/2, \lambda)$ with $(\varphi_{i+1} + \varphi_i)/2$.

The linear differential equation (8) is also solved by the implicit midpoint rule:

$$\left(\mathrm{I} - \frac{h}{2}\partial_1 f(\varphi(u, \frac{t_{i+1} + t_i}{2}, \lambda), \lambda)\right) \xi_{i+1} = \left(\mathrm{I} + \frac{h}{2}\partial_1 f(\varphi(u, \frac{t_{i+1} + t_i}{2}, \lambda), \lambda)\right) \xi_i + h\partial_2 f(\varphi(u, \frac{t_{i+1} + t_i}{2}, \lambda), \lambda)$$

It is convenient to use the same iterative linear solver for all three cases, with the same preconditioner. We finally note that similar remarks would hold if one replaced the implicit midpoint rule with a higher order implicit solver for (6)–(8).

5. Calculating Special Points

When tracing a solution branch $s \mapsto (u(s), T(s), \lambda(s)) =: x(s)$ of (5), one is particularly interested in locating special points on the branch. These can be of various types. Our cases of interest are covered by requiring that a certain test function $\tau : \mathbb{R}^{n+2} \to \mathbb{R}$ changes sign. Hence we seek a point $x^* \in \mathbb{R}^{n+2}$ such that

$$\begin{bmatrix} H(x^*) \\ \tau(x^*) \end{bmatrix} = 0 \ . \tag{10}$$

The following Lemma is easy to see:

Lemma 2: *Let $x^* = x(s^*)$ be a regular zero point of H, i.e., the Jacobian $H'(x^*)$ has maximal rank. Then the following statements are equivalent:*

(1) $\tau(x(s))$ has a simple zero at $s = s^*$.
(2) x^* is a regular zero point of (10).

Once an approximation of x^* is found, one could, of course, use an inexact Newton's method directly on (10) to obtain a better approximation, i.e., without continuing to follow a path. Note, however, that this places a calculation of τ into the innermost loop of the method, i.e., while evaluating the functions in (10).

In the context of bifurcation analysis an evaluation of τ may be rather costly, e.g., in the case of torus bifurcation, see Theorem 4 below. Therefore, in our path following context, we use a somewhat different approach, which places an evaluation of τ into the outermost loop: During the numerical continuation, we monitor the sign of τ. Assume that a situation $\tau(x_-)\tau(x_+) < 0$ is encountered for two subsequent points x_-, x_+ on the solution curve. Then we introduce the approximate tangent

$$S := (x_+ - x_-)/||x_+ - x_-||$$

and the linear approximations $p(s) = x_- + s\,S \approx x(s)$.

Now let $q(s)$ be the solution of $\begin{bmatrix} H(q) \\ S^T(q - p(s)) \end{bmatrix} = 0$. Hence, $q(s)$ can be viewed as the corrector-point to the predictor point $p(s)$. Clearly, $q(s)$ can be approximated via an iterative nonlinear solver using a matrix-free double loop, the outer loop consisting of a Newton iteration, and the inner loop being a transpose-free iterative linear solver. We now find a zero of the function $s \mapsto \tau(q(s))$ via a secant-like method such as Brent's Method.

The resulting method for calculating x^* is implemented as a matrix-free triple loop, the outer loop being the secant-like method. Note, however, that the iterative methods representing the two inner loops can be started with increasingly improved values. Alternatively, a modification of this approach can be implemented into the numerical continuation method as a steplength strategy, see [3], Section 8.1. for details. This modification permits a matrix-free implementation consisting of a double loop. We apply these ideas to test functions τ that signal certain types of bifurcation points.

6. Simple Bifurcation Points

Bordered matrices are an important tool for a numerical unfolding of singularities. For example, this is one of the principal themes of the book [13]. A consequence of [13], Proposition 3.2.1, is Keller's Lemma [14]:

Let $A \in \mathbb{R}^{n\times n}$ have rank $n-k$, and let $B, C \in \mathbb{R}^{n\times k}$. Then

$$\begin{bmatrix} A & B \\ C^T & 0 \end{bmatrix}$$

is nonsingular if and only if $\begin{bmatrix} A & B \end{bmatrix}$ and $\begin{bmatrix} A \\ C^T \end{bmatrix}$ have full rank.

Since the set of invertible matrices is open in the space of square matrices, the choice of matrices B, C such that the above matrices have full rank is usually easy to fulfill. However, for numerical purposes, one needs to take issues of condition into account, see Remark 5.

The following is a well-known fact, see, e.g., [10,13]. A simple bifurcation point $x^* = x(s^*)$ is characterized by the fact that the determinant

$$\det \begin{bmatrix} H'(x(s)) \\ S^T \end{bmatrix}$$

changes sign at $s = s^*$. Here S has to be some approximate tangent, i.e., $S \approx \dot{x}(s^*)$. However, this is not a numerically suitable choice of a test function. A better choice is to consider the following bordered system:

$$\begin{bmatrix} \begin{bmatrix} S^T \end{bmatrix} & a \\ b^T & 0 \end{bmatrix} \begin{bmatrix} \xi(s) \\ \tau(s) \end{bmatrix} = \begin{bmatrix} 0 \\ 1 \end{bmatrix}$$

If $a, b \in \mathbb{R}^{n+2}$ are chosen such that

$$\begin{bmatrix} \begin{bmatrix} H'(x(s^*)) \\ S^T \end{bmatrix} & a \end{bmatrix} \quad \text{and} \quad \begin{bmatrix} \begin{bmatrix} H'(x(s^*)) \\ S^T \end{bmatrix} \\ b^T \end{bmatrix}$$

have full rank, then the matrix of the bordered system is non-singular for $s \approx s^*$ where $x^* = x(s^*)$ is a simple bifurcation point. It is easy to see (e.g., via Cramer's Rule) that τ is a test-function for the simple bifurcation point x^*, i.e., τ has a simple zero at $s = s^*$.

Hence, in principle, one could use the method described in Section 5 to detect and approximate simple bifurcation points. However, the approximation cannot be executed very accurately since the Jacobian

$$\begin{bmatrix} H'(x(s)) \\ S^T \end{bmatrix}$$

becomes singular at $s = s^*$, and hence the numerical tracing of $x(s)$ becomes unstable for $s \approx s^*$. It is, however, possible to obtain a matrix-free stable method [10].

7. Period-Doubling Bifurcation

If u is a T-periodic orbit of (1), i.e., $H(u, T, \lambda) = 0$, then $\partial_1 \varphi(u, T, \lambda)$ is called the monodromy matrix. Note that we already characterized its action in (7). A simple period-doubling bifurcation point $x^* = (u^*, T^*, \lambda^*)$ is characterized in the following way, see, e.g., [19]: For an algebraically simple real eigenvalue $\nu(s)$ of $s \mapsto \partial_1 \varphi(u(s), T(s), \lambda(s))$ there holds $\nu(s^*) = -1$ and $\nu'(s^*) \neq 0$. Note that this implies that the determinant of

$$\partial_1 \varphi\big(u(s), T(s), \lambda(s)\big) + \mathrm{I}$$

changes sign at $s = s^*$.

In analogy to Section 6 we obtain the following numerical test function for detecting such a point while numerically following the curve $s \mapsto x(s)$.

Theorem 3: Let $x^* = x(s^*)$ be a simple period-doubling bifurcation point as defined above. Suppose $b, c \in \mathbb{R}^n$ are chosen such that the bordered matrix in the folling system is invertible

$$\begin{bmatrix} \partial_1 \varphi\big(u(s), T(s), \lambda(s)\big) + \mathrm{I} & b \\ c^T & 0 \end{bmatrix} \begin{bmatrix} \xi(s) \\ \tau(s) \end{bmatrix} = \begin{bmatrix} 0 \\ 1 \end{bmatrix}. \tag{11}$$

Then the system is non-singular for $s \approx s^*$, and $\tau(s)$ has a simple zero at $s = s^*$.

Contrary to a simple bifurcation point, however, a simple period-doubling bifurcation point is not a singular point on the curve $H^{-1}(0)$, and hence the method described in Section 5 applies. The only additional complexity of the problem comes from the calculation of τ which requires one matrix-free loop.

8. Torus Bifurcation

We consider again a local (i.e., for $s \approx s^*$) parametrization $s \mapsto (u(s), T(s), \lambda(s))$ of $H^{-1}(0)$, and define the monodromy matrix $A(s) := \partial_1\varphi(u(s), T(s), \lambda(s))$. A simple torus bifurcation point $(u^*, T^*, \lambda^*) = (u(s^*), T(s^*), \lambda(s^*))$ is characterized by an algebraically simple complex eigenvalue $\lambda(s) = \nu(s) + i\omega(s)$ of $A(s)$ crossing the unit circle. Suppose that $v_1(s) + iv_2(s)$ is a corresponding eigenvector. Let $E(s) = \text{span}\{v_1(s), v_2(s)\}$. The following theorem describes a test function for a simple torus bifurcation point which has been implemented numerically. The introduction of the system (12) below was motivated by similar systems for Hopf bifurcation, such as [6, 21].

Theorem 4: Let $(u(s^*), T(s^*), \lambda(s^*))$ be a simple torus bifurcation point. Assume that $c, d \in \mathbb{R}^n$ are chosen so that

$$\begin{bmatrix} A(s^*)^2 - 2\nu(s^*)A(s^*) + \mathrm{I} \\ c^T \\ d^T \end{bmatrix} \quad \text{and} \quad \begin{bmatrix} A(s^*) - (\nu(s^*) \pm i\omega(s^*))\,\mathrm{I} \\ d^T \end{bmatrix}$$

have full rank. Then there exists a unique $e(s) \in E(s)$ with $c^T e(s) = 1, d^T e(s) = 0$ for $s \approx s^*$. Furthermore assume that $a, b \in \mathbb{R}^n$ are chosen so that
$\left[A(s^*)^2 - 2\nu(s^*)A(s^*) + \mathrm{I}\ a\ b\right]$ and $\left[A(s^*)^2 - 2\nu(s^*)A(s^*) + \mathrm{I}\ a\ A(s^*)e(s^*)\right]$
have full rank. Then the bordered matrix in the linear system

$$\begin{bmatrix} A(s)^2 - 2\mu A(s) + \mathrm{I} & a & b \\ c^T & 0 & 0 \\ d^T & 0 & 0 \end{bmatrix} \begin{bmatrix} \xi(\mu, s) \\ \alpha(\mu, s) \\ \beta(\mu, s) \end{bmatrix} = \begin{bmatrix} 0 \\ 1 \\ 0 \end{bmatrix} \tag{12}$$

is non-singular for $s \approx s^*$ and $\mu \approx \nu(s^*)$. Hence $\begin{bmatrix} \xi(\mu, s) \\ \alpha(\mu, s) \\ \beta(\mu, s) \end{bmatrix}$ is well-defined. Furthermore, $\tau(s) := \alpha(\mu(s), s)$ has a simple zero at $s = s^*$ and can be used as a test function for torus bifurcation.

A more precise version of Theorem 4 has been proven in [2].

Remark 5: Note that the theorem gives a local result. We therefore propose to use it in conjunction with an Arnoldi iteration: While following a branch of periodic solutions, we occasionally apply an Arnoldi-type iteration (we used ARPACK [15]) to obtain snapshots of the dominant eigenvalues (Floquet multipliers) of the monodromy matrix $\partial_1\varphi(u, T, \lambda)$. The snapshot will show when a pair of conjugate complex eigenvalues passes through the unit circle. We then use the above test-function to approximate the torus bifurcation point more accurately. Note that the Arnoldi-like method not only gives us guesses for μ via the real part of the approximate eigenvalue, but also for a and b via the real and imaginary part of the approximate eigenvector. This proved to be an effective and efficient strategy.

9. Cellular Exclusion Algorithms

This section will outline a cellular exclusion algorithm for finding all real zeros of a map $F : \prod_{i=1}^{n}[a_i, b_i] \to R^n$. The idea here is to apply such an algorithm to a system of defining equations arising for example, when a Lyapunov-Schmidt reduction has been performed. Exclusion algorithms are a well-known tool in the area of interval analysis for finding all solutions of a system of nonlinear equations. They also have been introduced in [22, 23] from a slightly different viewpoint. In particular, such algorithms seem to be very useful for finding all solutions of low-dimensional, but highly nonlinear systems which have many solutions.

In $\mathbb{R}^n$ and $\mathbb{R}^{m\times n}$ we use the component-wise '$\leq$' as a partial ordering, and $|\cdot|$ is the component-wise absolute value. We only use the max norm '$||\cdot||$'. For example, for two matrices $A, B \in \mathbb{R}^{n\times n}$ the symbol $A \leq B$ means that $A(i,j) \leq B(i,j)$ for $i, j = 1 : n$.

An interval σ in $\mathbb{R}^n$ is a rectangular box, i.e., there are two vectors $m_\sigma, d_\sigma \in \mathbb{R}^n$ with $d_\sigma(i) > 0$, $i = 1 : n$, such that

$$\sigma = [m_\sigma - d_\sigma, m_\sigma + d_\sigma] = \{\, x \in \mathbb{R}^n \ : \ m_\sigma - d_\sigma \leq x \leq m_\sigma + d_\sigma \,\} \ .$$

We call $m_\sigma - d_\sigma$ the lower corner, $m_\sigma + d_\sigma$ the upper corner, m_σ the center, and d_σ the mesh vector of σ.

Let $\sigma \subset \mathbb{R}^n$ be an interval and $F : \sigma \to \mathbb{R}^n$ a function defined on σ. We call a test

$$T_F(\sigma) \in \{\, 0 \, , \, 1 \,\} \quad \text{where} \quad 0 \equiv \text{yes} \quad \text{and} \quad 1 \equiv \text{no}$$

an *exclusion test* for F on σ iff $T_F(\sigma) = 0$ implies that F has no zero point in σ. Hence, $T_F(\sigma) = 1$ is a *necessary* condition for F to have a zero point in σ.

If an exclusion test is given, then one can recursively bisect intervals and discard the ones which are testing negatively. This leads to the following recursive *Exclusion Algorithm* which starts from some initial interval Λ on which F is defined. Let us assume that an exclusion test $T_F(\sigma)$ is available for all subintervals $\sigma \subset \Lambda$.

Algorithm 2: **(Exclusion Algorithm)**

$\Gamma \leftarrow \{\Lambda\}$ (initial interval)
for $\ell = 1$: maximal_level
  for $a = 1 : n$
    let $\tilde{\Gamma}$ be obtained by bisecting each $\sigma \in \Gamma$ along the axis a
    for $\sigma \in \tilde{\Gamma}$
      if $T_F(\sigma) = 0$
        drop σ from $\tilde{\Gamma}$ (σ is excluded)
    $\Gamma \leftarrow \tilde{\Gamma}$
  $\Gamma_\ell \leftarrow \Gamma$ (for later reference)

To simplify the efficiency study, we examine the strategy of cyclic bisections of the intervals along successive co-ordinate axes. Whenever one cycle of bisections is accomplished, we say that a new level of bisection has been reached and regard an exclusion algorithm as performing a fixed number of levels of bisections. The intervals which have not been discarded after ℓ levels of bisection will be considered as the intervals which the algorithm generates on the ℓ-th bisection level, see Figure 1 for an illustration. The list of these intervals is denoted by Γ_ℓ in the algorithm. Obviously, if $\Gamma_\ell = \emptyset$ for some level ℓ, then the algorithm has shown that there are no zero points of F in the initial interval Λ.

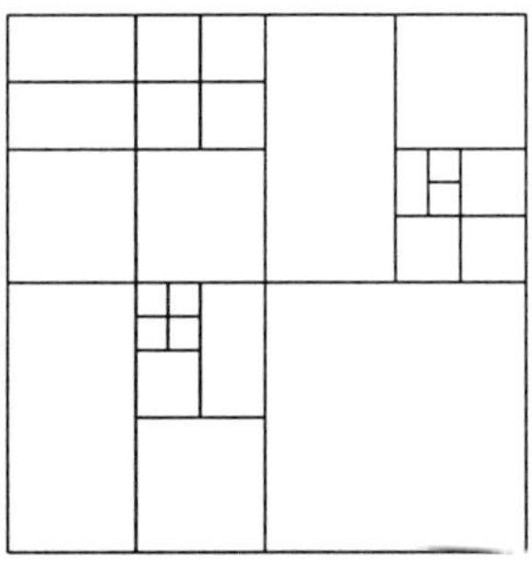

$\#(\Gamma_\ell)$	ℓ
1	0
4	1
8	2
10	3
6	4

Fig. 1. Illustration of Bisection Levels

All the exclusion tests that we will discuss are applied component-wise on the vector-valued function F. Hence, we only need to consider an exclusion test for a scalar-valued function $f : \sigma \to \mathbb{R}$, and then we can combine such (possibly differing types of) exclusion tests to obtain an exclusion test for a vector-valued function $F = \{ f_i \}_{i=1:n} : \sigma \to \mathbb{R}^n$ by setting

$$T_F(\sigma) := \prod_{i=1}^{n} T_{f_i}(\sigma) .$$

Thus, in the following, we will mainly restrict our attention to scalar functions $f : \sigma \to \mathbb{R}$ when designing exclusion tests.

It is clear that the efficiency of exclusion algorithms hinges mainly on the construction of a good exclusion test which is computationally inexpensive but relatively tight. Otherwise, too many intervals remain undiscarded on each bisection level, and this leads to significant numerical inefficiency.

Two exclusion tests which are easy to verify were given in [23]:

- Let $L > 0$ be a Lipschitz constant for f on the interval σ. Then

$$|f(m_\sigma)| \le L \, ||d_\sigma|| \tag{13}$$

  is an exclusion test for f on σ.
- If $f = g - h$ is the difference of two increasing functions on σ, then

$$h(m_\sigma - d_\sigma) \le g(m_\sigma + d_\sigma) \quad \textbf{and} \quad h(m_\sigma + d_\sigma) \ge g(m_\sigma - d_\sigma) \tag{14}$$

  is an exclusion test for f on σ.

Although such tests may be cheap and simple to apply, they often are not sufficiently tight to be efficiently useful. The aim of the next section is to generate and analyze refined tests in such a way that they lead to much more efficient exclusion algorithms.

10. Construction of Dominant Functions

Denote by $\mathbb{Z}_+$ the set of nonnegative integers. For a multi-index $\alpha = (\alpha_1, \ldots, \alpha_n) \in \mathbb{Z}_+^n$ we make the following standard definitions:

(1) The length of α is defined by $|\alpha| := \sum_i \alpha_i$.
(2) The factorial of α is defined by $\alpha! := \prod_i \alpha_i!$.
(3) If $x \in \mathbb{R}^n$, then we define $x^\alpha := \prod_i x_i^{\alpha_i}$
(4) We define the partial derivatives $\partial^\alpha = (\alpha!)^{-1} \prod_i \partial_i^{\alpha_i}$.

We also introduce the probability measures $\omega_k(dt) = k(1-t)^{k-1}\, dt$ on the interval $[0,1]$.

These definitions are given in such a way that for an integer $k > 0$ Taylor's formula with integral remainder is easy to write:

$$f(m+h) = f(m) + \sum_{0<|\alpha|<k} \partial^\alpha f(m) h^\alpha + \sum_{|\beta|=k} \int_0^1 \partial^\beta f(m+th)\, \omega_k(dt)\, h^\beta \tag{15}$$

Definition 6: Let $\sigma \subset \mathbb{R}^n$ be an interval. By $\mathcal{A}_k(\sigma)$ we denote the space of continuous functions $f : \sigma \to \mathbb{R}$ such that $\partial^\alpha f$ is absolutely continuous for $|\alpha| < k$. Note that for $f \in \mathcal{A}_k$ the Taylor formula (15) holds. In $\mathcal{A}_k(\sigma)$ we introduce the cone

$$\mathcal{K}_k(\sigma) = \{\, g \in \mathcal{A}_k(\sigma) \;:\; 0 \le \partial^\alpha g(x) \le \partial^\alpha g(y) \text{ for } 0 \le x \le y,\ |\alpha| \le k.\,\}$$

We also set $\mathcal{A}_\infty(\sigma) := \bigcap_{k=1}^\infty \mathcal{A}_k(\sigma)$ and $\mathcal{K}_\infty(\sigma) := \bigcap_{k=1}^\infty \mathcal{K}_k(\sigma)$.

We now introduce the notion of a dominant function which will be the basis for the estimates which follow:

Definition 7: Let $f \in \mathcal{A}_k(\sigma)$ and $g \in \mathcal{K}_k(\sigma)$. Then $f(x) \prec_k g(x)$ for $x \in \sigma$ (g dominates f with order k on σ) iff the estimates

$$|\partial^\alpha f(x)| \le \partial^\alpha g(|x|)$$

hold for all $x \in \sigma$ and $|\alpha| \le k$. If $f \in \mathcal{A}_\infty(\sigma)$ and $g \in \mathcal{K}_\infty(\sigma)$, then $f(x) \prec_\infty g(x)$ for $x \in \sigma$ means that $f \prec_k g$ for $x \in \sigma$ and all $k \ge 0$.

Note that $f(x) \prec_k g(x)$ for $x \in \sigma$ obviously implies that $f(x) \prec_q g(x)$ for $x \in \tau$, provided that $q \le k$ and $\tau \subset \sigma$. We will frequently use the notations $f \prec_k g$ or $f(x) \prec_k g(x)$ if not mentioning a specific interval does not lead to any ambiguity. The following example points out the differences between the various estimates:

Example 8: $\cos x \prec_\infty \cosh x$, but $\cos x \prec_1 1 + x$, $\cos x \prec_2 1 + \frac{1}{2}x^2$, and $\cos x \prec_3 1 + \frac{1}{2}x^2 + \frac{1}{6}x^3$.

11. Local Expansions to Obtain Exclusion Tests

The following theorem summarizes possible choices of exclusion tests:

Theorem 9: Let $\sigma \subset \mathbb{R}^n$ be an interval, and let $q > 0$ be an integer. Let $f(m_\sigma + x) \prec_q g(x)$ for $|x| \le d_\sigma$. Then

$$|f(m_\sigma)| \le g(d_\sigma) - g(0) - \sum_{0<|\alpha|<q} \underbrace{\left(\partial^\alpha g(0) - |\partial^\alpha f(m_\sigma)|\right)}_{\ge 0} d_\sigma^\alpha \tag{16}$$

is an exclusion test for f on σ.

Corollary 10: *Let $\sigma \subset \mathbb{R}^n$ be an interval, and let $q > 0$ be an integer. Let $f(x) \prec_q g(x)$ for $x \in \sigma$. Then*

$$|f(m_\sigma)| \le g(|m_\sigma| + d_\sigma) - g(|m_\sigma|) - \sum_{0<|\alpha|<q} \underbrace{(\partial^\alpha g(|m_\sigma|) - |\partial^\alpha f(m_\sigma)|)}_{\ge 0} d_\sigma^\alpha \tag{17}$$

is an exclusion test for f on σ.

The terms under the summation sign in (16) and (17) are nonnegative, and hence the test tightens with increasing q. To increase the efficiency of implementations, one would successively apply the test for $q = 1 : q_0$ and discard the interval as soon as the test fails. To describe a complexity result for the exclusion algorithm, we introduce the following definition.

Definition 11: We say that a zero point ξ of F has uniform order p if

(1) $\partial^\alpha F(\xi) = 0$ for $|\alpha| < p$.
(2) There exists an $\varepsilon > 0$ such that $\varepsilon\,||m - \xi||^p \le ||F(m)||$ for $||m - \xi|| \le \varepsilon$.

Theorem 12: Let each zero point ξ of F be of some uniform order which is at most q. Then there is a constant $C > 0$ such that the exclusion algorithm generates no more than C intervals on each bisection level, i.e., $\#\Gamma_k \le C$ independently of k.

An application of the exclusion algorithm typically consists of the following steps:

(1) Given the problem $F(x) = 0$, construct a G such that $F \prec_q G$. The above results and those in [11] are used in this step.
(2) Implement the exclusion test (16) or (17) for the given q. Note that for $q > 1$ many partial derivatives are involved, so one must formulate a script that actually writes these tests once F and G are given.
(3) Run the exclusion algorithm based on the test constructed in step 2.
(4) A typical feature of an exclusion algorithm is that each zero point causes the generation of several intervals, and therefore in a final step one needs to sort out which intervals represent the same zero point.

12. Exclusion Algorithms for Unconstrained Optimization

In this section we present a cellular exclusion algorithm for finding the global minimum of a continuous map $f : \prod_{i=1}^n [a_i, b_i] \to \mathbb{R}$. The algorithm has many features which are similar to the algorithm for finding all real solutions.

Let $\sigma \subset \mathbb{R}^n$ be an interval, and $f : \sigma \to \mathbb{R}$ a function defined on σ. We call a number $E_f(\sigma) \in \mathbb{R}$ a *lower bound* of f on σ iff $\inf_{x\in\sigma} f(x) \geq E_f(\sigma)$. Note that $E_f(\sigma)$ is also a lower bound (possibly not a good one) of f on τ if $\tau \subset \sigma$ is a subinterval.

If such bounds $E_f(\sigma)$ are available, then we can recursively bisect intervals and discard the ones whose lower bounds are larger than a currently found low value M of f. This leads to the following recursive *exclusion algorithm* which starts from some initial interval $\Lambda \subset \mathbb{R}^n$ on which f is defined. We assume that a lower bound $E_f(\sigma)$ is available for all subintervals $\sigma \subset \Lambda$.

Algorithm 3: **(Exclusion Algorithm for Minimization)**

```
Γ ← { Λ } (initial interval)
M ← f(m_Λ) (initial low value of f)
for ℓ = 1 : max_level
    for a = 1 : n
        let Γ̃ be obtained by bisecting each σ ∈ Γ along the axis a
        for σ ∈ Γ̃
            if E_f(σ) > M
                drop σ from Γ̃
            else
                M ← min (M, f(m_σ)) (update)
        Γ ← Γ̃
    Γ_ℓ ← Γ, M_ℓ ← M (for later reference)
```

The following Lemma is easy to show.

Lemma 13: *If f is continuous on Λ, then* $\lim_{\ell\to\infty} M_\ell = \min_{x\in\Lambda} f(x)$.

Again, it is clear that the efficiency of Algorithm 3 hinges mainly on the construction of good lower bounds which are computationally inexpensive, but relatively tight. The following theorem summarizes the possible choices of lower bound estimates which we will consider in this paper for Algorithm 3:

Theorem 14: Let $\sigma \subset \mathbb{R}^n$ be an interval, and let $q > 0$ be an integer. Let $f(m_\sigma + x) \prec_q g(x)$ for $|x| \le d_\sigma$. Then

$$E_f(\sigma) := f(m_\sigma) - \underbrace{(g(d_\sigma) - g(0))}_{\ge 0} + \sum_{0<|\alpha|<q} \underbrace{(\partial^\alpha g(0) - |\partial^\alpha f(m_\sigma)|)\, d_\sigma^\alpha}_{\ge 0} \tag{18}$$

is a lower bound for f on σ.

Corollary 15: *Let $\sigma \subset \mathbb{R}^n$ be an interval, and let $q > 0$ be an integer. Let $f(x) \prec_q g(x)$ for $x \in \sigma$. Then*

$$E_f(\sigma) := f(m_\sigma) - \underbrace{(g(|m_\sigma| + d_\sigma) - g(|m_\sigma|))}_{\ge 0} + \sum_{0<|\alpha|<q} \underbrace{(\partial^\alpha g(|m_\sigma|) - |\partial^\alpha f(m_\sigma)|)\, d_\sigma^\alpha}_{\ge 0} \tag{19}$$

is a lower bound for f on σ.

Proof: Note that $f(m_\sigma + x) \prec_q g(|m_\sigma| + x)$ for $|x| \le d_\sigma$ and apply the theorem. □

The terms under the summation sign in (18) and (19) are nonnegative, and hence the estimate tightens with increasing q, i.e., the lower bound estimate increases. To increase the efficiency of implementations, one would successively apply the estimate for $q = 1 : q_0$ in Algorithm 3 and discard the interval as soon as possible.

We recall that a point $\xi \in \mathcal{M}_f(\Lambda)$ which is also in the interior of Λ satisfies the (necessary) optimality conditions

(1) $\partial^\alpha F(\xi) = 0$ for $|\alpha| = 1$.
(2) The quadratic form $\sum_{|\beta|=2} \partial^\beta f(\xi) x^\beta$ is positive semidefinite.

If furthermore ξ is a regular point of the gradient of f (the generic case) then the quadratic form is positive definite, and Taylor's formula implies immediately that ξ has uniform order 1. Hence the previous definition (for $p > 1$) allows some degeneracy in the minimality of the point ξ. A Complexity Theorem analogous to Theorem 12 was proven in [1, 8]

13. Application to Nematic Electroconvection

In [4] the cellular exclusion algorithm for a global minimum has been applied to a surface of Hopf points resulting from a bifurcation analysis of the weak electrolyte model for electroconvection in a planar layer of nematic liquid crystals (NLC).

Electroconvection in NLC is a paradigm for pattern formation in anisotropic systems, exhibiting a rich variety of dynamical structures [5,18]. NLC are electrically conducting incompressible fluids that exhibit long range microscopic molecular alignment leading to anisotropic macroscopic properties. When an electric potential is applied across a thin NLC layer confined between two electrode plates, an electrohydrodynamic instability can occur above a critical field strength [5].

The traditionally used mathematical model to describe this type of instability is the so-called standard model (SM) and it involves equations for the velocity field, the director (a field of unit vectors describing the local orientation of the molecules), and the electric potential and charge, derived from the generalized Navier Stokes equations for anisotropic fluids, conservation of charge, and Poisson's and Ohm's laws. The SM explains quantitatively the phenomena observed in the conduction range, however it predicts always a stationary bifurcation. The recently introduced weak electrolyte model (WEM) [20], which incorporates molecular dissociation-recombination reactions and their effects on the conductivity into the SM, provides a basis for the understanding of the Hopf bifurcation observed quite frequently at threshold. The bifurcation analysis of the WEM has been discussed in [7].

The WEM–equations are too lengthy to be written down explicitly. Symbolically, they are of the form

$$\partial_t U = \mathcal{L}(R)U + \mathcal{N}(R, U),$$

where $U = U(t, x, y, z)$ represents the field variables, R is the bifurcation parameter, proportional to the strength of the external electric field, and $\mathcal{L}$ and $\mathcal{N}$ are linear and nonlinear operators. The NLC layer is sandwiched between two parallel plates which are idealized to be infinitely extended, hence the system is translation invariant in the horizontal directions. Owing to this invariance, the linearized system, $\partial_t U = \mathcal{L}U$, can be studied using the plane wave ansatz $U = e^{i(px+qy)}V(t, z)$. The operator L in the resulting linear system

$$\partial_t V = L(p, q, R)V$$

for V is an integro–differential operator in the z–direction. We use a Galerkin approximation of V by $3N$ vertical modes and accordingly represent L by a $3N \times 3N$–matrix. The Hopf bifurcation points are located on a surface $R = R(p,q)$ in (p,q,R)–space which is computed numerically using Werner's [21] bordered matrices algorithm.

Electroconvection sets in when R is increased above the global minimum, R_c, of the Hopf surface. To approximate this minimum and the associated critical wave numbers (p_c, q_c), in [4] the cell exclusion algorithm has been applied to $R(p,q)$. The number of modes has been successively increased until numerical convergence was observed. In Table 1 the approximated values of (p_c, q_c, R_c) are listed for $N = 5, 7, 9, 10$, for a certain choice of the other parameters of the WEM and reproduce those computed in [7] via a searching method. The level sets of $R(p,q)$ and the refinement of the domain is shown in Figure 2 for $N = 5$.

Table 1. Approximation of the global minimum of the Hopf surface.

N	p_c	q_c	R_c
5	1.130151	0.7065545	7.04302807
7	1.130146	0.7065543	7.04433204968870
9	1.1301459	0.706555429	7.04502799195270
10	1.1301459	0.706555429	7.04593846328180

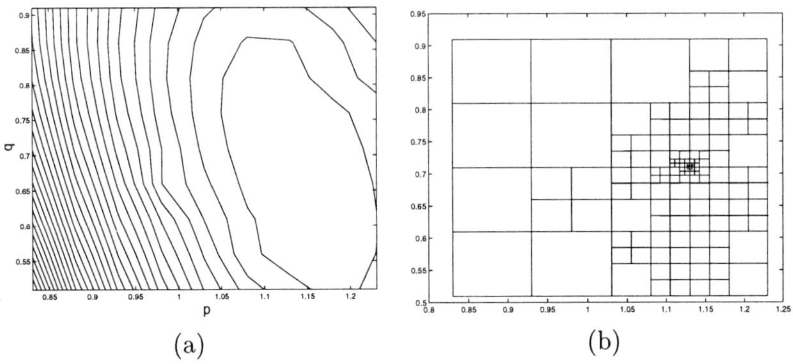

Fig. 2. (a): Level sets of $R(p,q)$. (b): Refinement of the domain.

References

[1] E. L. Allgower, M. Erdmann and K. Georg, *Journal of Complexity* **18**, 573 (2002).
[2] E. L. Allgower, U. Garbotz and K. Georg, *Rendiconti di Matematica*,**24**, (2004).
[3] E. L. Allgower and K. Georg. *Handbook of Numerical Analysis.* **5**, 3. North-Holland, (1997).
[4] I. Alolyan, *PhD Thesis Colorado State University* Fort Collins, (2004).
[5] S. Chandrasekhar. *Liquid Crystals.*, University Press, Cambridge, 1977.
[6] K.-W. E. Chu, W. Govaerts, and A. Spence, *SIAM J. Num.Anal.* **31**, 524 (1994).
[7] G. Dangelmayr and I. Oprea, *Mollec. Cryst. Liq. Cryst.*, **413**, 305 (2004).
[8] M. Erdmann, *PhD thesis, Colorado State University*, (2001).
[9] U. Garbotz, *PhD thesis, University of Marburg*, (2001).
[10] K. Georg, *Numerical Funct. Anal. and Optimization* **22**, 303 (2001).
[11] K. Georg, *Advances in Geometry* **1**, 193 (2001).
[12] K. Georg, *J. Comput. and Appl. Math.* **152**, 147 (2003).
[13] W. Govaerts, *Numerical Methods for Bifurcations of Dynamical Equilibria*, SIAM, Philadelphia 2000.
[14] H. B. Keller, *Applications of Bifurcation Theory*, 359, Academic Press 1977.
[15] R. B. Lehoucq, D. C. Sorensen, and C. Yang, *ARPACK Users' Guide*, SIAM, Philadelphia, 1998.
[16] K. Lust and D. Roose, *Numerical Methods for Bifurcation Problems and Large-Scale Dynamical Systems.* Springer-Verlag, 2000.
[17] K. Lust, D. Roose, A. Spence, and A. R. Champneys, *SIAM J. on Sci. Comput.* **19**, 1188 (1998).
[18] W. Pesch and U. Behn. In *Evolution of Spontaneous Structures in Dissipative Continous Systems*, F. Busse and S. Müller, eds., Springer, 1998, pp. 335.
[19] R. Seydel. *From Equilibrium to Chaos. Practical Bifurcation and Stability Analysis.* Elsevier, New York, 1988.
[20] M. Treiber and L. Kramer. *Mol. Cryst. Liqu. Cryst.*, 261:311, 1995.
[21] B. Werner, *SIAM J. Numer. Anal.*, 33:435–455, 1996.
[22] Z.-B. Xu, J.-S. Zhang, and Y.-W. Leung, *Appl. Math. Comput.* **86**, 235 (1997).
[23] Z.-B. Xu, J.-S. Zhang, and W. Wang, *Appl. Math. Comput.* **80**,181 (1996).

CHAPTER 3

VALIDITY OF THE GINZBURG-LANDAU APPROXIMATION IN PATTERN FORMING SYSTEMS WITH TIME PERIODIC FORCING

Norbert Breindl, Guido Schneider, and Hannes Uecker

Mathematisches Institut I, Universität Karlsruhe, 76128 Karlsruhe, Germany

We consider the validity of the Ginzburg-Landau equation in pattern forming systems with time periodic forcing. Beside the proof of an approximation result for a model problem we extend the possibility for the derivation of the Ginzburg-Landau equation to arbitrary frequencies in time by a modified ansatz.

1. Introduction

Our investigations are motivated by electro-convection in nematic liquid crystals, the paradigm for pattern formation in anisotropic systems [6, 10, 27]. In this experiment [4] nematic liquid crystals with negative or only mildly positive dielectric anisotropy are sandwiched between two glass plates with transparent electrodes subject to some external time-periodic electric field, see Figure 1. Liquid crystals are often called the fourth state of matter, because they combine properties of a liquid, like the flow behavior, with such of solids, especially the anisotropy. In nematic liquid crystals the rod-like molecules point at an average in the same direction and can be influenced by an electric field. If the amplitude of the applied alternate current voltage is above a certain threshold the trivial spatially homogenous time periodic solution gets unstable and bifurcates into spatially periodic patterns.

The mathematical analysis of the creation and interactions of such patterns is based very often on the reduction of the governing partial differential equations to finite or infinite-dimensional amplitude equations which are expected to capture the essential dynamics near the bifurcation point. The most famous amplitude equation occurring in such a setup is the so called Ginzburg-Landau equation. It is derived by multiple scaling analysis and describes slow modulations in time and space of the amplitude of

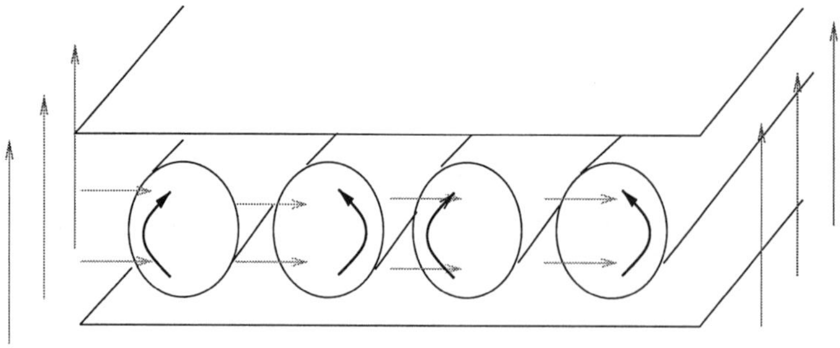

Fig. 1. Roll solutions in nematic crystals. The director field is almost parallel to the plates. The external time-periodic electric field is perpendicular to the plates.

the linearly most unstable modes. The Ginzburg-Landau equation has been derived for example for reaction-diffusion systems and hydrodynamical stability problems, as the Bénard and the Taylor-Couette problem. For these examples the Ginzburg-Landau equation has been justified as an amplitude equation by a number of mathematical theorems [3,8,11,12,19,21,29], also including approximation and attractivity results. For an overview see [16]. Hence, the Ginzburg-Landau equation really gives a proper description of these original systems near the bifurcation point.

The Ginzburg-Landau equation has also been used extensively to describe pattern formation in nematic liquid crystals [1, 17, 27, 30]. However, the literature cited above about the mathematical justification of the Ginzburg-Landau equation is restricted so far to autonomous systems and is not covering the situation of nematic liquid crystals due to the time-periodic forcing. Therefore, it is the purpose of this paper to justify the Ginzburg-Landau equation also in case of a time periodic setup.

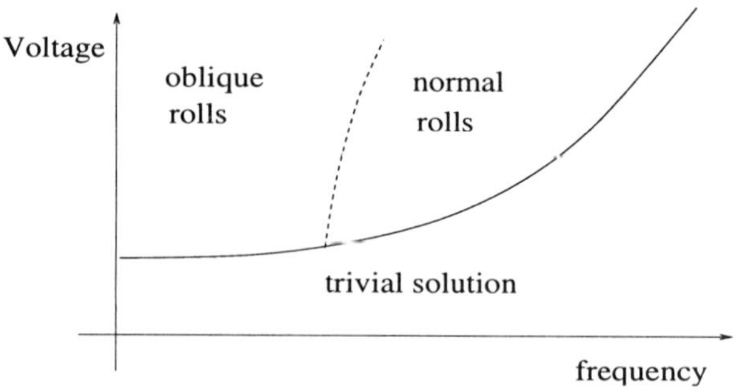

Fig. 2. The bifurcation diagram.

The Ginzburg-Landau equation occurs in the parameter region indicated in Figure 2 as "normal rolls". For the derivation of the Ginzburg-Landau equation in this region in the existing literature it is assumed implicitly that the external electric field oscillates with a sufficiently high frequency such that the governing partial differential equations can be replaced by an effective autonomous system such that the usual derivation of the Ginzburg-Landau equation applies.

Our approach is different. By a modified ansatz we remove the assumption of a highly oscillating external electric field, i.e. we extend the possibility for the derivation of a Ginzburg-Landau equation to arbitrary frequencies in time. However, in contrast to the autonomous case the solutions of the Ginzburg-Landau equation then have to be analytic. This is no serious restriction, since this is true for every $t > 0$ by the smoothing properties of the Ginzburg-Landau equation, but it has to be assumed at $t = 0$.

Due to the limited number of pages we restrict ourselves in the discussion of the validity question to a model problem which has the essential features of the nematic liquid crystal problem which are relevant for our purposes. We refer to a forthcoming paper for the application of the theory to electro-convection in nematic liquid crystals.

The plan of the paper is as follows. In Section 2 we present our result in Theorem 1 for a scalar valued partial differential equation with time periodic forcing on the real line. We prove that solutions of this original system can be approximated via the solutions of the associated Ginzburg-Landau equation. This is the first time that the Ginzburg-Landau formalism is justified in a non-autonomous situation. The proof of this approximation result is given in a number of steps from Section 3 to Section 7. The general situation is discussed in a number of remarks in Section 2. The proof of Theorem 1 goes along the lines of the autonomous case and reviews the basic concepts.

Notation: Throughout this paper many different constants are denoted by the same symbol C if they can be chosen independent of the small perturbation parameter $0 < \varepsilon \ll 1$. Fourier transform is denoted by

$$(\mathcal{F}u)(k) = \hat{u}(k) = \frac{1}{2\pi} \int_{\mathbb{R}} u(x) e^{-ikx} dx.$$

2. The Model

We consider

$$\partial_t u = -u + \beta \cos(\omega t) - \gamma \partial_x^8 u - u^2 (4\partial_x^2 + \partial_x^4) u + u \partial_x u, \tag{1}$$

with $u(x,t) \in \mathbb{R}$, $x \in \mathbb{R}$, and $t \geq 0$. The three parameters satisfy $\beta \geq 0$, $\omega \neq 0$, and $\gamma > 0$, but small. This model problem has the essential features of the nematic liquid crystal problem which are relevant for our purposes. We have

a) a linear damping $-u$,

b) a time-periodic forcing $\beta \cos(\omega t)$, in which β is the control parameter for the amplitude and $\omega \neq 0$ the fixed temporal wave number,

c) an 8th-order derivative term $\gamma \partial_x^8 u$ in which $\gamma > 0$ is constant and small. This term makes the problem semilinear, and is only added to avoid functional analytic difficulties which are not related to our purposes.

d) the nonlinear term $-u^2(4\partial_x^2 + \partial_x^4)u$ roughly corresponds to nonlinear viscous stress in liquid crystals. This term is responsible for the Turing instability of the trivial solution. Like for the electro-convection problem it will be proportional to β^2.

e) $u\partial_x u$ models the convection term in the Navier-Stokes equations.

For the model problem we find the trivial, spatially homogenous, time-periodic solution

$$u_0(t) = (1+\omega^2)^{-1}\beta(\cos(\omega t) + \omega \sin(\omega t)).$$

We set $u_0^2(t) = \bar{c} + c(t)$, where $\bar{c} = \dfrac{\omega}{2\pi}\displaystyle\int_0^{\frac{2\pi}{\omega}} u_0^2(t)dt$ is the mean value which is proportional to β^2. Note that $u_0(t)$ and $c(t)$ have a vanishing mean value. We do this to separate the equation for the perturbation $v(t) = u(t) - u_0(t)$ into an autonomous and into an oscillating part. We get

$$\begin{aligned} \partial_t v = {} & -v - \gamma\partial_x^8 v - \bar{c}(4\partial_x^2 + \partial_x^4)v \\ & -v^2(4\partial_x^2 + \partial_x^4)v + v\partial_x v \\ & -c(t)(4\partial_x^2 + \partial_x^4)v + u_0(t)\partial_x v \\ & -2u_0(t)v(4\partial_x^2 + \partial_x^4)v. \end{aligned} \tag{2}$$

The first line contains the autonomous linear terms, the second line the autonomous nonlinear terms, the third line the non autonomous linear terms, and the fourth line the non autonomous nonlinear terms. To determine the stability of the trivial solution we first consider the linearized autonomous part

$$\partial_t v = -v - \gamma\partial_x^8 v + \bar{c}(4\partial_x^2 + \partial_x^4)v. \tag{3}$$

We find the solutions

$$v(x,t) = e^{ikx+\lambda(k,\bar{c})t},$$

where

$$\lambda(k,\bar{c}) = -1 + 4\bar{c}k^2 - \bar{c}k^4 - \gamma k^8.$$

There are a critical $\bar{c}_{\text{crit}}$ and a critical wave number $k_c > 0$, such that $\lambda(k_c, \bar{c}_{\text{crit}}) = 0$, where $\lambda(k,\bar{c}) < 0$, if $\bar{c} < \bar{c}_{\text{crit}}$. We are interested in the case $\bar{c} > \bar{c}_{\text{crit}}$ and introduce the small perturbation parameter $\varepsilon > 0$ by

$$\bar{c} = \bar{c}_{\text{crit}} + \varepsilon^2 .$$

For $\varepsilon > 0$ sufficiently small, we have the typical situation for deriving a Ginzburg-Landau equation (see Figure 3). Adding the non autonomous

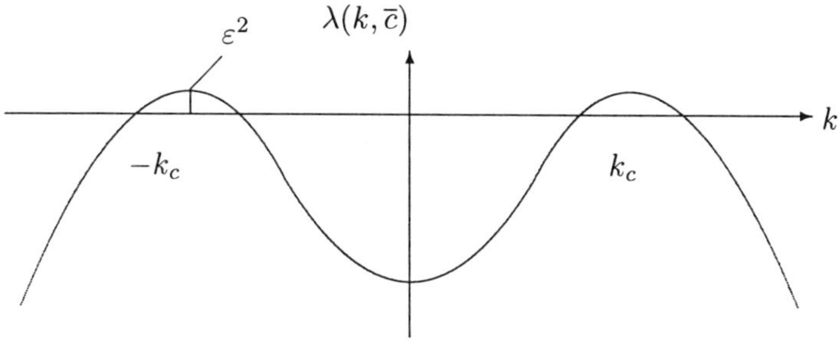

Fig. 3. The curve $k \mapsto \lambda(k,\bar{c})$ for $\varepsilon > 0$

linear terms from (2) to (3) we find solutions (see below)

$$v(x,t) = e^{ikx+\lambda(k,\bar{c})t}\hat{v}_k(t),$$

with $\hat{v}_k(t) = \hat{v}_k\left(t + \dfrac{2\pi}{\omega}\right)$, i.e. the Floquet exponents of the linearization about the trivial time periodic solution $u_0(t)$ are also given by $\lambda = \lambda(k,\bar{c})$.

In order to handle arbitrary but fixed ω we modify the usual ansatz for the derivation of the Ginzburg-Landau equation to

$$\psi_A(x,t,\varepsilon) = (\varepsilon A(X,T)e^{ik_c x + \varphi_0(t)} + c.c.) + \mathcal{O}(\varepsilon^2), \tag{4}$$

where

$$X = \varepsilon(x + \varphi_1(t)) \quad \text{and} \quad T = \varepsilon^2 t.$$

Because of the oscillating terms we have changed the usual ansatz by introducing functions $\varphi_0 = \varphi_0(t)$ and $\varphi_1 = \varphi_1(t)$. Inserting (4) into (1) and equating the coefficients in front of the $\varepsilon^m e^{ni(k_c x+\varphi_0(t))}$ to zero shows

$$\varphi_0(t) = (4k_c^2 - k_c^4)\int_0^t c(\tau)\,d\tau + ik_c\int_0^t u_0(\tau)\,d\tau,$$

$$\varphi_1(t) = (8k_c^3 - 4k_c)i\int_0^t c(\tau)\,d\tau + \int_0^t u_0(\tau)\,d\tau,$$

and that A has to satisfy a Ginzburg-Landau equation (7) with highly oscillating coefficients which can be simplified further to an autonomous Ginzburg-Landau equation

$$\partial_T A = c_1 A + c_2\partial_X^2 A - c_3 A|A|^2, \tag{5}$$

with coefficients $c_j \in \mathbb{C}$.

Note that φ_0 and φ_1 stay bounded, but possess non vanishing imaginary parts. Hence A which is evaluated at $X = \varepsilon(x + \varphi_1(t))$ has to be analytic in a strip in the complex plane. Our approximation theorem is as follows:

Theorem 1: Let $\delta_{GL} > 0$ and $A = A(X,T)$ be a solution of the Ginzburg-Landau equation (5) for $T \in [0, T_0]$, analytic in X in a strip $S_{\delta_{GL}} = \{z \in \mathbb{C} \mid |\mathrm{Im} z| < \delta_{GL}\}$ in the complex plane satisfying

$$\sup_{T\in[0,T_0]}\ \sup_{z\in S_{\delta_{GL}}} |A(z,T)| < \infty.$$

Then there are $\varepsilon_0 > 0$ and $C > 0$, such that for all $\varepsilon \in (0, \varepsilon_0)$ we have solutions v of (2) satisfying

$$\sup_{t\in[0,T_0/\varepsilon^2]}\ \sup_{x\in\mathbb{R}} |v(x,t) - \psi_A(x,t,\varepsilon)| \le C\varepsilon^2.$$

Remark 2: As a consequence of Theorem 1 the dynamics known for (5) can be found approximately in the original system (2), too. The error of order $\mathcal{O}(\varepsilon^2)$ is much smaller than the approximation ψ_A and the solution v which are both of order $\mathcal{O}(\varepsilon)$ for all $T \in [0, T_0]$ or $t \in [0, T_0/\varepsilon^2]$, respectively. This fact should not be taken for granted: there are modulation equations (for an example see [22]) which although derived by reasonable formal arguments do not reflect the true dynamics of the original equations.

Remark 3: Like in the autonomous case [23] the approximation theorem as stated above can be improved in a number of directions. However, the proof of an optimized result would be very technical and beyond the scope of this paper.

Remark 4: In the autonomous case approximation and attractivity results have been established by a number of authors, cf. [3, 8, 18, 19, 21, 26, 29] for model problems, but also for the general situation including the Navier-Stokes equation. Nowadays the theory is a well established mathematical tool which can be used to prove stability results [25, 28], upper semi-continuity of attractors [15, 24] and global existence results [20, 23].

Remark 5: To prove Theorem 1 we cannot directly use energy estimates because they do not work on the long timescale of order $\mathcal{O}(1/\varepsilon^2)$ due to quadratic terms in (2). We have to use mode filters as in the autonomous case [18, 19] to separate the error function into critical and stable modes.

Surprisingly there are two unexpected points.

Remark 6: First, in contrast to the existing literature by the modified ansatz (4) we are able to remove the assumption of a highly oscillating external electric field, i.e. we gain the possibility of deriving the Ginzburg-Landau equation for arbitrary frequencies.

Remark 7: Secondly, as a consequence of our approach in contrast to the autonomous case the solutions of the Ginzburg-Landau equation have to be analytic. However, this is no serious restriction, since this is true for every $t > 0$ by the smoothing properties of the Ginzburg-Landau equation [26], but it has to be assumed for $t = 0$.

Remark 8: The electro-convection problem in liquid crystals possesses two unbounded directions. Due to the anisotropy of the problem the instability takes place at two non zero wave vectors $\pm k_c \in \mathbb{R}^2$. The amplitude equation is then given by a Ginzburg-Landau equation [1]

$$\partial_T A = c_1 A + c_{2,X}\partial_X^2 A + c_{2,Y}\partial_Y^2 A - c_3 A|A|^2,$$

with solutions $A = A(X, Y, T) \in \mathbb{C}$ and coefficients $c_{2,X}, c_{2,Y} \in \mathbb{C}$. Hence the restriction to one unbounded direction is no restriction with respect to our purposes. For the classical isotropic Bénard convection instability occurs at a ring of wave vectors $k \in \mathbb{R}^2$ satisfying $|k| = k_c \in \mathbb{R}$. Hence in two unbounded directions in the isotropic case no longer a Ginzburg-Landau equation occurs.

Remark 9: For non small values of ε, i.e. away from the bifurcation point other amplitude equations take the role of the Ginzburg-Landau equation. In general the locally preferred patterns do not fit together globally, and so there will be some phase shifts in the pattern which will be transported or transformed by dispersion and diffusion. For the description of the evolution of the local wavenumber q of these pattern phase diffusion equations, conservation laws and Burgers equation can be derived. Recently, approximation results in the above sense have been proved in [7, 13, 14]. However, there are a number of restrictions. The estimates only hold locally in space and there is a global phase shift which cannot be estimated to be small on the time scale under consideration, i.e. only the form of the solution, but not its position can be approximated by these amplitude equations. These restrictions do not apply for patterns which are perfect for $x \to \pm\infty$.

The rest of the paper contains the proof of Theorem 1. It goes along the lines of the autonomous case and reviews the basic concepts.

3. Some Preparations

The solution $t \mapsto v(\cdot, t)$ of (2) defines a curve in an infinite-dimensional phase space. This means that $x \mapsto v(x,t)$ for fixed t lies in a suitable function space. In order to prove Theorem 1 we have to compare the distance between the curve $t \mapsto v(\cdot, t)$ and the associated approximation $t \mapsto \psi_A(\cdot, t, \varepsilon)$ for each fixed t in the norm of the phase space.

Sobolev spaces H^m equipped with the norm

$$\|u\|_{H^m} = \sum_{j=0}^{m} \|\partial_x^j u\|_{L^2}, \qquad \text{with} \qquad \|u\|_{L^2}^2 = \int_{\mathbb{R}} |u(x)|^2 dx,$$

which turned out to be a good choice for the handling of partial differential equations on finite domains are too small for our purposes. Fundamental solutions like constant functions, spatially periodic functions, and fronts are not contained in H^m. It turned out that it is advantageous [15, 19] to work with the $H^m_{l,u}$-space of uniform local Sobolev functions equipped with the norm

$$\|u\|_{H^m_{l,u}} = \sup_{x \in \mathbb{R}} \|u\|_{H^m(x,x+1)}$$

satisfying

$$\|T_y u - u\|_{H^m_{l,u}} \to 0 \quad \text{for} \quad y \to 0,$$

where $(T_y u)(x) = u(x+y)$. This space contains the missing functions and easily allows to use Fourier transform. In Fourier space linear differential operators and fundamental solutions of linear partial differential equations are multiplication operators. In order to control these operators in the $H^m_{l,u}$-spaces we use the following multiplier theorem.

Lemma 10: *Let W_1, W_2 be some Hilbert spaces, $m \in \mathbb{Z}$ and $k \mapsto (1+k^2)^{m/2}\hat{M}(k) \in C_b^2(\mathbb{R}, L(W_1, W_2))$. Then the linear operator $M_{l,u}$: $H^q_{l,u}(W_1) \mapsto H^{q+m}_{l,u}(W_2)$ is bounded for all $q \in \mathbb{N}_0$ with $q+m \geq 0$ by*

$$c(q,m)\|(1+|\cdot|^2)^{m/2}\hat{M}\|_{C_b^2(\mathbb{R},L(W_1,W_2))} ,$$

where $c(q,m)$ is independent of $\hat{M}$.

Proof. See page 441/442 of [19]. □

We mainly use multiplier theory to separate the critical and non critical modes by so called mode filters.

Fix $\delta > 0$ smaller than $k_c/8$ independent of $0 < \varepsilon \ll 1$. Let χ_c be a C_0^∞ cut off function with values in $[0,1]$ and

$$\chi_c(k) = \begin{cases} 1, & \text{for } k \in I_c = [-k_c - \delta\,, -k_c + \delta] \bigcup [k_c - \delta\,, k_c + \delta], \\ 0, & \text{for } k \in \mathbb{R} \backslash ([-k_c - 2\delta\,, -k_c + 2\delta] \bigcup [k_c - 2\delta\,, k_c + 2\delta])\,. \end{cases}$$

Then we define the mode filter for the critical modes by

$$E_c v = \mathcal{F}^{-1}\chi_c \mathcal{F} v\,.$$

The mode filter for the stable modes is defined by

$$E_s = 1 - E_c\,.$$

According to the fact that E_c and E_s are not projections we define auxiliary mode filters E_c^h and E_s^h satisfying $E_c^h E_c = E_c$ and $E_s^h E_s = E_s$ by

$$E_c^h v = \mathcal{F}^{-1}\chi_c^h \mathcal{F} v,$$

where χ_c^h is a C_0^∞-cut off function with values in $[0,1]$ and

$$\chi_c^h(k) = \begin{cases} 1, & \text{for } k \in I_c = [-k_c - 2\delta\,, -k_c + 2\delta] \bigcup [k_c - 2\delta\,, k_c + 2\delta], \\ 0, & \text{for } k \in \mathbb{R} \backslash ([-k_c - 3\delta\,, -k_c + 3\delta] \bigcup [k_c - 3\delta\,, k_c + 3\delta]) \end{cases}$$

and by

$$E_s^h v = \mathcal{F}^{-1}(1 - \chi_s^h)\mathcal{F} v,$$

where χ_s^h is a C_0^∞-cut off function with values in $[0,1]$ and

$$\chi_s^h(k) = \begin{cases} 1, & \text{for } k \in I_c = [-k_c - \delta/2\,, -k_c + \delta/2] \bigcup [k_c - \delta/2\,, k_c + \delta/2], \\ 0, & \text{for } k \in \mathbb{R} \backslash ([-k_c - \delta\,, -k_c + \delta] \bigcup [k_c - \delta\,, k_c + \delta])\,. \end{cases}$$

We will use that the critical part of the quadratic interaction of critical modes vanishes identically due to disjoint supports in Fourier space, i.e.

$$E_c((E_c u) \cdot (E_c v)) = 0\,. \tag{6}$$

With Lemma 10 we conclude

Lemma 11: *The operators E_c and E_c^h are linear bounded mappings from $L^2_{l,u}$ into $H^m_{l,u}$, i.e. for all $m \geq 0$ there exist $C_m \geq 0$, such that*

$$\|E_c u\|_{H^m_{l,u}} + \|E_c^h u\|_{H^m_{l,u}} \leq C_m \|u\|_{L^2_{l,u}} .$$

Proof: Using Lemma 10 shows

$$\begin{aligned}\|E_c u\|_{H^m_{l,u}} &\leq C\|(1+|\cdot|^2)^{m/2}\hat{E}_c(\cdot)\|_{C_b^2}\|u\|_{L^2_{l,u}} \\ &\leq C\|u\|_{L^2_{l,u}}\end{aligned}$$

using the compact support of E_c. □

As a direct consequence we have

Lemma 12: *The operators E_s and E_s^h are bounded mappings in $H^m_{l,u}$, i.e. for all $m \geq 0$ there exist $C_m \geq 0$ such that*

$$\|E_s u\|_{H^m_{l,u}} + \|E_s^h u\|_{H^m_{l,u}} \leq C_m \|u\|_{H^m_{l,u}} .$$

Our model (2) and the associated Ginzburg-Landau equation (5) are semilinear equations, i.e. the local existence and uniqueness of the solutions follows by a standard fixed point argument with the help of the variation of constant formula [9]. Thus, for given $A|_{T=0} \in H^m_{l,u}$ there exists a $T_1 > 0$ and a solution $A \in C([0,T_1], H^m_{l,u})$ of the Ginzburg-Landau equation (5) for all $m \geq 1$. Moreover, for given $v|_{t=0} \in H^m_{l,u}$ there exists a $t_1 > 0$ and a solution $v \in C([0,t_1], H^m_{l,u})$ of the model (2) for all $m \geq 4$. As a consequence the solutions of the Ginzburg-Landau equation (5) exist as a long as they can be bounded in $H^m_{l,u}$ for a $m \geq 1$ and the solutions of the model (2) exist as a long as they can be bounded in $H^m_{l,u}$ for a $m \geq 4$. By the smoothing properties of the linearized system the solutions are analytic for every $T > 0$ and $t > 0$, respectively.

Finally, we introduce the space

$$C_a^\omega = \{u : S_a \mapsto \mathbb{C} \mid u \text{ analytic in } S_a\},$$

where $S_a = \{z \in \mathbb{C} \mid |\mathrm{Im} z| < a\}$ equipped with the norm

$$\|u\|_{C_a^\omega} = \sup_{z \in S_a} |u(z)| < \infty.$$

4. Derivation of the Ginzburg-Landau Equation

The so called residual

$$\begin{aligned}\operatorname{Res}(v) = & -\partial_t v - v - \gamma\partial_x^8 v - \overline{c}(4\partial_x^2 + \partial_x^4)v \\ & -v^2(4\partial_x^2 + \partial_x^4)v + v\partial_x v \\ & -c(t)(4\partial_x^2 + \partial_x^4)v + u_0(t)\partial_x v \\ & -2u_0(t)v(2\partial_x^2 + \partial_x^4)v\end{aligned}$$

contains all terms which do not cancel after inserting the approximation into the model (2). Hence, $\operatorname{Res}(v) = 0$ if and only if v is a solution of (2). In this section we mainly estimate the residual for the approximation (4). We recall $X = \varepsilon(x + \varphi_1(t))$ and $T = \varepsilon^2 t$ and refine the ansatz to

$$\begin{aligned}\psi_A(x,t,\varepsilon) = & \ \varepsilon A_1(X,T)e^{ik_c x + \varphi_0(t)} + c.c. \\ & + \frac{\varepsilon^2}{2}A_0(X,T) + \varepsilon^2 A_2(X,T)e^{2(ik_c x + \varphi_0(t))} + c.c. \ .\end{aligned}$$

Equating the coefficient in front of $\varepsilon e^{ik_c x}$ to zero shows

$$\varphi_0(t) = (4k_c^2 - k_c^4)\int_0^t c(\tau)d\tau + ik_c\int_0^t u_0(\tau)d\tau.$$

The integrals stay bounded due to the vanishing mean values of u_0 and c. Equating the coefficient in front of $\varepsilon^2 e^{ik_c x}$ to zero shows

$$\varphi_1(t) = (4k_c^3 - 8k_c)\, i\int_0^t c(\tau)d\tau + \int_0^t u_0(\tau)d\tau.$$

Again the integrals stay bounded due to the vanishing mean values of u_0 and c. Especially, we have the $\mathcal{O}(1)$-boundedness of the imaginary part of φ_1.

Remark 13: The solutions of the Ginzburg-Landau-equation are analytic in a strip of width δ_{GL}. Since $|\operatorname{Im}(X)| \leq \mathcal{O}(\varepsilon) \ll \delta_{GL}$ the solution can be evaluated at the position $X = \varepsilon(X + \varphi_1(t))$.

Equating the coefficient in front of $\varepsilon^3 e^{ik_c x}$ to zero gives an equation of the form

$$\begin{aligned}\partial_T A_1 = & \ d_1(\frac{T}{\varepsilon^2})A_1 + d_2(\frac{T}{\varepsilon^2})\partial_X^2 A_1 + d_3(\frac{T}{\varepsilon^2})A_1 A_0 \\ & + d_4(\frac{T}{\varepsilon^2})A_2 A_{-1} + d_5(\frac{T}{\varepsilon^2})A_1 A_1 A_{-1},\end{aligned}$$

with time-dependent coefficients $d_j = d_j(\frac{T}{\varepsilon^2})$. In order to get an equation for A_0 we equate the coefficient in front of $\varepsilon^2 e^{0k_c x}$ to zero. We obtain an equation of the form

$$d_6(\frac{T}{\varepsilon^2})A_0 = d_7(\frac{T}{\varepsilon^2})A_{-1}A_1,$$

with

$$d_6(\frac{T}{\varepsilon^2}) = \lambda(0, \bar{c}_{\text{crit}}) = -1 \neq 0.$$

In order to get on equation for A_2 we equate the coefficient in front of $\varepsilon^2 e^{2ik_c x}$ to zero. We obtain an equation of the form

$$d_8(\frac{T}{\varepsilon^2})A_2 = d_9(\frac{T}{\varepsilon^2})A_1 A_1,$$

with

$$d_6(\frac{T}{\varepsilon^2}) = \lambda(2k_c) - c(t)(-16k_c^2 + 16k_c^4) + 2u_0(t)ik_c \neq 0,$$

which can be computed explicitly. Eliminating A_0 and A_2 in the equation of A_1 shows that A_1 satisfies an equation of the form

$$\partial_T A_1 = d_1(\frac{T}{\varepsilon^2})A_1 + d_2(\frac{T}{\varepsilon^2})\partial_x^2 A_1 + d_{10}(\frac{T}{\varepsilon^2})A_1|A_1|^2. \tag{7}$$

Then for a given solution A of (7), we can construct A_0 and A_2 such that the residual for the critical modes is of order $\mathcal{O}(\varepsilon^4)$ and for the stable modes of order $\mathcal{O}(\varepsilon^3)$. This can be made rigorous with the help of Lemma 10, cf. [19].

Notation. If the approximation is constructed via the solutions of the autonomous Ginzburg-Landau equation (5) it will be denoted with $\varepsilon\psi_A$ in the following. If the approximation is constructed via the solutions of the non autonomous Ginzburg-Landau equation (7) it will be denoted with $\varepsilon\psi_B$ in the following.

Lemma 14: *Let $\delta_1 > 0$ and $C_1 > 0$. For all $\varepsilon \in (0,1)$ let $A_1 = A_1(X,T) \in C([0,T_0], C^\omega_{\delta_1})$ be a solution of (7) with $\sup_{T\in[0,T_0]} \|A_1(\cdot,T)\|_{C^\omega_{\delta_1}} \leq C_1$. Then we have a $C_2 > 0$ such that for all $\varepsilon \in (0,1)$ we have*

$$\sup_{t\in[0,T_0/\varepsilon^2]} \|E_s(\text{Res}(\psi_B))\|_{H^m_{l,u}} = \mathcal{O}(\varepsilon^3),$$

$$\sup_{t\in[0,T_0/\varepsilon^2]} \|E_c(\text{Res}(\psi_B))\|_{H^m_{l,u}} = \mathcal{O}(\varepsilon^4).$$

5. The Non Autonomous Case

As a major step of the proof of Theorem 1 we show here that the solutions of (2) can be approximated via the solutions of the non autonomous Ginzburg-Landau equation (7).

Theorem 15: Let $\delta_1 > 0$ and $C_1 > 0$. For all $\varepsilon \in (0,1)$ let $A_1 = A_1(X,T) \in C([0,T_0], C^{\omega}_{\delta_1})$ be a solution of (7) with $\sup_{T\in[0,T_0]} \|A_1(\cdot,T)\|_{C^{\omega}_{\delta_1}} \leq C_1$. Then there are $\varepsilon_0 > 0$ and $C > 0$ such that for all $\varepsilon \in (0,\varepsilon_0)$ we have solutions v of (2) satisfying

$$\sup_{t\in[0,T_0/\varepsilon^2]} \|v(\cdot,t) - \psi_B(\cdot,t,\varepsilon)\|_{H^m_{l,u}} \leq C\varepsilon^2 .$$

Proof. We write (2) as

$$\partial_t v = \Lambda v + B(v,v) + C(v,v,v), \tag{8}$$

with

$$\begin{aligned} \Lambda v &= -v - \gamma\partial_x^8 v - \overline{c}(4\partial_x^2 + \partial_x^4)v - c(t)(4\partial_x^2 + \partial_x^4)v + u_0(t)\partial_x v, \\ B(v,v) &= v\partial_x v - 2u_0(t)v(4\partial_x^2 + \partial_x^4)v, \\ C(v,v,v) &= -v^2(4\partial_x^2 + \partial_x^4)v. \end{aligned}$$

Inserting

$$v = \varepsilon\psi_c + \varepsilon^2\psi_s + \varepsilon^2 R_c + \varepsilon^3 R_s \ ,$$

with $R_c = E_c^h R_c$, $R_s = E_s^h R_s$, $\psi_c = E_c^h\psi_c$, and $\psi_s = E_c^h\psi_s$ gives, by using (6),

$$\begin{aligned} \partial_t R_c &= \Lambda R_c + \varepsilon^2 L_c(R) + \varepsilon^3 N_c(R) + \varepsilon^2 \mathrm{Res}_c \ , \\ \partial_t R_s &= \Lambda R_s + L_s(R_c) + \varepsilon N_s(R) + \mathrm{Res}_s \ , \end{aligned} \tag{9}$$

where

$$\begin{aligned} \mathrm{Res}_c &= \varepsilon^{-4} E_c(\mathrm{Res}(\psi_B)) \ , \\ \mathrm{Res}_s &= \varepsilon^{-3} E_s(\mathrm{Res}(\psi_B)) \ , \\ L_c(R) &= 2E_c(B(R_s,\psi_c) + B(R_c,\psi_s)) \ , \\ L_s(R_c) &= 2E_s B(R_c,\psi_c) \ , \end{aligned}$$

and where $N_c(R)$ and $N_s(R)$ satisfy

$$\begin{aligned} \|N_c(R)\|_{H^m_{l,u}} &\leq C(D_c,D_s)(\|R_c\|_{H^m_{l,u}} + \|R_s\|_{H^m_{l,u}})^2, \\ \|N_s(R)\|_{H^{m-4}_{l,u}} &\leq C\|R_s\|_{H^m_{l,u}} + C(D_c,D_s)(\|R_c\|_{H^m_{l,u}} + \|R_s\|_{H^m_{l,u}})^2 \end{aligned}$$

as long as

$$\|R_c\|_{H^m_{l,u}} \leq D_c \quad \text{and} \quad \|R_s\|_{H^m_{l,u}} \leq D_s \,, \tag{10}$$

where $C(D_c, D_s)$ is a constant depending on D_c and D_s independent of $0 \leq \varepsilon \ll 1$. The constants D_c and D_s will be chosen later on independent of $0 \leq \varepsilon \leq 1$. This system is solved with initial datum $(R_c(0), R_s(0)) = (0,0)$. The solution of

$$\partial_t R = \Lambda R, \quad R|_{t=\tau} = R_0$$

is denoted with $R(t) = \mathcal{K}(t,\tau)R_0$ which defines a linear evolution operator $\mathcal{K}(t,\tau)$ satisfying $\mathcal{K}(t,\tau) = \mathcal{K}(t+\frac{2\pi}{\omega}, \tau+\frac{2\pi}{\omega})$. In Fourier space we have

$$\partial_t \hat{R}(k) = \Lambda(k)R(k) = \lambda(k)\hat{R}(k) - d_1(k)c(t)\hat{R}(k) + d_2(k)u_0(t)\hat{R}(k),$$

with constants $d_j = d_j(k)$ which is solved by

$$\begin{aligned}\hat{v}(k,t) &= e^{\int_0^t (\lambda(k)+d_1(k)c(\tau)+d_2(k)u_0(\tau))\, d\tau}\hat{v}(k,0) \\ &= e^{\lambda(k)t}e^{d_1(k)\int_0^t c(\tau)d\tau}e^{d_2(k)\int_0^t u_0(\tau)d\tau}\hat{v}(k,0)\end{aligned}$$

leading to the Floquet multipliers $e^{\lambda(k)\frac{2\pi}{\omega}}$ due to the vanishing mean values of c and u_0. As a direct consequence it follows [19]:

Lemma 16: *There exist C, $\sigma > 0$ independent of $0 < \varepsilon \ll 1$ such that we have for the stable part*

$$\|\mathcal{K}(t,\tau)E_s^h\|_{L(H^m_{l,u}, H^m_{l,u})} \leq Ce^{-\sigma(t-\tau)}$$

and

$$\|\mathcal{K}(t,\tau)E_s^h\|_{L(H^{m-4}_{l,u}, H^m_{l,u})} \leq C\max(1,(t-\tau)^{-1/2})e^{-\sigma(t-\tau)}.$$

Moreover, we have for the critical part

$$\|\mathcal{K}(t,\tau)E_c^h\|_{L(H^m_{l,u}, H^m_{l,u})} \leq Ce^{C\varepsilon^2(t-\tau)}.$$

We apply the variation of constant formula to (10) and obtain

$$\begin{aligned}R_c(t) &= \int_0^t \mathcal{K}(t,\tau)E_c^h(\varepsilon^2 L_c(R) + \varepsilon^3 N_c(R) + \varepsilon^2\mathrm{Res}_c)(\tau)d\tau \,, \\ R_s(t) &= \int_0^t \mathcal{K}(t,\tau)E_s^h(L_s(R_c) + \varepsilon N_s(R) + \mathrm{Res}_s)(\tau)d\tau \,.\end{aligned}$$

With $S_i(s) := \sup_{0\leq t\leq s}\|R_i(t)\|_{H^m_{l,u}}$, $(i = s,c)$ we find, by using (9) and

$$\Big(\int_0^t C\max(1,\tau^{-1/2})e^{-\sigma\tau}d\tau\Big) = \mathcal{O}(1)$$

independent of $t > 0$, that

$$\begin{aligned} S_s(t) &\le CS_c(t) + \varepsilon(CS_s(t) + C_s(D_c, D_s)(S_c(t) + S_s(t))^2) + C_{Res}, \\ &\le CS_c(t) + 1 + C_{Res}, \end{aligned}$$

if

$$\varepsilon(CD_s + C_s(D_c, D_s)(D_c + D_s)^2) \le 1. \tag{11}$$

Similarly, we find

$$\begin{aligned} S_c(t) &\le \varepsilon^2 \int_0^t (C(S_c(\tau) + S_s(\tau)) + \varepsilon C_s(D_c, D_s)(S_c(\tau) + S_s(\tau))^2 + C_{Res}) d\tau, \\ &\le \varepsilon^2 \int_0^t (C(S_c(\tau) + S_s(\tau)) + 1 + C_{Res}) d\tau, \end{aligned}$$

if

$$\varepsilon C_s(D_c, D_s)(D_c + D_s)^2 \le 1. \tag{12}$$

Using the above estimate for $S_s(t)$ finally shows

$$S_c(t) \le \varepsilon^2 \int_0^t C(S_c(\tau) + 1 + C_{Res}) d\tau.$$

Gronwall's inequality yields

$$S_c(t) \le C(1 + C_{Res}) T_0 e^{CT_0} =: D_c$$

for all $t \in [0, T_0/\varepsilon^2]$. Then by the estimate for $S_s(t)$

$$S_s(t) \le CD_c + 1 + C_{Res} =: D_s.$$

We are done, if we choose $\varepsilon_0 > 0$ so small that for all $\varepsilon \in (0, \varepsilon_0)$ the conditions (11) and (12) are satisfied. □

6. Comparison of the Ginzburg-Landau Equations

Obviously the Ginzburg-Landau equation (7) is useless for constructing approximations for the solutions of (2) since it still contains the small parameter ε in a singular way. We expect that for $\varepsilon \to 0$ only the average of the d_j will play a role and that it is sufficient to consider an autonomous Ginzburg-Landau equation. Therefore, it is the purpose of this section to compare the solutions of the autonomous Ginzburg-Landau-equation

$$\partial_T A = c_1 \partial_X^2 A + c_2 A + c_3 A|A|^2, \tag{13}$$

with the solutions of the non-autonomous Ginzburg-Landau-equation

$$\partial_T B = (c_1 + c_4(\frac{T}{\varepsilon^2}))\partial_X^2 B + (c_2 + c_5(\frac{T}{\varepsilon^2}))B + (c_3 + c_6(\frac{T}{\varepsilon^2}))B|B|^2, \quad (14)$$

with highly oscillating coefficients c_j for $j = 4, 5, 6$ satisfying

$$c_j(t) = c_j(t + 2\pi/\omega) \qquad \text{and} \qquad \int_0^{2\pi/\omega} c_j(t)dt = 0.$$

We prove

Theorem 17: Let $\delta > 0$ and let $A \in C([0, T_0], C_{2\delta}^{\omega})$ be a solution of (13). Then there exist $C, \varepsilon_0 > 0$ such that for all $\varepsilon \in (0, \varepsilon_0)$ equation (14) possesses solutions B satisfying

$$\sup_{T\in[0,T_0]} \|A(\cdot, T) - B(\cdot, T)\|_{C_\delta^\omega} \leq C\varepsilon^2 \, .$$

Proof. We introduce the error function R by $B = A + \varepsilon^2 R$ which satisfies

$$\begin{aligned}\partial_T R = &\Lambda R + 3C(A, A, R) + 3\varepsilon^2 C(A, R, R)\\ &+\varepsilon^4 C(R, R, R) + \varepsilon^{-2}\text{Res}(A) \, ,\end{aligned}$$

where

$$\begin{aligned}\Lambda R &= (c_1 + c_4(\frac{T}{\varepsilon^2}))\partial_X^2 R + (c_2 + c_5(\frac{T}{\varepsilon^2}))R \, ,\\ 3C(R_1, R_2, R_3) &= (c_3 + c_6(\frac{T}{\varepsilon^2}))(R_1 R_2 \bar{R}_3 + R_1 \bar{R}_2 R_3 + \bar{R}_1 R_2 R_3) \, ,\\ \text{Res}(A) &= c_4(\frac{T}{\varepsilon^2})\partial_X^2 A + c_5(\frac{T}{\varepsilon^2})A + c_6(\frac{T}{\varepsilon^2})A|A|^2 \, .\end{aligned}$$

We define the linear evolution operator $S(T, \tau)$ by $R(T) = S(T, \tau)R(\tau)$, where $R(t)$ solves

$$\partial_T R = \Lambda R \quad , \quad R|_{T=\tau} = R(\tau) \, .$$

Then, we consider

$$\begin{aligned}R(T) = \int_0^T &S(T, \tau)(3C(A, A, R) + 3\varepsilon^2 C(A, R, R)\\ &+ \varepsilon^4 C(R, R, R) + \varepsilon^{-2}\text{Res}(A))(\tau)d\tau \, .\end{aligned}$$

We estimate

$$\sup_{T\in[0,T_0]} \left\| \int_0^T S(T, \tau)\varepsilon^{-2}\text{Res}(A)(\tau)d\tau \right\|_{C_\delta^\omega} \leq C_{Res},$$

with $C_{Res} > 0$ a constant independent of $0 < \varepsilon \ll 1$. This follows for instance from

$$\begin{aligned}
&\| \int_0^T S(T,\tau) c_4(\tfrac{\tau}{\varepsilon^2}) \partial_X^2 A(\tau) d\tau \|_{C_\delta^\omega} \\
&\leq \| \varepsilon^2 S(T,\tau) \tilde{c}_4(\tfrac{\tau}{\varepsilon^2}) \partial_X^2 A(\tau)|_{\tau=0}^T \|_{C_\delta^\omega} \\
&\quad + \| \int_0^T \varepsilon^2 S(T,\tau) \tilde{c}_4(\tfrac{\tau}{\varepsilon^2}) \partial_X^2 \partial_\tau A d\tau \|_{C_\delta^\omega} \\
&\quad + \| \int_0^T \varepsilon^2 (-S(T,\tau)\Lambda(\tau)) \tilde{c}_4(\tfrac{\tau}{\varepsilon^2}) \partial_X^2 A \, d\tau \|_{C_\delta^\omega} \\
&\leq C\varepsilon^2 ,
\end{aligned}$$

where $\partial_\tau \tilde{c}_4 = c_4$ and $\partial_\tau S(T,\tau) = -S(T,\tau)\Lambda(\tau)$. The function $\tilde{c}_4$ stays $\mathcal{O}(1)$-bounded due to the vanishing mean value of c_4. The rest of the theorem then follows by a simple application of Gronwall's inequality. □

Remark 18: It is easy to see that for every $0 \leq \delta' < \delta$ we have $\|\partial_T B\|_{C_{\delta'}^\omega} = \mathcal{O}(1)$ by expressing $\partial_T B$ by the right hand side of (14), but $\partial_T^2 B = \mathcal{O}(\varepsilon^{-2})$.

7. The Final Step

It remains to conclude Theorem 1 from Theorem 15 and Theorem 17. Let ψ_A be the approximation constructed via the solution A of the autonomous Ginzburg-Landau equation (5). Let ψ_B be the approximation constructed via the solution $B = A_1$ of the non-autonomous Ginzburg-Landau equation (7). Moreover, let v be the solution of the model (2). From Theorem 15 we have

$$\sup_{t\in[0,T_0/\varepsilon^2]} \|\psi_B(\cdot,t,\varepsilon) - v(\cdot,t)\|_{H_{l,u}^m} = \mathcal{O}(\varepsilon^2).$$

From Theorem 17 and $C_\delta^\omega \subset H_{l,u}^m$ we have

$$\sup_{T\in[0,T_0]} \|A(\cdot,T) - B(\cdot,T)\|_{H_{l,u}^m} = \mathcal{O}(\varepsilon^2)$$

which implies

$$\sup_{t\in[0,T_0/\varepsilon^2]} \|\psi_B(\cdot,t,\varepsilon) - \psi_A(\cdot,t,\varepsilon)\|_{H_{l,u}^m} = \mathcal{O}(\varepsilon^2).$$

Hence, by the triangle inequality and Sobolev's embedding theorem we have

$$\begin{aligned}
&\sup_{t\in[0,T_0/\varepsilon^2]} \sup_{x\in\mathbb{R}} |\psi_A(x,t,\varepsilon) - v(x,t)| \\
&\leq C \sup_{t\in[0,T_0/\varepsilon^2]} \|\psi_A(\cdot,t,\varepsilon) - v(\cdot,t)\|_{H^m_{l,u}} \\
&\leq C(\sup_{t\in[0,T_0/\varepsilon^2]} \|\psi_A(\cdot,t,\varepsilon) - \psi_B(\cdot,t,\varepsilon)\|_{H^m_{l,u}} \\
&\qquad + \sup_{t\in[0,T_0/\varepsilon^2]} \|\psi_B(\cdot,t,\varepsilon) - v(\cdot,t)\|_{H^m_{l,u}}) \\
&= \mathcal{O}(\varepsilon^2).
\end{aligned}$$

Therefore, we are done. □

Acknowledgments

The paper is partially supported by the Deutsche Forschungsgemeinschaft DFG under the grant Kr 690/18-1/2. The authors are grateful for discussions with Gerhard Dangelmayr and Lorenz Kramer.

References

[1] I. Aranson, L. Kramer, *Rev. Modern Phys.* **74**, no. 1, 99-143 (2002).
[2] P. Bollerman, G. Schneider, A. van Harten, *In Nonlinear Dynamics and Pattern Formation in Natural Environment* (eds: A. Doelman and A. van Harten) Pitman Research Notes **335**, Longman, 20-36 (1995).
[3] P. Collet, J.-P. Eckmann, *Comm. Math. Phys.* **132**, 139-153 (1990).
[4] S. Chandrashekar, *Liquid Crystals.* University Press, Cambridge, 1977.
[5] M.C. Cross, P.L. Hohenberg,*Rev. Mod. Phys.* **65**, 851-1090 (1993).
[6] G. Dangelmayr, I. Oprea, *Mollec. Cryst. Liq. Cryst.*, **413**, 305 (2004).
[7] A. Doelman, B. Sandstede, G. Schneider, A. Scheel, The dynamics of modulated wave trains. In preparation.
[8] W. Eckhaus, *J. Nonlinear Science* **3**, 329-348 (1993).
[9] D. Henry, *Geometric Theory of Semilinear Parabolic Equations*, LNM **840**, Springer, Berlin-New York 1981.
[10] L. Kramer, W. Pesch, *Annual review of fluid mechanics*, **27**, 515-541, Annual Reviews, Palo Alto, CA, 1995.
[11] I. Melbourne, *J. Nonlinear Sci.* **8**, 1-15 (1998).
[12] I. Melbourne, *Trans. Amer. Math. Soc.* **351**, 1575-1603 (1999).
[13] I. Melbourne, G. Schneider, *J. Differential Equations* **199**, no. 1, 22-46 (2004).
[14] I. Melbourne, G. Schneider, *Math. Nachr.* **263/264**, 171-180 (2004).
[15] A. Mielke, G. Schneider, *Nonlinearity* **8**, 743-768 (1995).
[16] A. Mielke, The Ginzburg-Landau equation in its role as a modulation equation. In: *Handbook of Dynamical Systems II* (B. Fiedler, ed.). North-Holland, Amsterdam, 759-834 (2002).

[17] W. Pesch, U. Behn. Electrohydrodynamic Convection in Nematics in *Evolution of Spontaneous Structures in Dissipative Continuous Systems.* F. H. Busse and S. Ç. Mueller, eds., Springer, 1998.
[18] G. Schneider, *Journal of Nonlinear Science* **4**, 23-34 (1994).
[19] G. Schneider, *J. Appl. Math. Physics (ZAMP)* **45**, 433-457 (1994).
[20] G. Schneider, *Comm. Math. Phys.* **164**, 157-179 (1994).
[21] G. Schneider, *J. Diff. Eqns.* **121**, 233-257 (1995).
[22] G. Schneider, *Mathematische Nachrichten* **176**, 249-263 (1995).
[23] G. Schneider, *Journal de Mathématiques Pures et Appliquées* **78**, 265-312 (1999).
[24] G. Schneider, *Integral and Differential Equations* **12**, 913-926 (1999).
[25] G. Schneider, H. Uecker, *Z. Angew. Math. Phys.* **54**, no. 4, 677-712 (2003).
[26] P. Takáč, P. Bollerman, A. Doelman, A. van Harten and E.S. Titi, *SIAM J. Math. Anal.* **27**, 424-448 (1996).
[27] M. Treiber, L. Kramer, *Phys. Rev.* ***E*** **58**, 1973 (1998).
[28] H. Uecker, *J. Nonlinear Sci.* **11**, no. 2, 89-121 (2001).
[29] A. van Harten, *J. Nonlinear Science* **1**, 397-422 (1991).
[30] H. Zhao, L. Kramer, *Phys. Rev.* ***E*** **62**, 5092 (2000).

CHAPTER 4

STABILITY AND BIFURCATION FROM RELATIVE EQUILIBRIA AND RELATIVE PERIODIC ORBITS

Pascal Chossat

I.N.L.N. (CNRS) Sophia Antipolis, 06560 Valbonne, FRANCE
and
Embassy of France in New Delhi

Continuous symmetries in dynamical systems are responsible for non-generic regimes which are frequently observed in experiments, like rotating and travelling waves in fluids or meandering spiral waves in spatially extended systems of reaction-diffusion type. These behaviors can only be resolved if the symmetries are incorporated in the analysis of the dynamics. This problem has been extensively studied in the past 15 years, in view of the frequent occurrence of continuous symmetries in mathematical models of evolution systems in Nature. The aim of this paper is to introduce the subject in a comprehensive and elementary way.

1. Introduction

A symmetry for a dynamical system or more generally for an equation defined in some vector space V is an invertible map of V into itself which does not modify the form of the equation. For example the laws of Newton's mechanics are invariant under the isometries (transformations which do not modify distances) of the usual Euclidean space: translations, reflections, rotations. When combining such transformations, we get Euclidean transformations again. The set of all these transformations is called the group of Euclidean symmetries which we denote by $E(3)$. In an n-dimensional Euclidean space, the group of all isometries is $E(n)$. It is generated by the group of $n \times n$ orthogonal matrices $O(n)$ and by the group of translations $\mathbb{R}^n$. We write $SO(n)$ for the subgroup of $O(n)$ of matrices with determinant equal to $+1$ (pure rotations). The largest group of symmetries of classical laws of mechanics is the group of Galilean transformations which incorporates the invariance under time translations. In other words these laws are autonomous with respect to time. In this paper we shall mainly be interested with Euclidean symmetries of autonomous systems, although certain results would apply to more general groups.

Specific systems which are governed by Newton's mechanics do not in general possess full Galilean symmetry, because they are subjected to constraints. For example a pendulum is anchored at a ceiling and therefore its equation of motion cannot be translation invariant. In addition it is subjected to gravity, which implies that its equation cannot be invariant under all rotations in space. It may however be invariant under the rotations around the vertical axis passing through the point of fixation of the pendulum at the ceiling, if no additional constraint is imposed to the system. These rotations form a subgroup of $E(3)$, and more precisely of the group $SO(3)$ of rotations in space, which we write, by a slight abuse of notation, $SO(2)$. If θ is the angle of rotation, $SO(2)$ is defined as the set of matrices of the form

$$R_\theta = \begin{pmatrix} \cos\theta & -\sin\theta & 0 \\ \sin\theta & \cos\theta & 0 \\ 0 & 0 & 1 \end{pmatrix} \tag{1}$$

Therefore the group $SO(2)$ is parameterized by the angle θ, which itself is defined modulo $2Q$. We write $\theta \in \mathbb{R}/2Q\mathbb{Z}$ and we usually set $\mathbb{R}/2Q\mathbb{Z} = S^1$. It follows that $SO(2)$ can be endowed with a differentiable manifold structure (it is identified with the oriented circle). Groups which possess this property are called Lie groups. $SO(2)$ is a one dimensional Lie group. Finite groups are 0-dimensional Lie groups, but we don't call them Lie groups because their manifold structure is trivial. If reflections through the vertical planes are also considered, then the symmetry group becomes $O(2)$, which we may identify with the full group of symmetries of the circle. The full symmetry group of the pendulum is therefore $O(2)$. The group $SO(2)$ can also be seen as the connected component of the identity matrix in $O(2)$. The fundamental property of a Lie group G is that its tangent space at the neutral element, which we denote by $\mathfrak{g}$ and which is well defined since the Lie group is a differentiable manifold, contains most of the information about the Lie group structure. In particular given any ξ in $\mathfrak{g}$, $\exp\xi = \sum \frac{1}{n!}\xi^n$ is an element of G. In the case of $SO(2)$ as a subgroup of $SO(3)$, the tangent space is obtained by differentiating R_θ at $\theta = 0$, which gives

$$\xi = \begin{pmatrix} 0 & -1 & 0 \\ 1 & 0 & 0 \\ 0 & 0 & 0 \end{pmatrix} \tag{2}$$

The element ξ is called the infinitesimal generator of the rotations around the vertical axis. We have $R_\theta = \exp(\theta\xi)$. For any Lie group G, the expo-

nential map exists and defines an isomorphism between a neighborhood of the identity in G and a neighborhood of 0 in $\mathfrak{g}$.
Lie groups whose elements are matrices are particularly important for applications. They are called linear Lie groups. The Euclidean groups can always be seen as linear Lie groups (see [13]).

Any rotation belonging to the group $SO(3)$ can be identified with three angles, two of them defining the position in space of an axis and one angle of rotation around that axis. From this one can show that $SO(3)$ is a 3 dimensional Lie group. The tangent space to $SO(3)$ at the identity matrix is therefore a vector space of dimension 3 which is spanned by the infinitesimal generators ξ_j, $j = 1, 2, 3$, of the rotations around the three coordinate axes (the matrix (2) is one of them). The group acts on its own tangent space at the identity by conjugation: this can be easily checked. By differentiating this action, we can see that the infinitesimal generators satisfy certain commutation relations which endow the tangent space at the identity with the structure of a Lie algebra. This is a general property of Lie groups and we will usually call *Lie algebra* the tangent space $\mathfrak{g}$ of a group G at the identity.

Taking into account translations in space, which we can identify with vectors in $\mathbb{R}^3$, we see that $E(3)$ is a Lie group of dimension 6. There is a very important difference between rotations and translations. The former are bounded transformations while the latter are unbounded, meaning that translations can have length as large as we wish, while rotations, being orthogonal matrices, have "length" (norm) equal to 1. In fact, as a topological set, $SO(3)$ is compact, while $E(3)$ is not.

A comprehensive introduction to Lie groups in the perspective of equivariant dynamical systems and bifurcation theory can be found in [4], and a physics oriented, however a rigorous and rather complete exposition of linear Lie groups and their representations can be found in [13].

Let us come back to the pendulum. Only the two states of rest (with the pendulum in vertical position) are invariant under the symmetry $O(2)$. Whenever a motion sets in, the symmetry is broken. We may also consider the invariance of the system with respect to time. The states of rest are invariant under time translations $t_0 \mapsto t_0 + t$, $t \in \mathbb{R}$. Motion breaks this time invariance too. Non stationary solutions can still possess a certain degree of symmetry, which forms a subgroup of the group of spatio-temporal symmetries generated by $O(2)$ and $\mathbb{R}$. For example oscillations in a plane are time periodic solutions which in addition are invariant under the reflection through that plane. Their spatio-temporal symmetries are therefore generated by this reflection and by the multiples of the period (which form a discrete subgroup of $\mathbb{R}$ isomorphic to the group of integer numbers $\mathbb{Z}$). It is well-known that the spherical pendulum has two kinds of periodic solutions which are the planar oscillations and the uniform rotations around the vertical axis with a constant angle of deflection. In the latter case, for an observer moving around the vertical axis at the same (constant) speed as the rotating pendulum, the system looks *stationary*. For that reason, this type of solution is called a *relative equilibrium*(RE). For solutions of the former type, note that any vertical plane passing through the point of fixation of the pendulum is a plane of periodic oscillations. Moreover, if an impetus is given to such a planar oscillation transversally to its plane, then a drift will be superposed to the oscillation and the motion will undergo the well-known figure of oscillations with a precession of constant angle. Motions of this type would look *time periodic* to an observer moving around the vertical axis at a constant speed which is synchronized with the drift. For that reason, this type of solution is called a *relative periodic orbit*(RPO).

RE's and RPO's occur in many systems with continuous symmetries. Fluid mechanics provides a large number of examples, such as travelling water waves (which are RE's). A paradigm to such instabilities in hydrodynamics is the Couette-Taylor problem, see [3]. Physics, chemistry and even life science provide many other examples, I refer to [4] and to [8] for more information on examples and bibliography. Systems of reaction-diffusion equations defined in the plane or in higher dimension provide an important class of examples, in which the symmetry group is unbounded ($E(2)$ or $E(3)$) but the states which are investigated are "localized" in space, meaning that they are not invariant under any subgroup of the group of translations. There has been an extensive analysis of these systems in the past 10 years, the basic difficulty being the non compactness of the symmetry group. Bifurcations from relative equilibria or relative periodic orbits are interesting because they can give rise to rich spatio-temporal structures which are observed in experiments. In the case of isolated equilibria or periodic orbits undergoing a change of stability, hence a bifurcation as a parameter is varied, the usual tools to analyze this singularity and to compute the bifurcated solutions are the Lyapunov-Schmidt decomposition and the center manifold reduction (see [4], [11], [8]). The Lyapunov-Schmidt decomposition allows for the computation of the steady-state or periodic solutions which may arise at the bifurcation point. It is a very simple and effective method, however it does not (in general) provide a description of the changes in the dynamics and of the stability of the branching solutions. In contrast, the center manifold reduction is a procedure which consists in restricting the analysis in phase space to a finite dimensional manifold which contains all the relevant information about the asymptotic dynamics. In particular it contains the bounded solutions which are close to the bifurcation point and since it is (locally) attracting, it also contains the possibly unstable manifolds which may arise from these solutions.

In this paper, I will present a simple method of reduction which permits to lift the degeneracy induced by the continuous part of the symmetry in order to analyze the dynamics near a relative equilibrium or relative periodic orbit. This method is based on an idea which was initially developed by Iooss [10], [3] for the Couette-Taylor problem, then extended to systems with a compact symmetry group in [4]. The geometrical theory for relative equilibria and relative periodic orbits was set up by Krupa [12] and Field [6]. In the case of a non compact symmetry group, the problem is more involved. A rigorous theory of the Lyapunov-Schmidt decomposition has been given by [19] in this case, while a center manifold reduction has been developed by Sandstede, Scheel and Wulff [15], [16]. The method which I present here is also valid for certain non compact group actions. It also allows for a computation of the center manifold even when the group is non compact. In the non compact case, a more geometrical approach has been followed by Golubitsky, Leblanc and Melbourne [7], assuming the existence of the center manifold (see also [5]).

Relative equilibria and relative periodic orbits are examined respectively in Sections 1 and 2. In each of these sections, the first subsection introduces the concept while the second subsection exposes the reduction method.

2. Relative Equilibria

2.1. *Definition and Properties*

We consider an evolution equation in a Banach (often a Hilbert) space $\mathcal{Y}$:

$$\frac{dx}{dt} = \mathcal{F}(x) \tag{3}$$

where $\mathcal{F}$ is a smooth operator with domain $\mathcal{X}$ dense in $\mathcal{Y}$ (by smooth I mean k times continuously differentiable for k sufficiently large). We assume that $\mathcal{F}$ generates a local semi-flow $x \mapsto \Phi_t(x)$. In other words, any initial condition $x_0 \in \mathcal{X}$ generates a solution $\Phi_t(x_0)$ for $0 \le t \le c$, c a positive constant which may depend on the initial condition, and moreover Φ_t is a smooth semigroup acting in $\mathcal{Y}$. If $\mathcal{Y}$ has finite dimension or if $\mathcal{F}$ is smooth from $\mathcal{Y}$ to itself, then Φ_t is a one-parameter group of transformations in $\mathcal{Y}$, as it follows from the classical theory of ordinary differential equations. More generally the framework which we adopt here is suited for partial differential equations of parabolic type, such as reaction-diffusion equations, Navier-Stokes equations etc.

We assume that the Lie group G acts in $\mathcal{Y}$ by a homomorphism

$$\rho : G \longrightarrow GL(\mathcal{Y})$$

where $GL(\mathcal{Y})$ is the group of linear invertible operators in $\mathcal{Y}$. We also assume that G is a subgroup of $E(n)$ for some n. We assume in addition that ρ is *strongly continuous* in $\mathcal{Y}$: for any $x \in \mathcal{Y}$,

$$\lim_{g \to \mathrm{id}} (\rho(g) - I) = 0.$$

This defines a *representation* of G in $\mathcal{Y}$. The image $\rho(G)$ is a subgroup Γ of $GL(\mathcal{Y})$. We also assume that $\mathcal{X}$ is invariant by this action.
There are many ways in which a symmetry group can act on a Banach space. For example if $\mathcal{Y}$ is the space $C_b(\mathbb{R}^2)$ of uniformly continuous, bounded real functions on $\mathbb{R}^2$, then we may define a representation of the planar Euclidean group $E(2)$ on $\mathcal{Y}$ by

$$\rho(g, \vec{a})f(x) = f(g^{-1}(x - \vec{a})), \quad g \in O(2), \ \vec{a} \in \mathbb{R}^2 \tag{4}$$

The operator ρ is obviously continuous with respect to $\vec{a}$, however this is not true with respect to rotations in $SO(2)$ because rotations by angles as small as we wish can induce deviations as large as we wish "far enough" from the center of the rotation. Therefore the smoothness of the representation of the group depends on the choice of the Banach space in which it is defined. On the other hand Wulff [18] showed that the functions in $C_b(\mathbb{R}^2)$ on which rotations act continuously form a dense Banach subspace of $C_b(\mathbb{R}^2)$, and moreover the Cauchy problem for (3) is well-posed in this space.

It follows from classical semigroup theory [14] that, given $\xi \in \mathfrak{g}$ in the Lie algebra of G, the one-parameter group $\rho(\exp \xi s)$, $s \in \mathbb{R}$, admits a generator κ_ξ which is a closed linear operator in $\mathcal{Y}$ with dense domain

$$D(\xi) = \{x \in Y \ / \ \kappa_\xi x := \lim_{t \to 0} \frac{1}{t}(\rho(\exp \xi t)x - x) \text{ exists}\}.$$

Note that κ_ξ is also linear with respect to ξ. The following has been proved in [15]:

(i) the intersection $\mathcal{Y}_1$ of all the domains $D(\xi)$ for $\xi \in \mathfrak{g}$ is dense in $\mathcal{Y}$.

(ii) $\mathcal{Y}_1$ is a Banach space for the norm $\|x\|_1 := \|x\| + \sup_{j=1,\dots,k} \|\kappa(\xi_j)x\|$ where $\xi_1, \dots, \xi_k$ is a basis of $\mathfrak{g}$.

(iii) $\rho(g)x$ is continuously differentiable as a function from G into $\mathcal{Y}$ for any $x \in \mathcal{Y}_1$.

We now assume that the operator $\mathcal{F}$ is *equivariant* under the action of G:

$$\mathcal{F}(\rho(g)x) = \rho(g)\mathcal{F}(x), \quad \text{all } g \in G,\ x \in \mathcal{Y}.$$

This expresses the *symmetry* of the problem (3) with respect to the action of G.

Definition 1: A relative equilibrium (RE) for the evolution equation (3) is a solution which satisfies the relation $\varphi_t(x_0) = \rho(\exp(t\xi_0))x_0$, for a fixed element $\xi_0 \in \mathfrak{g}$ and an initial condition $x_0 \in \mathcal{Y}$ (therefore, $x_0 \in \mathcal{Y}_1$).

Therefore the semiflow $\varphi_t(x_0)$ is in fact a flow: it is defined for any $t \in \mathbb{R}$. Moreover the group associated with this flow coincides with the action of G. Let us analyze the consequences and give an interpretation of this fact. The set $\mathcal{O}_{\xi_0} := \{\exp(t\xi_0) \ / \ t \in \mathbb{R}\}$ is a subgroup of G. Let H be the *isotropy subgroup* of x_0 in G: $H = \{h \in G \ / \ \rho(h)x_0 = x_0\}$. Let Fix$(H)$ be the subspace of $\mathcal{Y}$ consisting of all the points which are fixed under the action of H. It is well-known (and easy to check thanks to the G-invariance of the equation) that any solution with initial condition in Fix(H) must lie entirely in Fix(H). It is also easy to check that the largest subgroup of G which acts in this subspace is $N(H)$, the *normalizer*[a] of H in G. Therefore $\mathcal{O}_{\xi_0}$ is in fact a subgroup of $N(H)$. Suppose that $N(H)$ is compact, then the closure of $\mathcal{O}_{\xi_0}$ (an abelian connected Lie group by definition) is a torus in $N(H)$, *i.e.* an abelian group which is isomorphic to $\mathbb{R}^d/\mathbb{Z}^d$ (d is the dimension of the torus). In general (in a sense which I do not make precise here, see [4]) d is equal to the maximal dimension a torus can have in $N(H)/H$. If the quotient group $N(H)/H$ is finite, the relative equilibrium is actually an equilibrium.

The interpretation of a RE is very simple: it follows from its definition and from the G invariance of the equation that $x(t) = \rho(\exp(t\xi_0))x_0$ is a RE for (3) if and only if

$$\kappa_{\xi_0} x_0 = \mathcal{F}(x_0), \tag{5}$$

an equation which expresses the fact that x_0 is a steady-state when observed in a frame which is moving along the group orbit at the velocity ξ_0. In order

[a]that is the group $\{g \in G \mid gHg^{-1} = H\}$

to compute a RE, one must therefore determine two unknown quantities: the element $x_0 \in \mathcal{Y}$ and the "velocity" ξ_0 in G. I refer to [4] for details and examples.

For completeness, I mention another definition of RE's which is valid when the group is compact. Let $\mathcal{Y}/G$ denote the *orbit space* of $\mathcal{Y}$ by the group action, i.e. the set of equivalence classes for the relation "to be in the same G-orbit". This space, although not a manifold in general, possesses nice properties when G is compact, and the canonical projection $Q : \mathcal{Y} \to \mathcal{Y}/G$ allows to smoothly project the dynamics of (3) onto the orbit space. We may therefore define relative equilibria as solutions of (3) which are projected on equilibria in the orbit space. Techniques to compute the orbit space and the projected flow when $\dim(\mathcal{Y}) < \infty$ are often used for Hamiltonian systems. They apply too when a a finite dimensional center manifold exists, see [4] for details.

The pendulum can serve as an example to illustrate these properties of RE's, in a case when the group is compact. The rotating periodic solutions are RE's with respect to the symmetry group $O(2)$. The points on these trajectories have trivial isotropy, therefore $N(H)/H = O(2)$: the trajectories are indeed isomorphic to circles (1 dimensional tori). Now consider the quasiperiodic solutions which show oscillations with precession and which we defined as relative periodic orbits in the introduction. Suppose in addition that the equations of the pendulum have been put in normal form. This means that after a suitable change of variables, which is valid only in a neighborhood of the trivial equilibrium, the equations have a leading polynomial part which is invariant under the action of the group S^1 generated by the quadratic part of the Hamiltonian function (averaging method to compute small oscillations), and the rest is neglected. Then it can be shown that these solutions are RE's for the action of $O(2) \times S^1$ on the normal form. In this case, $N(H)/H = SO(2) \times S^1$, implying that in general the trajectories fill tori of dimension 2 (hence the quasiperiodic behavior). Of course the S^1 symmetry is not physical and the steady behavior in a rotating frame is seen in phase space of the normal form, not directly in physical space. If we now look at the planar periodic oscillations, those are again RE's for the action of $O(2) \times S^1$. However, for a solution of this kind, H is the group generated by the reflection through its plane of oscillation. In this case $N(H)$ is spanned by this reflection, by the rotation of angle π in $SO(2)$, and by S^1 itself. The group $N(H)/H$ is therefore one dimensional and these RE's are periodic orbits.

Examples in hydrodynamics involving similar symmetries can be found in

the Couette-Taylor problem of a fluid flow filling the space between two coaxial rotating cylinders, see [3].
A class of examples of a different kind but which has focused a lot of attention in the recent past, is the occurrence of localized states in spatially extended systems such as convective fluid layers, chemical reactions and catalysis on films, population dynamics of bacteria in Petri boxes, among others. In certain cases, these patterns have the form of spirals which exhibit the properties of RE's with respect to the group of planar rotation. It is an experimental as well as computational fact that when the domain is very extended in two spatial coordinates, the system is well approximated by assuming that the domain has infinite planar extension. This is the reason for the occurrence of $E(2)$ symmetry in these models. The fact that $E(2)$ be a non compact group is a serious obstacle to the computation of solutions which are not periodic in the space variables. If spatially periodic solutions are seeked, then the problem can be restricted to the class of states which are spatially periodic, allowing to reduce the symmetries to those of a periodic lattice in the plane. Such symmetries generate a compact subgroup of $E(2)$. Then, pattern selection principles like the equivariant branching lemma for bifurcation problems, do apply [8]. In the non compact case, the basic difficulty is that there is no finite dimensional center manifold, see [4] for details. The computation of spiral solutions involves other techniques, see e.g. [9] and [17]. The spiral states can be observed experimentally as well as numerically, for example in systems of reaction-diffusion.

The interest for RE's has grown sharply when it was realized that their instabilities can lead to surprising and interesting dynamics. This was notably observed in the case of spirals in spatially extended systems. These spirals can develop instabilities which lead them to "meander" in the plane. The first convincing tentative to understand this behavior was made by Barkley [1]. It led to a considerable literature and developments.

2.2. *Dynamics near a Relative Equilibrium*

In order to study the dynamics near a relative equilibrium $x(t) = \rho(\exp(t\xi))x_0$, we will show that, under some suitable hypotheses, the flow (or semiflow) can be decomposed into a component along the G-orbit of the RE and a component transverse to it. We will see that all the relevant information is contained in the transverse component to the G-orbit. By (5), x_0 satisfies the equation

$$\mathcal{F}(x) - \kappa_{\xi_0} x = 0 \tag{6}$$

Let $\mathcal{L}_*$ be the linearization of $\mathcal{F} - \kappa_{\xi_0}$ at x_0: $\mathcal{L}_* = DF(x_0) - \kappa_{\xi_0}$. This is an (in general) unbounded operator in $\mathcal{Y}$, with domain $\mathcal{X}$. We assume that the spectrum of $\mathcal{L}_*$ consists of a "stable" part σ_s which is included into the half plane $\{z \mid Re(z) \leq \nu < 0\}$ (with $\nu < 0$) and a center part σ_c which consists of a finite number of eigenvalues of finite multiplicity which are located on the imaginary axis. We denote by E_c the finite-dimensional subspace of $\mathcal{Y}$ which is the sum of the generalized eigenspaces for the eigenvalues in σ_c.
Let $\mathcal{T}_0$ be the tangent space to the G-orbit of x_0 at x_0. Here we are assuming that the $\rho(G)x_0$ is a smooth manifold, which is the case when the group is compact and in the applications in general. There exists a basis $\xi_1, \ldots, \xi_q$ of $\mathfrak{g}$ such that if $p \leq q$ is the dimension of $\mathcal{T}_0$, then a basis of $\mathcal{T}_0$ is given by $\kappa_{\xi_1}, \ldots, \kappa_{\xi_p}$.
When the symmetry group G is a subgroup of an Euclidean group, $\mathcal{T}_0$ belongs to E_c. Indeed, let $e^{\xi s}$ be the one parameter subgroup of G generated by an element $\xi \in \mathfrak{g}$. By equivariance of $\mathcal{F}$ we have

$$\mathcal{F}(\rho(e^{\xi s})x_0) - \rho(e^{\xi s})\kappa_{\xi_0}x_0 = 0.$$

By differentiating at $s = 0$ and after some elementary algebra, we obtain

$$\mathcal{L}_*\kappa_\xi x_0 = \kappa_{[\xi_0,\xi]}x_0$$

where $[\xi_0, \xi] = \xi_0\xi - \xi\xi_0$ is the Lie bracket of the elements ξ_0 and ξ in $\mathfrak{g}$. It turns out that for Lie subgroups of $E(n)$, the eigenvalues of the map $\xi \mapsto [\xi_0, \xi]$, which is defined in $\mathfrak{g}$, are located on the imaginary axis (see [18]). The result follows. This means that the dynamics in a neighborhood of the RE is degenerate along its group orbit.
We also assume that the isotropy subgroup H of x_0 is compact. I claim that $\mathcal{L}_*$ is H-equivariant. This is clear for $DF(x_0)$, it is therefore enough to prove that κ_{ξ_0} commutes with $\rho(h)$ for all $h \in H$, or, equivalently, that ξ_0 commutes with h for all $h \in H$. By equivariance of the flow we know that $\exp(\xi_0 t) \in N(H)$ (the normalizer of H in G) for all t. Therefore, for all $h \in H$, there exists $\tilde{h}(t) \in H$ such that

$$\exp(\xi_0 t)h\exp(-\xi_0 t) = h\tilde{h}(t),$$

and $\tilde{h}$ depends differentiably on t. By differentiating at $t = 0$, we obtain

$$h^{-1}\xi_0 h - \xi_0 = \tilde{h}'(0). \tag{7}$$

Since $\tilde{h}(0) = \mathrm{id}$, $\tilde{h}'(0)$ is an element of the Lie algebra $\mathfrak{h}$ of H. Now remark that the choice of the velocity ξ_0 of the RE is not unique in general. Indeed, if $\exp(\eta t)$ is a subgroup of H, then clearly $\exp(\xi_0 t)\exp(\eta t)x_0 = \exp(\xi_0 t)x_0$

for all t, and therefore $\xi + \eta$ is also a velocity for the RE. If however we choose ξ_0 in the orthogonal complement $\mathfrak{h}^\perp$ of the Lie algebra $\mathfrak{h}$ of H (for an H-invariant scalar product in $\mathfrak{h}$), then ξ_0 is unique. Then in (7), the left hand side lies in $\mathfrak{h}^\perp$ and the right hand side lies in $\mathfrak{h}$, which implies the claim. It also follows that E_c is invariant under the action of H.

These hypotheses are generally satisfied when the symmetry group G is compact and the check is not difficult, see for example [3]. When $G = E(2)$ and x_0 is a rotating spiral wave for a system of reaction-diffusion equations, they are also satisfied [18].

In order to decompose the problem as announced, we need first to define a projection on $\mathcal{T}_0$. This can be formally achieved as follows. Let Q be the projection onto E_c associated with the spectral decomposition $\sigma_c \cup \sigma_s$. Specifically, let Γ be a C^1 closed curve surrounding σ_c and not containing a part of σ_s. Then we can set (Dunford calculus)

$$Q = \frac{1}{2Q} \int_\Gamma (I - \gamma L_*)^{-1} d\gamma \ .$$

Q is clearly H-equivariant. Then in E_c, which is a finite dimensional vector space, we define the orthogonal projection P_0 on $\mathcal{T}_0$. Since H is compact, one can choose in E_c an H-invariant inner product $\langle \cdot, \cdot \rangle$ by averaging any given inner product over the group action by the mean of its Haar measure [4]. Now $P = QP_0$ defines a projection onto $\mathcal{T}_0$ in $\mathcal{Y}$.

We set $\mathcal{N}_0 = (1 - P)\mathcal{Y}$. By the H-equivariance of Q and P_0, this space is clearly H-invariant. We call it a transverse section to the G-orbit of the RE at x_0.

Let us now look for solutions of (3) of the following form:

$$x(t) = \rho(g(t))[x_0 + w(t)] \tag{8}$$

where $w(t) \in \mathcal{N}_0$ and $g(t)$ is a smooth trajectory in G. We are looking for solutions of (3) of the form (8) such that $g(t) = \exp(\xi_0 t)$ when $w = 0$. Formally, thanks to the G-equivariance of the operators, (3) gives

$$\frac{dw}{dt} + \rho(g)^{-1} D\rho(g) g'(t)(x_0 + w) = \mathcal{F}(x_0 + w) \tag{9}$$

I claim that $\rho(g)^{-1} D\rho(g) g'(t) = \kappa_{\xi(t)}$ where $\xi(t)$ is a curve in $\mathfrak{g}$ such that $\xi(0) = 0$ and $\kappa_\xi x := \lim_{t\to 0} \frac{1}{t}(\rho(\exp \xi(t))x - x)$. Indeed, let

$$\gamma(s) = \rho(g(t))^{-1} \rho(g(t + s))$$

for fixed t. Then of course $\gamma(s) \in \rho(G)$, $\gamma(0) = \mathrm{id}$ and $\gamma'(0) = \rho(g(t))^{-1} D\rho(g(t)) g'(t)$ which proves the claim.

We may now rewrite (9) as

$$\frac{dw}{dt} + (\kappa_{\xi_0} + \kappa_{\eta(t)})(x_0 + w) = \mathcal{F}(x_0 + w)$$

where $\eta(t)$ defines a curve in $\mathfrak{g}$ such that $\eta = 0$ at $w = 0$, and applying (5) we can further put this equation into the form

$$\frac{dw}{dt} + \kappa_{\eta(t)}(x_0 + w) = \mathcal{F}(x_0 + w) - \mathcal{F}(x_0) - \kappa_{\xi_0} w \tag{10}$$

However the representation ρ may not act differentiably on $w \in \mathcal{N}_0$, and therefore $\kappa_\xi w$ may not be defined, unless $w \in \mathcal{Y}_1$. In many applications however (it happens for example in the Couette-Taylor problem [3]), $\mathcal{X} \subset \mathcal{Y}_1$. In this case, the equation (9) is meaningful because the operators κ_ξ are bounded relatively to $D\mathcal{F}(x_0)$ and the Cauchy problem for (10) is well-posed. Otherwise, Sandstede, Scheel and Wulff [15] have shown that under the additional hypothesis that any G-orbit intersecting the finite dimensional space E_c is smooth, a center manifold exists for the RE and the G-orbits on this center manifold are smooth. Moreover, when the isotropy subgroup H is compact (which we always assume here), they show that the flow can be decomposed into a component in $\mathcal{T}_0$ and a component in $\mathcal{N}_0$. Then the above decomposition and subsequent analysis are valid for the computation of the center manifold. This procedure applies notably to the bifurcation analysis from spiral waves in spatially extended media.

Let us now decompose (10) according to (8). We obtain one equation in E_c

$$P\kappa_{\eta(t)}(x_0 + w) = P[\mathcal{F}(x_0 + w) - \mathcal{F}(x_0) - \kappa_{\xi_0} w] \tag{11}$$

and another equation in $\mathcal{N}_0$

$$\frac{dw}{dt} + (1 - P)\kappa_{\eta(t)} w = (1 - P)[\mathcal{F}(x_0 + w) - \mathcal{F}(x_0) - \kappa_{\xi_0} w] \tag{12}$$

The first equation can be formally solved with respect to η as follows. Let us set $\Xi_j = \kappa_{\xi_j} x_0$, $j = 1, \ldots, p$. We can choose the ξ_j's so that the Ξ_j's form an orthonormal basis for the H-invariant inner product in E_c. We also set

$$\eta = \alpha_1 \xi_1 + \cdots + \alpha_p \xi_p.$$

Then we have

$$P\kappa_{\eta(t)} x_0 = \kappa_{\eta(t)} x_0 = \alpha_1(t)\Xi_1 + \cdots + \alpha_q(t)\Xi_p \tag{13}$$

Similarly, the component of $P\kappa_{\eta(t)} w$ along Ξ_j, $j = 1, \ldots, q$, can be written

$$(P\kappa_{\eta(t)} w)_j = \alpha_1(t)\ell_{j1}(w) + \cdots + \alpha_q(t)\ell_{jq}(w) \tag{14}$$

where the ℓ_{jk} are linear functionals of w. When $\mathcal{Y}$ is a Hilbert space with scalar product $\langle \cdot, \cdot \rangle$, we can directly define

$$Px = \langle x, \Xi_1 \rangle \Xi_1 + \cdots + \langle x, \Xi_q \rangle \Xi_q$$

in which case

$$\ell_{jk} = \left\langle \kappa_{\xi_j} w, \Xi_k \right\rangle$$

Now, replacing (13) and (14) in (11), we obtain a system of p scalar equations. If the operators defining these equations are smooth, in particular if κ_ξ is bounded relatively to $D\mathcal{F}(x_0)$, then we can solve this system of p scalar equations for the α_j's by the implicit function theorem. We therefore obtain the following proposition

Proposition 2: *Suppose that $X \subset \mathcal{Y}_1$. Then there exist smooth functions $h_1(w), \ldots, h_p(w)$ such that $h_j(0) = 0$ and setting $\alpha_j(t) = h_j(w(t))$ solves the equation (11) in $\mathcal{T}_0$.*

An example of these calculations for the Couette-Taylor flow can be found in [3].

We are left with one equation (12) in $\mathcal{N}_0$ for the only unknown w, hence we have been able to get rid of the flow component along the group orbits. Let us comment on this equation which contains all the relevant information about the dynamics in a neighborhood of the RE. First, we rewrite it, thanks to the resolution of (11), as

$$\frac{dw}{dt} = (1-P)[\mathcal{F}(x_0+w) - \mathcal{F}(x_0) - (\kappa_{\xi_0} + h_1(w)\kappa_{\xi_1} + \cdots + h_p(w)\kappa_{\xi_p})w] \quad (15)$$

The equilibrium at $w = 0$ corresponds to the RE for (3). The stability of this trivial state is governed by the spectrum of the linearized operator $(1-P)\mathcal{L}_*$ in $\mathcal{N}_0$. Because we have projected transversally to the group orbit of x_0, we have got rid of the eigenvalues of $\mathcal{L}_0*$ which are forced on the imaginary axis by the symmetry. If the spectrum of the operator $(1-P)\mathcal{L}_0 * (1-P)$ in $\mathcal{N}_0$ is entirely contained in the half-plane $\{Rez \leq \chi < 0\}$ where $\chi > 0$, then the solution $w = 0$ is asymptotically stable (therefore the RE is asymptotically stable). If instead it contains some eigenvalues on the imaginary axis, a center manifold reduction can be performed on (15) as in the usual case, see [11], [4].

It remains to find how much of the symmetry is transmitted to (15). Since all the operators which enter in the expression for this equation are H-equivariant, we can state the

Proposition 3: *The equation (15) is invariant under the action of H.*

As an application, if the system depends on some parameter and a bifurcation occurs, then we are back to a bifurcation problem with symmetry H in $\mathcal{N}_0$. Again, see the bibliography for examples of bifurcations and consequences.

Remark. If the hypothesis in proposition 2 is not satisfied but the hypotheses of [15] are met (in particular, any orbit intersecting E_c is smooth), then the above decomposition still applies to formally compute the center manifold. For this, we can set $w = w_0 + w_s$ where $w_0 \in (1-P)E_c$ (the center subspace in $\mathcal{N}_0$) and $w_s \in (1-Q)\mathcal{Y}$. We know that there exists a smooth map $\Psi : E_c \to E_s$ such that the center manifold is parameterized by $w_s = \Psi(w_0)$. Replacing into (11) and (12) we can use these equations to compute a Taylor expansion of the functions h_j and Ψ with respect to w_0.

Note that, a steady-state bifurcation from $w = 0$ for the reduced equation (15) would lead to the occurrence of new branches of RE's, while a Hopf bifurcation would lead to the occurrence of solutions of the form (8) with a - in general - periodic w. This periodic oscillating behavior is superposed to the initial drift of the RE. The result is what we call a *relative periodic orbit.* As I said in the Introduction, the oscillations with precession in the spherical pendulum (or in the spinning top) are an example of relative periodic orbits. A Hopf bifurcation from a RE is also responsible for the occurrence of *modulated wavy Taylor vortices* in the Couette-Taylor problem and of *meandering* for spiral waves moving in an extended planar domain (see bibliography). In the next section I shall briefly introduce relative periodic orbits and the method to analyze their stability and bifurcations.

3. Relative Periodic Orbits

3.1. *Definition and Properties*

Definition 4: A solution $x(t)$ of Equation (3) is called a relative periodic orbit (RPO) if it satisfies the following property: there exists a $T > 0$ and an element $\gamma_0 \in G$ such that $x(t+T) = \rho(\gamma_0)x(t)$ for all t. The smallest T satisfying this condition, which we write T_0, is called the relative period of the RPO.

Being a trajectory of a G-invariant equation, $x(t) \in \mathrm{Fix}(H)$ for all t, where H is the isotropy subgroup of the initial condition $x(0)$. It follows that $\gamma_0 \in N(H)$, moreover γ_0 is only defined up to an element in H. The definition of an RPO is easy to interpret: the trajectory consists of pieces which are duplicated from the piece between $x(0)$ and $x(T)$ by successive applications of the group elements γ_0, γ_0^2, etc... The RPO's possess therefore a *spatio-temporal symmetry group* in addition to the pure spatial one. We define σ to be the closure of the group generated by the element γ_0 and by the group of spatial isotropies H, and call it the group of *spatio-temporal symmetries* of the RPO.

An analysis and classification of RPO's with respect to these properties was made by Krupa [12] when the group is compact and by Field [6]. This was further generalized by Wulff, Lamb & Melbourne [20]. A Poincaré map and center manifold reduction for RPO's was given by [4] in the case of compact groups and by [16] in a more general setting. Here we only expose the part of the theory which is required to be able to perform stability and bifurcation analysis and we refer to the bibliography for a more complete view. I will expose the method of [4], which is also valid for noncompact groups under suitable hypotheses. In particular we assume throughout this section that the G-orbit of the RPO is a smooth manifold, and that H is a compact subgroup of G (conditions which are usually satisfied in the applications).

A first important remark is that when G is an algebraic group (this includes compact and Euclidean groups) there exists an integer k and a $\xi_0 \in \mathfrak{g}$ such that $\gamma_0^k = \exp(\xi_0)$ (see [20]). The proof for compact G is simple. First notice that G^0 is (always) a normal subgroup of G. We can therefore define a natural group structure on G/G^0. When G is compact, the quotient group G/G^0 is finite. Therefore we can write $\gamma_0 = \delta g$ with δ of finite order k and $g \in G^0$, and since $\delta g = g'\delta$ for a $g' \in G^0$, we conclude that $\gamma_0^k \in G^0$. Now, since G^0 is compact, there exists an isomorphism between G^0 and its Lie algebra $\mathfrak{g}$. Therefore there exists a $\xi_0 \in \mathfrak{g}$ which satisfies the claim.

Here as in the previous section, we define

$$\kappa_\xi x = \lim_{s\to 0} \frac{1}{s}[\rho(e^{\xi s})x - x] \text{ for } x \in \mathcal{Y}_1 \text{ and } \xi \in \mathfrak{g}.$$

Proposition 5: *Let $x(t)$ be a RPO for a system with compact or Euclidean symmetry. Then there exists an element ξ in the Lie algebra of $N(H)$ and a* periodic *solution $y(t)$ of the equation*

$$\dot{y} + \kappa_\xi y = \mathcal{F}(y) \tag{16}$$

such that $x(t) = \rho(e^{\xi t})y(t)$, all $t \in \mathbb{R}$.

Proof. Let $\xi \in \mathfrak{g}$ and $y(t) = \rho(e^{-\xi t})x(t)$. As stated above, there exists an integer k, a number T and a $g_0 \in [N(H)/H]^0$ such that $x(kT) = \rho(g_0)x(0)$.

Therefore

$$y(kT) = \rho(e^{-kT\xi}g_0)y(0).$$

Moreover we can choose ξ such that $g_0 = \exp(kT\xi)$. The fact that y be solution of the equation $\dot{y} + \kappa_\xi y = \mathcal{F}(y)$ is a straightforward consequence of the equivariance of $\mathcal{F}$. ∎

This proposition allows us to easily prove that the flow on a RPO can either be unbounded when σ is noncompact, or otherwise fill tori of dimension $p+1$ where p is the dimension of a torus group included in $N(H)/H$ [4]. However, in contrast to what happens for relative equilibria, p can be generically strictly smaller than the dimension of the maximal tori in $N(H)/H$. The reason is that the spatiotemporal symmetry imposes restrictions on the flow, see [20] for a precise statement and complete argument about this point.

3.2. *Construction of a Poincaré Map for Relative Periodic Orbits*

In order to analyze the flow near the RPO we proceed in the same spirit as for RE's. It can be seen as a generalization of the method which was introduced in [2] for the study of bifurcations from standing waves (periodic obits with reflection symmetry) in systems with $O2$ symmetry. To get rid of the dynamics along the group action, the idea is to define a Poincaré map in a slice which is transverse to the G-orbit at the initial condition x_0. Let us first formally define this Poincaré map.

Recall that $\Phi_t(x)$ denotes the semiflow of the evolution equation (3). When taking the initial condition x_0 on the RPO, we have by definition $\Phi_{T_0}(x_0) = \gamma_0 x_0$. Therefore x_0 is a fixed point for the map $\gamma_0^{-1}\Phi_{T_0}(x_0)$. We set $\mathcal{L}_* = \gamma_0^{-1}D_x\Phi_{T_0}(x_0)$ and we assume that the spectrum of $\mathcal{L}_*$ consists of a part σ_s lying inside the unit circle and another part σ_c lying on the unit circle and consisting of a finite number of eigenvalues with finite multiplicity. Let E_c denote the (generalized) eigenspace associated with the eigenvalues on the unit circle. We therefore have $\dim E_c < \infty$. The tangent space to the G-orbit of the RPO at $x(0) = x_0$ is spanned by $\dot{x}(0)$ (i.e. $\mathcal{F}(x_0)$) and by the elements $\kappa_{\xi_j}x_0$, $j = 1, \ldots, p$ where $\xi_j \in \mathfrak{g}$ can be chosen as an orthogonal complement to the Lie algebra of H in $\mathfrak{g}$. We will assume that this tangent space is a subspace of E_c, a property which is true for compact groups as well as for the Euclidean group, see [16].

As in Section 1.2, we can now define an H-equivariant projection Q on E_c, then an H-equivariant projection P_0 on the tangent space in E_c, and we set

$P = QP_0$. We set $\mathcal{N}_0 = (1-P)\mathcal{Y}$ and $x = x_0 + w$, $w \in \mathcal{N}_0$. $\mathcal{N}_0$ is invariant under the action of H.
Let us define

$$\Psi(T, \alpha_1, \ldots, \alpha_p, w) = \rho(e^{\alpha_1\xi_1+\cdots+\alpha_p\xi_P}\gamma_0^{-1})\Phi_T(x_0 + w) - x_0 \tag{17}$$

Note that $\Psi(T_0, 0, \ldots, 0, 0) = 0$. Our aim is to find functions $T(w)$ close to T_0 and $\alpha_j(w)$ close to 0 such that $P\Psi(T(w), \alpha_1(w), \ldots, \alpha_p(w), w) = 0$. Then the map

$$\Pi(w) = (1-P)\Psi(T(w), \alpha_1(w), \ldots, \alpha_p(w), w)$$

will be a first return map on the transverse section $\mathcal{N}_0$.
As for relative equilibria, we are forced to assume that the group representation ρ is differentiable in $\mathcal{X}$. Then we remark that evaluating the derivative of $P\Psi$ with respect to T and $\alpha_1, \ldots, \alpha_p$ at the point $T = T_0$, $\alpha_1 = \cdots = \alpha_p = 0$ and $w = 0$ gives

$$D_{T,\alpha_1,\ldots,\alpha_p} P\Psi(T_0, 0, \ldots, 0)(dT, d\alpha_1, \ldots, d\alpha_p) = \dot{x}(0)dT + \sum_{j=1}^{p} d\alpha_j \kappa_{\xi_j} x_0,$$

which is surjective since $\dot{x}_0(0)$ and the $\kappa\xi_j x_0$ are linearly independent. Therefore the implicit function theorem applies, which allows us to conclude that the map Π exists.
The map Π takes the value 0 at $w = 0$. Moreover its derivative at $w = 0$ is $(1-P)\mathcal{L}_*$. Therefore a center manifold reduction can be performed on Π, and if the problem depends on some parameter, a bifurcation analysis can be undertaken.
However we have assumed that the representation ρ is smooth. We have seen that it is not always the case, for example when the group G is the Euclidean group acting in a space of functions defined in the plane. Additional assumptions similar to those introduced in the case of relative equilibria are needed in order to give a meaning to the above computations. This problem has been thoroughly investigated by [16] and we refer to this paper for more details. In particular, it is proved that RPO's in systems of reaction-diffusion with Euclidean group symmetry satisfy these conditions.

The relative Poincaré map is equivariant under the action of the isotropy group H. The proof is similar to the proof of the H-invariance of the reduced equation (15) in the case of relative equilibria. I refer to [4] for an exposition of the bifurcation theory for equivariant maps.

We finish this section with the following remark:

Remark 6: The foregoing construction of the Poincaré map can be related to the Poincaré map for a *periodic orbit* of an explicit equation. Indeed, as we already noticed (Proposition 5), if $x_0(t)$ is a solution on a RPO, then there exists ξ_0 in $\mathfrak{g}$ such that $y_0(t) = \rho(\exp(-\xi_0 t))x_0(t)$ is T_0-periodic and satisfies the equation

$$\frac{dy}{dt} = \mathcal{F}(y) - \kappa_\xi y \tag{18}$$

at $\xi = \xi_0$. It is now a simple exercise to check that the Poincaré map for $y_0(t)$ at the point $y_0(0) = x_0$ is precisely equal to Q. As an immediate consequence $\mathcal{L}_*$ is the monodromy operator associated with the equation

$$\frac{dy}{dt} = D\mathcal{F}(x_0)y - \xi_0 y$$

4. Conclusions

I mentioned in the introduction that bifurcation analysis from relative equilibria and relative periodic orbits is a powerful tool to analyze the occurrence of complex spatio-temporal patterns. Pattern formation in the Couette-Taylor system is a paradigm for this analysis. I briefly recall the problem: the gap between two coaxial cylinders is filled with a fluid (typically water). The cylinders rotate at different but constant speeds. When the speed of rotation of the inner cylinder reaches a critical value, which depends on the speed of the outer cylinder, the laminar flow becomes unstable to either a steady flow which consists of a number of doughnut-like identical cells which are superposed along the cylinders but which do not break its rotational symmetry (in particular the boundaries between these cells are flat), or a periodic flow which assumes the form of a rotating spiral vortex which looks similar to the sign indicating a barber's shop. The former flow is called Taylor vortex flow and the latter is a spiral flow. Both are relative equilibria with respect to the translational symmetry of the system if it is assumed of infinite length (valid assumption if the apparatus is spatially extended in the axial direction). The spiral flow is in fact a relative equilibrium with respect to the group generated by the translational and the rotational symmetries. I refer the reader to [3] for the bifurcation analysis from these states, which leads to new states which are indeed observed in the experiment, such as wavy Taylor vortices, wavy spirals and other more complicated flows.

It should however be emphasized that this type of analysis does not explain all kinds of behavior which are observed in systems with continuous symmetry, especially when the symmetry group is not compact. This is even visible in the Couette-Taylor system with cylinders of infinite length. For example the bifurcation of Taylor vortex flow from the laminar state occurs, in this case, through a continuous spectrum crossing the imaginary axis. One can heuristically explain the continuous spectrum by the fact that the absence of boundaries (or their presence "far away") does not force the system to select a spatial wave length. In a first approximation, one can restrict the study to "waves" of the "most unstable" length. This allows for a reduction of the symmetries to translations up to the given period. Then the group of translational symmetry becomes compact ($\mathbb{R}/\mathbb{Z}$ instead of $\mathbb{R}$) and of course the spectrum of the linearized system becomes discrete. If however we want to take account of all possible unstable modes, then the center manifold reduction fails and other, more subtle phenomena may occur (such as the phase instability of the Taylor vortex flow). Another phenomenon which frequently occurs in spatially extended systems is the coexistence of several patterns which do not "match" each other. For example, spiral flows which are rotating in opposite directions, one forming in the upper part of the apparatus and the other forming in the lower part. These patterns connect to each other at a boundary which is called a "defect". The computation of these states requires different techniques, see [3]. The computation of spiral waves for reaction-diffusion systems in planar media is a similar type of problem. When however these states are identified as relative equilibria or RPO's, it is possible, using the methods exposed in this paper, to analyze their stability and dynamics. The dynamics of defects and of patterns in extended systems can be modelled by equations which take account of the slow time evolution of the amplitudes of the leading modes (Ginzburg-Landau equations), but this is another story, see Chapter 8.

References

[1] D. Barkley, *Phys. Rev. Lett.* **72**, 164-167 (1994).

[2] P. Chossat, M. Golubitsky, *SIAM J. Math. Anal.* **19**, 1259-1270 (1988).

[3] P. Chossat, G. Iooss, *The Couette-Taylor Problem*, Studies in Applied Math Series **102**, Springer Verlag, New York (1994).

[4] P. Chossat, R. Lauterbach, *Methods in equivariant bifurcation and dynamical systems*, Advanced series in nonlinear dynamics 15, World Scientific, Singapur (2000).

[5] B. Fiedler, B. Sandstede, A. Scheel, C. Wulff, *Doc. Math. J. DMV,* **1**, 479-505 (1996).

[6] M. Field, *Local structure of equivariant dynamics*, in M. Roberts & I. Stewart

ed., *Singularity Theory and its Applications, Warwick 1989*, Lecture Notes in Mathematics 1463, Springer Verlag, New-York(1996).

[7] M. Golubitsky, V.G. Leblanc, I. Melbourne, *J. Nonlin. Sc.* **7** (6), 557-586 (1997).

[8] M. Golubitsky, I. Stewart, D. Schaeffer, *Singularities and groups in bifurcation theory* Vol. 2, Appl. Math. Sci. **69**, Springer Verlag (1988).

[9] P.S. Hagan, *SIAM J. Appl. Math.* **42**, 762-786 (1982).

[10] G. Iooss, *J. Fluid Mech,* **173**, 273-288 (1986).

[11] G. Iooss, M. Adelmeyer, *Topics in bifurcation theory and applications*, Advanced Series in Nonlinear Dynamics 3, World Scientific, Singapur(1992).

[12] M. Krupa, *SIAM J. Math. Anal.* **21** (6), 1453-1486 (1990).

[13] W. Miller, *Symmetry Groups and their Applications*, Academic Press, New York (1972).

[14] A. Pazy, *Semigroups of linear operators and applications to partial differential equations*, Appl. Math. Sc. 44, Springer Verlag, New-York (1983).

[15] B. Sandstede, A. Scheel, C. Wulff, *J. Diff. Eq.* **141**, 122 - 149(1997).

[16] B. Sandstede, A. Scheel, C. Wulff, *J. Nonlin. Sci.* **9** (4), 439-478 (1999).

[17] A. Scheel, *SIAM J. Math. Anal.* **29** (6), 1399-1418 (1998).

[18] C. Wulff, *Doc. Math.* **5**, 227-274 (2000).

[19] C. Wulff, Dissertation: *Theory of Meandering and Drifting Spiral Waves in Reaction-Diffusion Systems*, Freie Universität Berlin (1996).

[20] C. Wulff, J. Lamb, I. Melbourne, *Ergodic Theory Dynam. Systems,* **21**, 605–635 (2001).

CHAPTER 5

ROTATING MAGNETOCONVECTION WITH MAGNETOSTROPHIC BALANCE

Keith Julien*, Edgar Knobloch†, and Steve M. Tobias†

* *Department of Applied Mathematics, University of Colorado, Boulder, CO 80309, USA*

† *Department of Applied Mathematics, Leeds University, Leeds LS2 9JT, UK*

The linear theory for the onset of convection in a rotating layer with an imposed vertical magnetic field is studied for large Taylor (Ta) and Chandrasekhar (Q) numbers. The velocity and magnetic field perturbations are in magnetostrophic balance. For small values of $\sigma = \nu/\kappa$ and $\zeta = \eta/\kappa$, where ν, κ and η are the kinematic viscosity, and thermal and ohmic diffusivities, three oscillatory modes are typically present, and these undergo a complex sequence of transitions as parameters are varied. Two of these modes are related to the MAC (Magneto-Archimedean-Coriolis) modes discussed by Braginskii, while the third is a magneto-convective mode. Three different asymptotic regimes are identified in the limit of large Ta and Q, depending on their relative magnitude, and the characteristics of the three modes in each regime are described in detail. Of greatest interest is the regime $Ta = \mathcal{O}(Q^{\frac{3}{2}})$ which captures all the transitions among the three modes including the transition to steady convection.

1. Introduction

In this chapter we revisit a classical problem, originally looked at by Chandrasekhar [1]: the linear theory for the onset of convection in a rotating layer with an imposed vertical magnetic field. Although the basic formulation of the problem is contained in Chandrasekhar's book, we have recently realized that his discussion of the governing dispersion relation is incomplete in certain regimes of astrophysical interest, and that it contains several surprises. In the present work we elucidate these, focusing on convection in rapidly rotating objects with strong magnetic fields, such as might be present in A-type stars with starspots at or near the poles [2]. We are therefore particularly interested in the regime in which both the Taylor number Ta and the Chandrasekhar number Q are large. Moreover, in contrast to

Chandrasekhar who focused on liquid metals for which the ohmic diffusivity is always large compared to thermal diffusivity, we explore the properties of the dispersion relation in the opposite and astrophysically relevant case of small ohmic diffusivity and kinematic viscosity. We investigate in detail the regime in which the velocity and magnetic field perturbations are in magnetostrophic balance, since this results in a highly nontrivial competition between the rotation and magnetic field, resulting under appropriate conditions in as many as three competing oscillatory modes. In the case of small Prandtl numbers overstability in the form of inertial waves is favored, while small ohmic diffusivities favor overstable magnetoconvection. Of course, the inertial waves will in general be affected by the presence of a strong magnetic field while magnetoconvection (both oscillatory and steady) will be affected by the strong rotation. In the following we refer to waves in which both ingredients are important as MAC (Magneto-Archimedean-Coriolis) waves, using the terminology first introduced by Braginskii [3], despite the fact that Braginskii was concerned with hydromagnetic waves in a horizontal field only. The competition between these two instability mechanisms is responsible for the particular richness of the present problem, both at the linear level and in the fully nonlinear regime. In this note we present detailed solutions of the linear stability problem in this regime and point out the presence of three essentially distinct asymptotic regimes, depending on the relative magnitude of Ta and Q. Since we are interested in the effects of rotation on magnetoconvection we choose to scale Ta relative to Q rather than vice versa, and define these regimes in terms of an exponent p such that the oscillation frequency $\omega = \mathcal{O}(Q^p)$. The first of these regimes, defined by $p > \frac{1}{2}$, is rotation-dominated, while the third ($p < \frac{1}{2}$) is dominated by the magnetic field. In each case, however, the competing restoring force can be retained in the theory. In the most interesting (and most complex!) case, $p = \frac{1}{2}$, the linearized Lorentz force enters at the same order as the Coriolis force and the behavior of the system is dominated by the requirements of magnetostrophic balance. The range of applicability of each of these asymptotic regimes is tested against a numerical solution of the full dispersion relation.

2. The Dispersion Relation

The dimensionless Boussinesq equations describing magnetoconvection in a plane horizontal layer of depth d rotating uniformly about the vertical

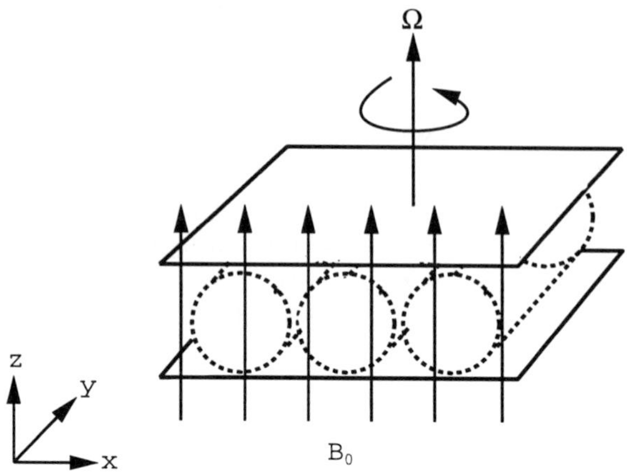

Fig. 1. The geometry of the rotating plane-parallel layer illustrating two-dimensional rolls permeated by a vertical magnetic field.

with angular velocity Ω (see Figure 1) are

$$\frac{1}{\sigma}\frac{D\mathbf{u}}{Dt}+Ta^{\frac{1}{2}}\hat{\mathbf{z}}\times\mathbf{u}-\zeta Q\mathbf{B}\cdot\nabla\mathbf{B}=-\nabla p+RaT\hat{\mathbf{z}}+\nabla^2\mathbf{u}, \tag{1}$$

$$\frac{DT}{Dt}=\nabla^2 T, \tag{2}$$

$$\frac{D\mathbf{B}}{Dt}=\mathbf{B}\cdot\nabla\mathbf{u}+\zeta\nabla^2\mathbf{B}, \tag{3}$$

$$\nabla\cdot\mathbf{u}=0, \qquad \nabla\cdot\mathbf{B}=0, \tag{4}$$

where $\mathbf{u}\equiv(u,v,w)$ is the velocity in Cartesian coordinates (x,y,z) with z vertically upwards, and T, $\mathbf{B}$ and p denote the dimensionless temperature, magnetic field and total pressure, respectively. The equations have been nondimensionalized with respect to the thermal diffusion time d^2/κ in the vertical. The resulting dimensionless parameters

$$Ta=\frac{4\Omega^2d^4}{\nu^2}, \quad Q=\frac{\mu B_0^2d^2}{4\pi\rho\nu^2}, \quad Ra=\frac{g\alpha\Delta Td^3}{\nu\kappa}, \quad \sigma=\frac{\nu}{\kappa}, \quad \zeta=\frac{\eta}{\kappa} \tag{5}$$

are the Taylor, Chandrasekhar, Rayleigh, and thermal and magnetic Prandtl numbers, respectively. Here B_0 is the strength of the uniform vertical magnetic field threading the layer (Figure 1), ΔT is the applied temperature difference, g is the gravitational acceleration and α the coefficient of thermal expansion; of the remaining quantities ν represents the kinematic viscosity, κ the thermal diffusivity, η the magnetic diffusivity, and ρ is the (constant) fluid density.

These equations are supplemented with periodic boundary conditions in the horizontal, and stress-free, perfectly (thermally) conducting boundary conditions at the top and bottom. If the magnetic field is assumed not to exert any stress on the horizontal boundaries the resulting problem, linearized about the conduction state, $\mathbf{u} = 0$, $T = 1 - z$, $\mathbf{B} = (0, 0, 1)$, is separable in the horizontal and vertical directions, and small amplitude perturbations of the conduction state satisfy the dispersion relation

$$\left(\frac{i\omega}{\sigma} + a^2 + \frac{\zeta Q \pi^2}{i\omega + \zeta a^2}\right)^2 a^2 + \pi^2 Ta = \left(\frac{i\omega}{\sigma} + a^2 + \frac{\zeta Q \pi^2}{i\omega + \zeta a^2}\right) \frac{k^2 Ra}{i\omega + a^2}. \quad (6)$$

Here $a^2 = k^2 + \pi^2$, k is the horizontal wavenumber of the perturbations, and ω is their frequency. This frequency is real along the neutral stability curve $Ra = Ra(k)$. If $\omega(k) \neq 0$ this neutral stability curve corresponds to the onset of overstability; if $\omega(k) \equiv 0$ the onset is steady. It should be mentioned that in addition to the modes derived from this dispersion relation there is a pair of slowly decaying long wavelength modes. In two dimensions these describe the decay of a large scale mean flow parallel to the rolls and of the magnetic flux function [4]. In the following we do not consider these modes further.

As in our previous work [5, 6] the details of the velocity and magnetic boundary conditions become immaterial in the limit of large Ta and Q. In this regime the boundary conditions manifest themselves in ever thinner and passive boundary layers along the top and bottom boundaries. As a result the idealized boundary conditions used above suffice to describe the linear stability properties of the conduction state even for other, more realistic choices of the boundary conditions, provided only that Ta and Q are both sufficiently large (cf. [7]).

3. Numerical Results: The Full Dispersion Relation

The results that follow are ordered according to whether $\sigma < 1$, favoring inertial wave instability, or $\zeta < 1$, favoring overstable magnetoconvection, or both. The last case is of course the most interesting. Figures 2-4 show the result of solving the dispersion relation (6) when $\zeta = 0.1$ and $Q = 10^{10}$, for various values of the Prandtl number σ and the quantity $\delta \equiv TaQ^{-\frac{3}{2}}$, which measures the rotation rate relative to the strength of the magnetic field. In all cases the results are presented in terms of a scaled

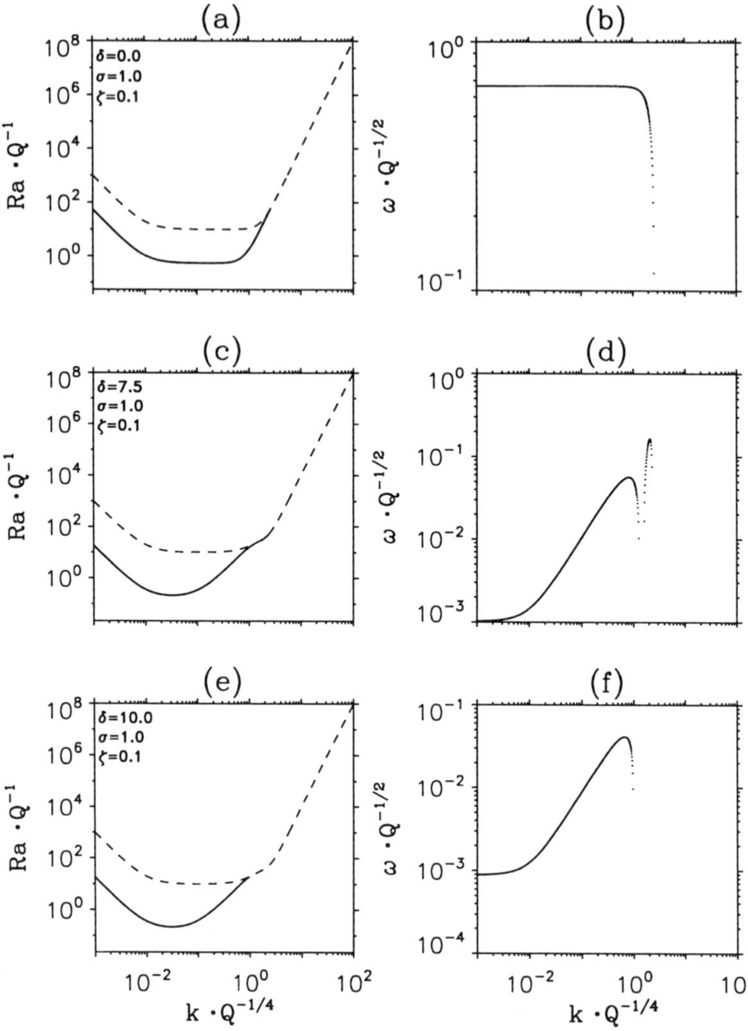

Fig. 2. Neutral stability curves $Ra(k) \cdot Q^{-1}$ for steady (dashed line) and oscillatory (solid line) convection, and the corresponding frequency $\omega \cdot Q^{-1/2}$ when $Q = 10^{10}$ and $\zeta = 0.1$, $\sigma = 1.0$. (a,b) $\delta = 0.0$; (c,d) $\delta = 7.5$; (e,f) $\delta = 10.0$. Here $\delta \equiv Ta \cdot Q^{-3/2}$.

Rayleigh number RaQ^{-1}, scaled wavenumber $kQ^{-\frac{1}{4}}$ and scaled frequency $\omega Q^{-\frac{1}{2}}$. These scalings are motivated by the asymptotic analysis described in Section 4. We find that the nature of an unstable oscillatory mode is best characterized by the dependence of its (scaled) frequency on the (scaled) wavenumber. Figure 2 indicates that the most interesting behavior occurs for $RaQ^{-1} = \mathcal{O}(1)$, $kQ^{-\frac{1}{4}} = \mathcal{O}(1)$. The values $\sigma = 1.0$, $\zeta = 0.1$ used in the figure favor overstability of magnetoconvective origin. The figure shows the neutral stability curves $Ra(k)$ for overstability (solid curve) and pure exponential growth (dashed curve) for different values of the rotation rate δ. For fixed k these curves indicate the presence of Hopf and steady state bifurcations, respectively. In the cases shown the marginal stability curve for overstability always falls below that for steady state instability and terminates on it at a Takens-Bogdanov (hereafter TB) point. At this point the oscillation frequency ω vanishes. The weak dependence of the frequency on k when $\delta = 0$ (Figure 2b) is characteristic of modes we refer to as magnetoconvective, and is a consequence of the small value of ζ used. However, as the rotation rate increases ($\delta > 0$) the frequency at small k drops as the mode acquires the characteristics of a slow MAC mode (see below). These include the small frequency as $k \to 0$ and the characteristic power law growth with k, before dissipative effects set in for higher k. In fact the frequencies of all the modes encountered below become independent of k as $k \to 0$ since dissipation has no effect on large scales. At the same time the neutral curve becomes more complicated. For $\delta = 7.5$ the Hopf curve breaks up into two parts with the creation of two additional Takens-Bogdanov points. If δ is increased further, the large wavenumber portion shrinks to zero so that by $\delta = 10.0$ only the lower wavenumber portion remains. However, despite these changes in the oscillatory neutral curve, there is only one type of overstable mode, a slow MAC mode that is continuously connected to the magnetoconvective mode as the parameter $\delta \to 0$; as a result the neutral curves resemble those familiar from the non-rotating problem [6].

For smaller σ the situation is more interesting since we can now have inertial oscillations as well as the magnetoconvective modes. Figure 3 shows an example for $\sigma = 0.01$. As shown in Figures 3a,b the non-rotating case ($\delta = 0$) behaves much like the case $\sigma = 1.0$. However, for sufficiently large rotation (e.g., the case $\delta = 0.1$ shown in Figures 3c,d) a pair of new neutral stability curves appears at small k; of these, the higher one is associated with lower frequency modes and vice versa. The net effect of $\delta \neq 0$ is therefore to split the magnetoconvective mode present when $\delta = 0$ into three modes, each characterized by different $\omega(k)$. In the following we refer to the high frequency mode as the fast MAC mode; its frequency decreases with increasing k. The low frequency mode is referred to as the slow MAC mode; its frequency increases with k. This is the mode already encountered in Figure 2 for $\sigma = 1.0$. Finally, there is a mode of intermediate frequency that is almost independent of k over a large range. In the following we shall refer

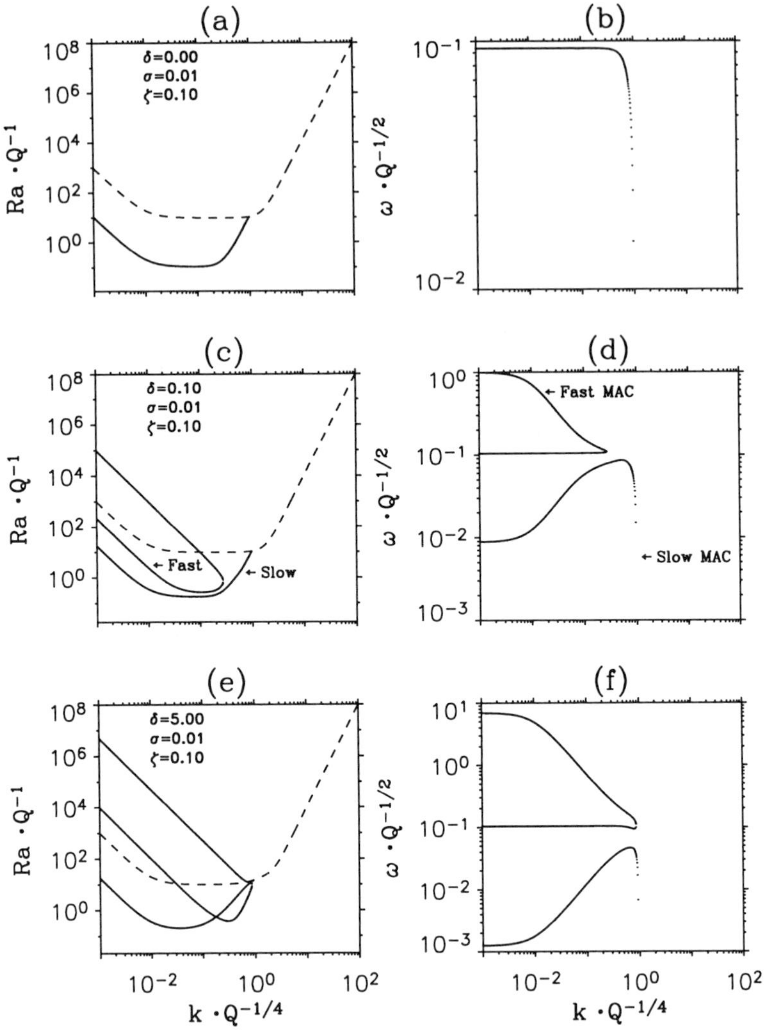

Fig. 3. As for Figure 2 but for $\zeta = 0.1$, $\sigma = 0.01$. (a,b) $\delta = 0.0$; (c,d) $\delta = 1.0$; (e,f) $\delta = 5.0$; (g,h) $\delta = 11.0$; (i,j) $\delta = 12.0$; (k,l) $\delta - 20.0$.

to this mode as the magnetoconvective mode since its frequency is almost independent of the rotation rate, and its dispersion relation resembles that for the magnetoconvective mode in the non-rotating case. However, it must be emphasized that this is so only once the rotation rate is large enough and that rotation is in fact essential for the presence of this mode. Observe that the frequencies of all three modes become comparable at $k = \mathcal{O}(1)$ (in our scaling) allowing them to interact. It is this interaction region that is

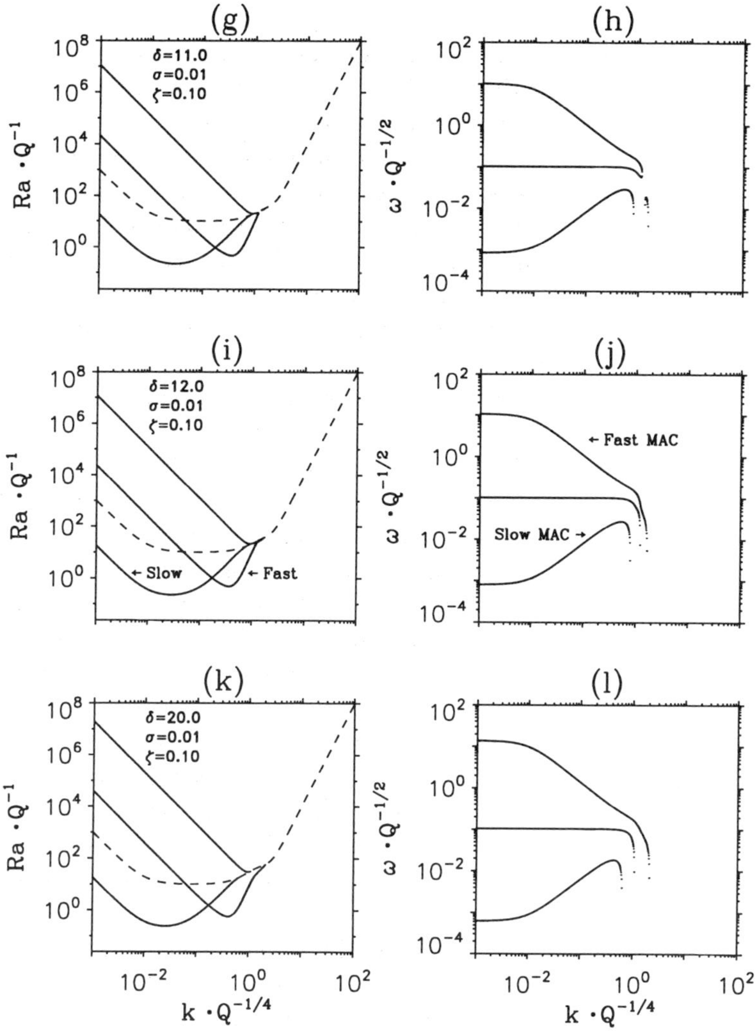

Fig. 3. continued.

of primary interest in this paper. Since the product of the frequencies of the slow and fast modes is almost independent of k (and of δ) and comparable to the square of the magnetoconvective frequency, it is appropriate to think of these two modes as the result of (strongly k-dependent) rotational splitting of the magnetoconvective mode. Note also that in the limit $k \to 0$ the frequencies of all three modes approach constant, k-independent values, while the neutral stability curves remain strongly k-dependent. The two higher neutral stability curves, corresponding to the magnetoconvective mode (upper curve) and the fast MAC mode (lower curve) both depend strongly on the rotation rate as measured by δ. In contrast the lowest neutral curve, corresponding to the slow MAC modes, is almost independent of δ. The fact that the neutral stability curve for the slow modes resembles, for small to moderate values of δ, the shape of the neutral stability curve for the magnetoconvective mode when $\delta = 0$ highlights the close relation between these two modes (see Figure 2). Because of its low frequency the slow mode is the mode most strongly affected by buoyancy, in contrast to the two remaining modes which require buoyancy for their destabilization but remain essentially Alfvénic (the magnetoconvective mode) or inertial (the fast mode).

For larger values of δ the frequency of the fast mode continues to increase while that of the slow mode drops, resulting in a gradual separation of their neutral curves. However, a wavenumber region remains in which the frequencies of these two modes remain comparable resulting in a localized interaction between them. The beginning of this interaction can be seen for $\delta = 5.0$ (Figures 3e,f). The interaction splits the slow mode neutral curve into two by the time $\delta = 11.0$ with the creation of two (additional) Takens-Bogdanov bifurcations, forming an isola of overstability at large k (cf. Figures 2c,d). This isola reconnects with the magnetoconvective and fast modes by $\delta = 12.0$ (Figures 3i,j), with the result that the neutral curves for the fast and magnetoconvective modes now terminate on the steady state neutral curve as well (see Figures 3k,l). However, despite these changes the slow mode remains the mode that sets in first as Ra increases.

Figure 4 shows the results of varying the Prandtl number σ for fixed rotation rate $\delta = 10.0$. This time a pair of new neutral stability curves appears as σ decreases. This is because the dominant damping mechanism for modes that are essentially inertial is provided by viscosity. Consequently one expects such modes to appear as σ decreases. These curves appear as a pair because of magnetic splitting of the inertial mode; the corresponding frequency curves indicate that the top neutral stability curve corresponds to modes we have called magnetoconvective while the lower curve corresponds to the fast MAC modes. Despite the large frequency splitting the thresholds are dominated by rotation and hence are very similar. With decreasing σ these neutral stability curves separate, and begin to interact with the steady neutral curve once $\sigma < 0.01$. The curves first touch and then connect with the steady neutral curve creating two new Takens-Bogdanov

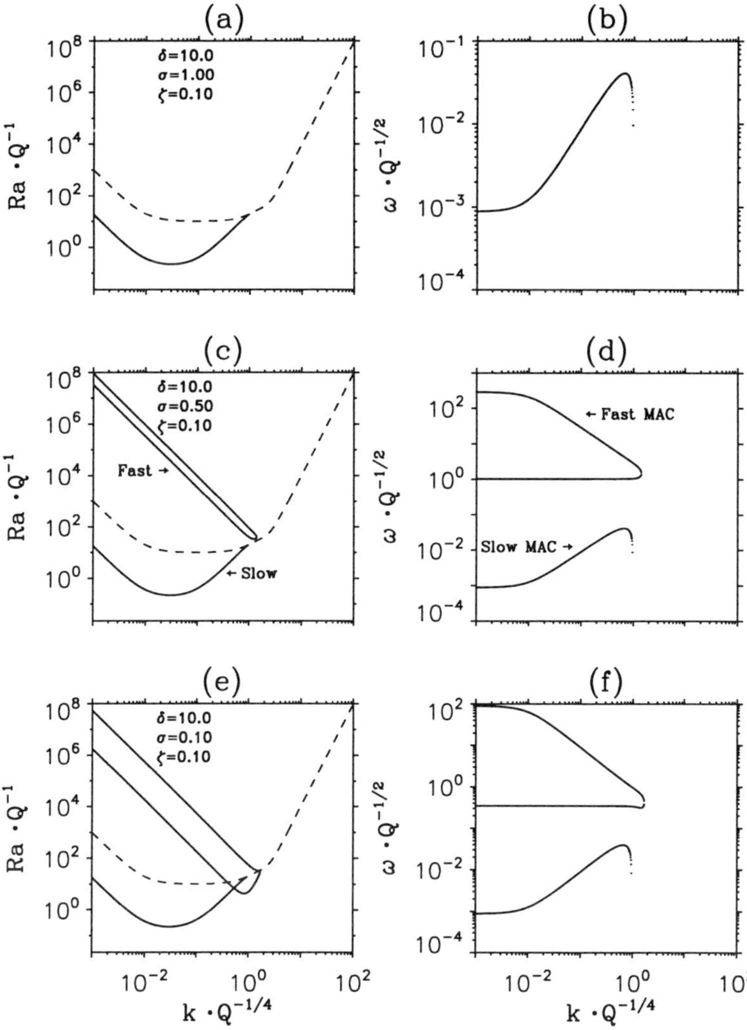

Fig. 4. As for Figure 2 but for $\zeta = 0.1$, $\delta = 10.0$. (a,b) $\sigma = 1.0$; (c,d) $\sigma = 0.5$; (e,f) $\sigma = 0.1$; (g,h) $\sigma = 0.01$; (i,j) $\sigma = 0.001$; (k,l) $\sigma = 0.0001$.

bifurcations, i.e., the frequencies of both modes drop to zero towards large k, much as in Figure 3. The frequency curves for the two modes then pinch off leaving behind an isola of overstability between two Takens-Bogdanov bifurcations (see Figure 4j for $\sigma = 0.001$) and a curve connecting the remaining magnetoconvective and fast modes. The net result is that the neutral stability curves for these modes shift towards smaller k with decreasing σ and hence toward lower Rayleigh numbers, allowing the fast MAC mode

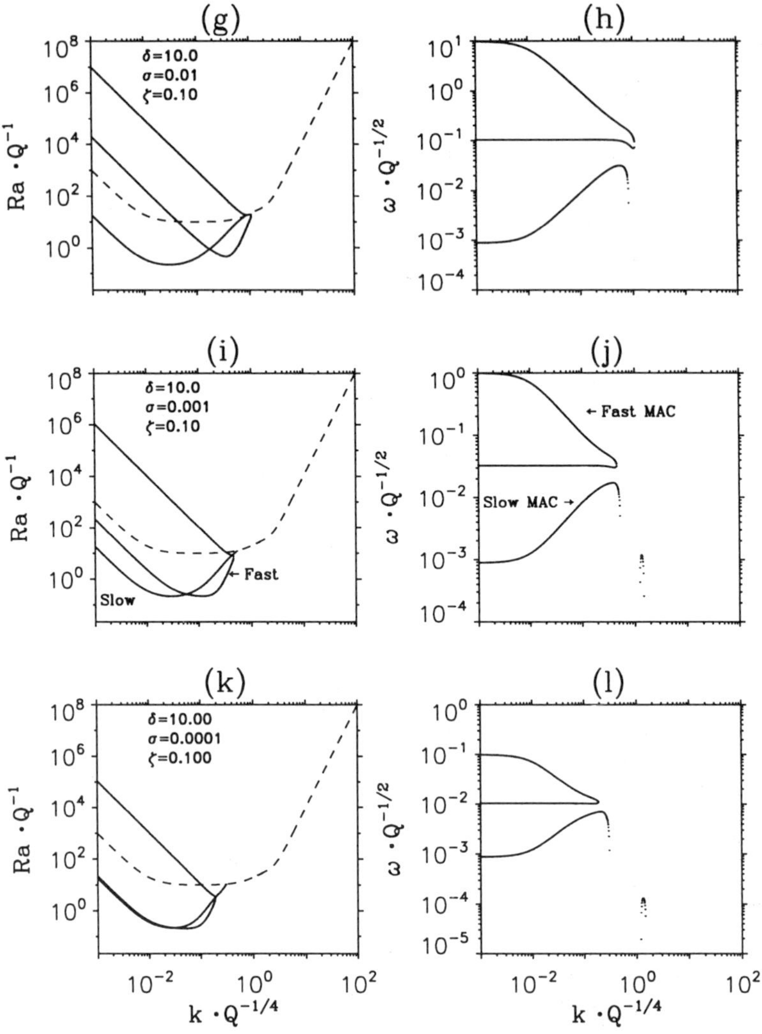

Fig. 4. continued.

to become the first unstable mode. Note that for $\sigma = 0.0001$ the slow and fast MAC modes have almost identical thresholds for small k; this is because the (scaled) frequency of the fast mode drops rapidly with decreasing σ towards the already small frequency of the slow mode, while the latter remains insensitive to the value of the Prandtl number. As a result the frequency difference between the two modes has almost no effect on their stability thresholds. This behavior persists down to $\sigma = 0$. We remark that because of our nondimensionalization small Prandtl numbers correspond to small *viscosity* and not to large thermal conductivity.

4. Asymptotic Analysis

This section is devoted to an analytical understanding of the dispersion relation (6) derived above. We consider various asymptotic regimes motivated by astrophysical considerations, followed by a study of the long wavelength properties of this relation from which a particularly simple picture of the various modes emerges.

4.1. *The Large Q, Ta Limit*

In this section we focus on large horizontal wavenumbers and frequencies, both of which are favored for large Q and/or Ta. We suppose therefore that $\omega = \mathcal{O}(Q^p)$. In order to retain at least some effects of horizontal diffusion in the analysis that follows we also suppose that k (and hence a) is $\mathcal{O}(Q^{p/2})$. It is necessary to distinguish between three cases: $p > \frac{1}{2}$, $p = \frac{1}{2}$ and $p < \frac{1}{2}$. The first of these cases we consider is $p > \frac{1}{2}$, and in the limit $Q \to \infty$ we set $\delta \equiv TaQ^{-3p}$ and take δ to be of order one. Here the rotation is so strong that magnetic field effects drop out. The problem then reduces simply to rapidly rotating convection, described by

$$\left(\frac{i\omega}{\sigma} + k^2\right)^2 k^2 + \pi^2\delta = \left(\frac{i\omega}{\sigma} + k^2\right)\frac{k^2\hat{R}}{i\omega + k^2}, \tag{7}$$

where $\hat{R} \equiv RaQ^{-2p}$, and both ω and k are the corresponding scaled quantities and hence of order one. Thus $Ra = \mathcal{O}(Ta^{\frac{2}{3}})$ for all $p > \frac{1}{2}$. Magnetic effects can be brought in at higher order by writing $\hat{R} = R_0 + R_1Q^{1-2p} + \mathcal{O}(Q^{2-4p})$ and similarly for the frequency. We refer to this case as Regime I. This regime is the least interesting (see remark at the end of Section 4.1) and we do not consider it further.

The case of greatest interest is the case $Ta = \mathcal{O}(Q^{\frac{3}{2}})$ corresponding to $p = \frac{1}{2}$, hereafter Regime II. This regime corresponds to the slowest rotation rate for fixed Q of all three regimes. Here we consider wavenumbers $k = \mathcal{O}(Q^{\frac{1}{4}}) = \mathcal{O}(Ta^{\frac{1}{6}})$ and frequencies $\omega = \mathcal{O}(Q^{\frac{1}{2}}) = \mathcal{O}(Ta^{\frac{1}{3}})$, and find that $Ra = \mathcal{O}(Q) = \mathcal{O}(Ta^{\frac{2}{3}})$. This particular scaling with Ta was identified in [5, 8] for the preferred mode in rapidly rotating convection with no magnetic field. The scaling for the critical Rayleigh numbers and frequencies also agrees with that obtained in [6, 9] for (nonrotating) magnetoconvection in a strong magnetic field, viz. $Ra = \mathcal{O}(Q)$ and $\omega = \mathcal{O}(Q^{\frac{1}{2}})$, even though the wavenumber scaling corresponds to that of the Takens-Bogdanov point rather than that for the preferred wavenumber which scales as $k = \mathcal{O}(Q^{\frac{1}{6}})$. However, in earlier work on (nonrotating) magnetoconvection we noted [6] that scaling the wavenumbers with $Q^{\frac{1}{4}}$ instead of $Q^{\frac{1}{6}}$ allows us to retain the effects of horizontal dissipation at leading order, and hence all the essential features of the linear problem, including wavenumber selection and the appearance of Takens-Bogdanov bifurcations as the wavenumber changes. Moreover, the scaled dispersion relation predicts the correct critical Rayleigh numbers and frequencies in the limit of (scaled) $k \to 0$, i.e., their values at the true minimum of the neutral stability curve at $k = \mathcal{O}(Q^{\frac{1}{6}})$. In this regime the linear frequencies of all three modes are $\mathcal{O}(Q^{\frac{1}{2}})$ for $\mathcal{O}(1)$ Prandtl numbers, and the dispersion relation that results,

$$\left(\frac{i\omega}{\sigma} + k^2 + \frac{\zeta\pi^2}{i\omega + \zeta k^2}\right)^2 k^2 + \pi^2\delta = \left(\frac{i\omega}{\sigma} + k^2 + \frac{\zeta\pi^2}{i\omega + \zeta k^2}\right)\frac{k^2\hat{R}}{i\omega + k^2}, \quad (8)$$

retains all the essential properties of the original dispersion relation (6), as discussed further below. Here $\delta \equiv TaQ^{-\frac{3}{2}}$ and $\hat{R} = RaQ^{-1}$.

The final regime (Regime III) corresponds to $p < \frac{1}{2}$ and retains the full influence of the magnetic field even though the rotation rate is larger (for fixed Q) than in Regime II. In this case the dispersion relation (6) becomes

$$\pi^2 + \frac{\delta}{\zeta^2 k^2}(i\omega + \zeta k^2)^2 = \frac{\hat{R}}{\zeta}\left(\frac{i\omega + \zeta k^2}{i\omega + k^2}\right), \quad (9)$$

where $\delta \equiv TaQ^{p-2}$ and $\hat{R} = RaQ^{-1}$. Observe that in this scaling the Prandtl number σ drops out. As a result the two MAC modes are filtered out. However, as suggested by Figure 4, we can restore the influence of the Prandtl number by scaling it appropriately with Q: $\sigma \equiv \hat{\sigma}Q^{2p-1}$, where $\hat{\sigma} = \mathcal{O}(1)$. In this case, which applies to asymptotically small Prandtl numbers, the dispersion relation (6) becomes

$$\left(\frac{i\omega}{\hat{\sigma}} + \frac{\zeta\pi^2}{i\omega + \zeta k^2}\right)^2 k^2 + \pi^2\delta = \left(\frac{i\omega}{\hat{\sigma}} + \frac{\zeta\pi^2}{i\omega + \zeta k^2}\right)\frac{k^2\hat{R}}{i\omega + k^2}, \quad (10)$$

where $\delta \equiv TaQ^{p-2}$ and $\hat{R} = RaQ^{-1}$ as before. In the following we refer to this regime as III'.

It is useful to list the order of magnitude of the perturbations in the different regimes identified above. For two-dimensional instabilities in the form of rolls with axes in the y direction, we write $\mathbf{B} = (a, b, 1 + c)$, and suppose that the temperature perturbation $\theta = \mathcal{O}(1)$. It follows that in all three regimes $w = \mathcal{O}(Q^p)$, while in regime I ($p > \frac{1}{2}$): $u = \mathcal{O}(Q^{\frac{p}{2}})$, $v =$

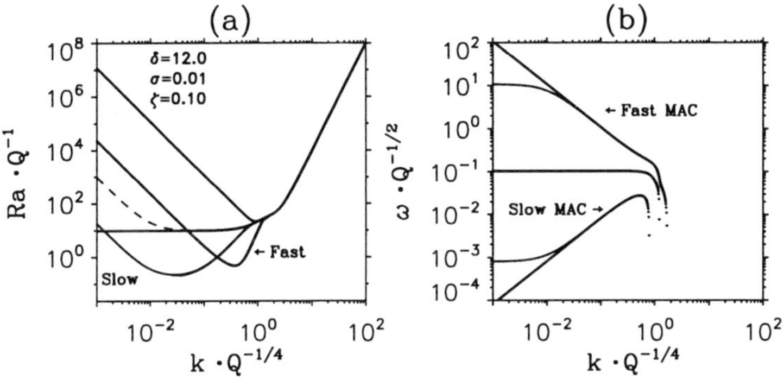

Fig. 5. Comparison of Figure 3(i,j) and the asymptotic dispersion relation (8). The two agree for $k > 3 \times 10^{-2} Q^{\frac{1}{4}}$.

$\mathcal{O}(Q^{3p-1}) \gg w$, and $a = \mathcal{O}(Q^{-\frac{p}{2}})$, $b = \mathcal{O}(Q^{2p-1})$, while in regime II ($p = \frac{1}{2}$): $u = \mathcal{O}(Q^{\frac{1}{4}})$, $v = \mathcal{O}(Q^{\frac{1}{2}}) = \mathcal{O}(w)$, and $a = \mathcal{O}(Q^{-\frac{1}{4}})$, $b = \mathcal{O}(1)$. Finally in regime III ($p < \frac{1}{2}$): $u = \mathcal{O}(Q^{\frac{p}{2}})$, $v = \mathcal{O}(Q^{p}) = \mathcal{O}(w)$, and $a = \mathcal{O}(Q^{-\frac{p}{2}})$, $b = \mathcal{O}(1)$. Note that the regime I and III results agree with those of regime II when $p = \frac{1}{2}$. One can check that in all three cases the perturbations in the y direction are in magnetostrophic balance, a consequence of the vanishing pressure gradient in this direction.

We now explore the adequacy of these simplified dispersion relations for the parameter values of interest. As already mentioned regime II focuses on wavenumbers $k = \mathcal{O}(Q^{\frac{1}{4}})$, picking out the power law growth and decrease of the frequencies of the slow and fast modes in Figures 2-4. Figure 5 compares the results obtained from the simplified dispersion relation (8) (thick lines) with those from the full relation (thin lines). Observe that the dispersion relation (8) captures accurately the behavior of all three modes for all $k > 3 \times 10^{-2} Q^{\frac{1}{4}}$, with discrepancies observable at smaller values of k only, i.e., the simplified dispersion relation focuses precisely on those properties that are characteristic of the three modes.

In contrast Figure 6 shows the results of solving the asymptotic dispersion relation (10), Regime III′, for $\hat{\sigma} = 1.0$, $\zeta = 0.1$ and several values of δ in the case $p = \frac{1}{3}$. Since this regime focuses on small Prandtl numbers it comes as no surprise that one finds in general three neutral stability curves for the onset of oscillations. However, as in the full problem (6), only one of these is present when $\delta = 0$, the magnetoconvective mode. Figure 6a shows that this mode sets in with $k = 0$, indicating a larger scale than assumed in our scaling. However, as soon as δ is nonzero this mode becomes the slow MAC mode and its neutral stability curve acquires a minimum at a finite value of the (scaled) wavenumber. At the same time two new neutral curves appear at small wavenumbers and move to the right with increasing δ, much as in the full dispersion relation (Figure 3) for small (unscaled) σ. The neutral stability curve for the fast mode introduced in this way remains nearby during this process, and one may expect both modes to be important near onset.

With increasing δ a new Takens-Bogdanov bifurcation appears at large wavenumbers (Figure 6e), bringing in a new neutral curve of fast modes extending to yet larger wavenumbers. For larger values of δ the two TB bifurcations on the steady state neutral curve collide and the combined fast mode neutral curve detaches from the steady state neutral curve and now extends across all (scaled) wavenumbers. Analogous behavior takes place with increasing $\hat{\sigma}$. Figure 6e should be compared with the region near $kQ^{-\frac{1}{4}} \approx 1$ in Figure 3g,h for $\sigma = 0.01$. Evidently, the rightmost Takens-Bogdanov bifurcation is off-scale in the scaled theory, and it is for this reason that the scaled problem possesses a fast mode extending to large wavenumbers. These transitions are similar to those seen for the full dispersion relation, as seen on comparing Figures 6j,l with Figures 3h,j. In both cases, the corresponding frequency curves show how well the scaled theory captures the behavior of the linear stability problem in the regime $k = \mathcal{O}(Q^{\frac{1}{6}})$. Any discrepancies can be attributed to the fact that the values of δ and σ in the two sets of plots are, in some cases, somewhat different. Figure 7 shows a blow up of some of the reconnections that take place between Figures 6j and 6l.

Figure 8 shows that when $\zeta = 1.0$ the situation is considerably simpler. The slow mode is now absent, leaving the fast and magnetoconvective modes, although these too undergo reconnections as δ varies. In particular, with increasing δ the magnetoconvective mode terminates in a TB bifurcation, and the two parts of the fast mode neutral curve reconnect (see Figures 6g,i). As a result, for large enough δ overstability is present at all k; this is not necessarily so for smaller rotation rates.

The simplicity of the dispersion relation (9) in regime III allows us to describe analytically many of the features of the numerical results in Figures 2-4 that otherwise appear somewhat mysterious. The relation (9) leads immediately to

$$\hat{R}_{osc} = \frac{2\zeta^2\pi^2}{1+\zeta} + 2\delta k^2(1+\zeta) \tag{11}$$

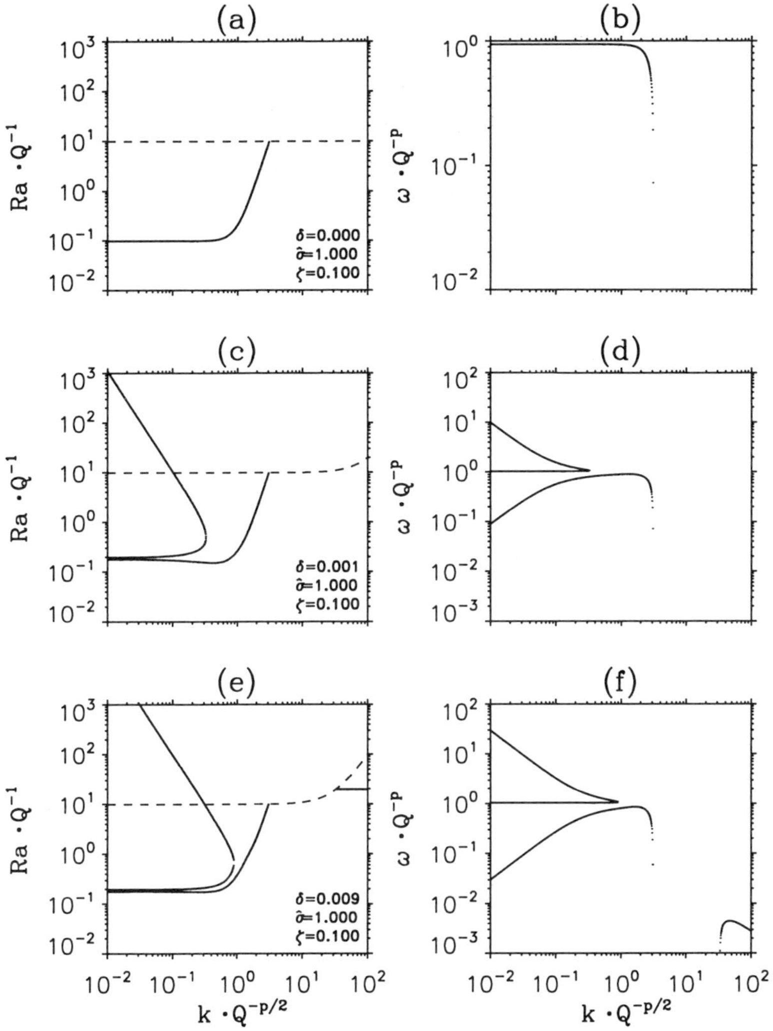

Fig. 6. Neutral stability curves $Ra(k)$ for steady (dashed line) and oscillatory (solid line) convection, and the corresponding frequency $\omega(k)$ in regime III′ ($Ta = \delta Q^{2-p}, p = 1/3$) with $\zeta = 0.1$, $\hat{\sigma} = 1.0$. (a,b) $\delta = 0.0$; (c,d) $\delta = 0.001$; (e,f) $\delta = 0.009$; (g,h) $\delta = 0.1$; (i,j) $\delta = 0.5$; (k,l) $\delta = 1.0$.

$$\omega^2 = -\zeta^2 k^4 + \frac{\zeta^2(1-\zeta)}{1+\zeta}\frac{k^2\pi^2}{\delta}. \qquad (12)$$

Note that this regime has filtered out two of the three oscillatory modes, leaving only the slow MAC mode ($\delta \neq 0$). The corresponding critical

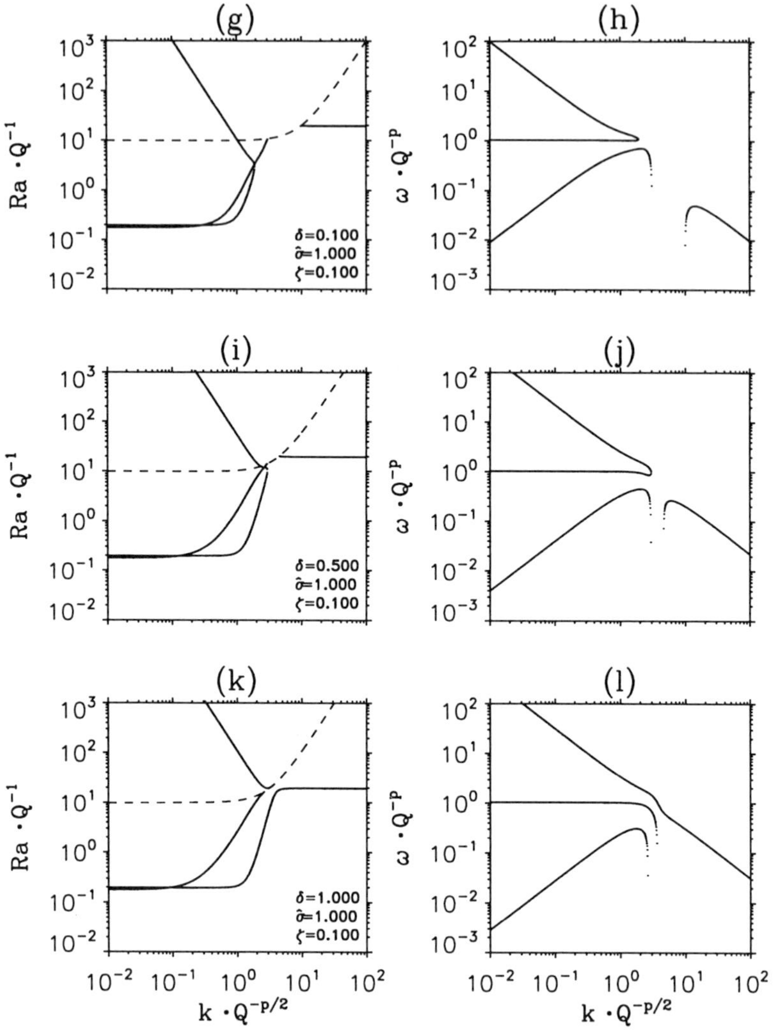

Fig. 6. continued.

Rayleigh number for the onset of steady convection is

$$\hat{R}_s = \pi^2 + \delta k^2. \tag{13}$$

Thus in either case convection sets in with zero wavenumber, formally violating the implicit assumption that (the scaled) k is $\mathcal{O}(1)$.

The linear theory results reveal an unusual property: when $\delta \equiv 0$ (i.e., in the "non-rotating" case) no oscillations are present and the primary instability sets in at $\hat{R}_* = \pi^2$, in apparent conflict with the limit $\delta \downarrow 0$

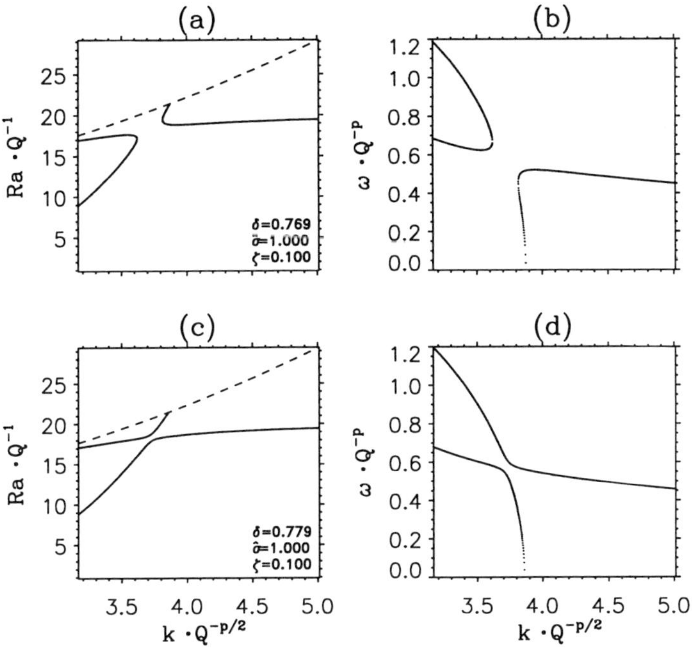

Fig. 7. Blowup of the reconnection of neutral stability curves in Figures 6i,k.

which predicts very high frequency oscillations (if $1-\zeta>0$) and a *lower* threshold Rayleigh number:

$$\hat{R}_0 = \frac{2\zeta^2\pi^2}{1+\zeta}. \tag{14}$$

Note that this is in fact the minimum of the Rayleigh number (11) as a function of k.

Figure 3 helps us understand what is happening. The figure shows that for $\delta = 0$ the magnetoconvective frequency is relatively large before dropping precipitously to $\omega = 0$ as k increases. The figure also shows that the corresponding frequency (the slow MAC frequency) drops dramatically as δ increases (Figures 3d,f). Moreover, for small δ the neutral stability curves for both steady and oscillatory modes are almost independent of the wavenumber k unless this is either very small or becomes comparable to its value at the Takens-Bogdanov point. These results are all consistent with the analytic predictions (11-14) in their range of applicability, viz. $k = \mathcal{O}(Q^{\frac{1}{6}})$, corresponding to $kQ^{-\frac{1}{4}}$ in the range 0.01 - 0.1 in Figure

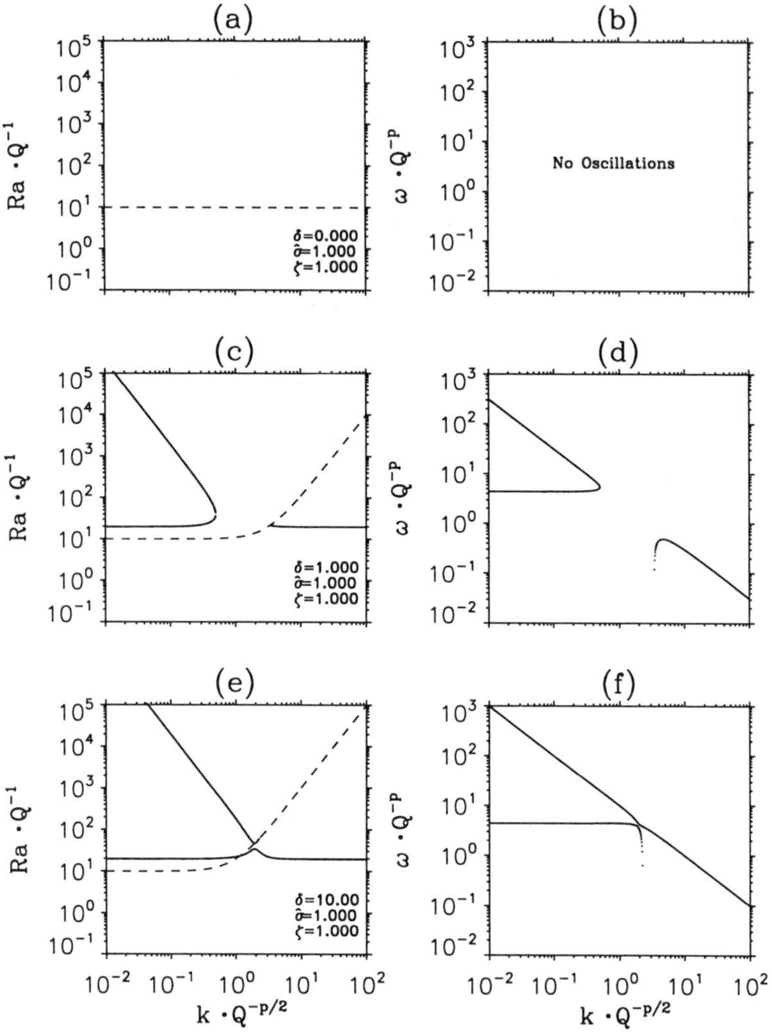

Fig. 8. Neutral stability curves $Ra(k)$ for steady (dashed line) and oscillatory (solid line) convection, and the corresponding frequency $\omega(k)$ in regime III′ ($Ta = \delta Q^{2-p}, p = 1/3$) with $\zeta = 1.0$, $\hat{\sigma} = 1.0$. (a,b) $\delta = 0.1$; (c,d) $\delta = 1.0$; (e,f) $\delta = 10.0$.

3. In the present scaling the TB wavenumber is given by

$$\delta k_{TB}^2 = \frac{1-\zeta}{1+\zeta}\,\pi^2; \tag{15}$$

at this point the critical Rayleigh numbers for oscillatory and steady convection coincide. Thus as $\delta \downarrow 0$ the TB bifurcation moves to infinite wavenumbers, and the wavenumber dependence drops out. Moreover, it is a simple matter to check that an $\mathcal{O}(1)$ frequency in the scaling of Figure 3 corresponds to an $\mathcal{O}(Q^{\frac{1}{2}-p})$ frequency in regime III, thereby explaining the divergence of the frequency with decreasing δ in equation (12). These facts reflect the presence of a crossover from regime II to regime III with increasing rotation rate (relative to the magnetic field).

We remark that we have presented Figures 6-8 for $p = \frac{1}{3}$ only. In this regime (Regimes III, III′) $Ta = \mathcal{O}(Q^{\frac{5}{3}})$ but magnetic effects are retained. We can, however, get the same rotation rate, $Ta = \mathcal{O}(Q^{\frac{5}{3}})$, in regime I as well. To this end we must take $p = 5/9$. Performing the scalings used for regime I we find that magnetic effects scale out. Thus for identical values of the external parameters different internal balances isolate different modes, and lead to a different picture of the dynamics of the system. Evidently the balances quantified by the scaling used in Regime III retain the essence of the system, while those used in Regime I do not.

4.2. *The Limit* $k \to 0$

In Section 2 we used the long wave (small k) behavior of the three modes present to identify and label the modes concerned. It is helpful to characterize this behavior precisely.

4.2.1. *The Full Dispersion Relation (6)*

In this section we examine the long wave limit of the dispersion relation (6). The key to understanding the behavior shown in Figures 2-4 is the observation that if $Ta = \mathcal{O}(Q^{1+\varepsilon})$ for any $\varepsilon > 0$, however small, rotation will dominate the mode structure of the problem. We focus here on this case since it contains Regime II, which as already explained is of greatest interest. In this case we can look for the magnetoconvective modes in the form $\omega^2 = \omega_0^2 + \mathcal{O}(k^2)$, $Ra = R_0/k^2 + \mathcal{O}(1)$, where $\omega_0 \gg 1$, $R_0 \gg 1$ because Q (and Ta) is large. We then find

$$\omega^2 = \sigma\zeta Q\pi^2 \frac{1+\zeta}{1-\sigma} + \mathcal{O}(k^2), \qquad Ra = \frac{\sigma(1+\zeta)}{\sigma+\zeta}\frac{\pi^2}{k^2}Ta + \mathcal{O}(1). \tag{16}$$

Thus the asymptotic frequency of magnetoconvective modes is independent of the rotation rate and of the wavenumber k, while the threshold for the excitation of this mode increases with Ta, but decreases with increasing k.

If, in contrast, one looks for modes with frequency larger than $\mathcal{O}(Q^{\frac{1}{2}})$ one finds the fast MAC mode:

$$\omega^2 = \frac{\sigma^2(1-\sigma)}{1+\sigma}Ta + \mathcal{O}(k^2), \qquad Ra = \frac{2\sigma^2}{1+\sigma}\frac{\pi^2}{k^2}Ta + \mathcal{O}(1). \tag{17}$$

Evidently, this mode is the inertial wave, and its long wave properties are entirely independent of the magnetic field. Note in particular that the critical Rayleigh number increases with Ta and decreases with increasing k.

Finally, we also need to locate the slow MAC mode. To this end we suppose that the frequency is smaller than $\mathcal{O}(Q^{\frac{1}{2}})$. We find that

$$\omega^2 = \frac{\zeta^2(1-\zeta)}{1+\zeta}\frac{Q^2\pi^4}{Ta} + \mathcal{O}(k^2), \qquad Ra = \frac{2\zeta^2}{1+\zeta}\frac{\pi^4}{k^2}Q + \mathcal{O}(1). \tag{18}$$

Evidently, this mode is a pure thermally excited Alfvén oscillation, and its critical Rayleigh number is independent of the rotation. Consequently the threshold for this mode is independent of the rotation rate δ, as noted in Figures 2-4.

Thus both of these modes only acquire their MAC character at finite k, and it is at these wavenumbers that the characteristic decrease of the fast frequency with k and the corresponding increase in the slow frequency first manifest themselves, a property that results in their interaction with one another as well as with the magnetoconvective mode. It is this behavior that is identified by the scaling leading to Regime III.

It should be emphasized that different asymptotic results hold in the case $Ta = \mathcal{O}(Q^{1-\varepsilon})$, $\varepsilon > 0$. In this case the magnetic field determines the main properties of the system, with the rotation playing a secondary role. For brevity we omit the results.

4.2.2. *Regime II: The Dispersion Relation (8)*

The dispersion relation (8) can be used to extend the long wave results derived from (6) into a more interesting regime in which some of the k dependence is retained. This is of course because this dispersion relation in fact only applies for $k \gg \pi$. Hence the long wave limit investigated in this section corresponds to (unscaled) $k \sim \pi$. The resulting frequency and the critical Rayleigh number are again determined from the real and imaginary parts of (8). The magnetoconvective mode is characterized by $\omega = \omega_0 + \mathcal{O}(k^2)$, $R = R_0/k^2 + \mathcal{O}(1)$. This time one finds

$$\omega^2 = \sigma\zeta\pi^2\frac{1+\zeta}{1-\sigma} + \mathcal{O}(k^2), \qquad Ra = \frac{\sigma(1+\zeta)}{\sigma+\zeta}\frac{\pi^2\delta}{k^2} + \mathcal{O}(1). \tag{19}$$

Thus as leading order the frequency of the magnetoconvective mode is independent of both k and δ, while the threshold decreases both with increasing k and at fixed k with decreasing δ.

The fast mode is characterized by the properties $\omega^2 = \lambda/k^2 + \mathcal{O}(1)$, $R = R_0/k^2 + \mathcal{O}(1)$, where λ and R_0 are $\mathcal{O}(1)$ quantities. Straightforward analysis now shows that

$$\omega^2 = \frac{1-\sigma}{1+\sigma}\frac{\sigma^2\pi^2\delta}{k^2} + \mathcal{O}(1), \qquad R = \frac{2\sigma^2}{1+\sigma}\frac{\pi^2\delta}{k^2} + \mathcal{O}(1). \tag{20}$$

These relations describe not only the decrease of the frequency of the fast mode with increasing k but indicate in addition that at fixed k the frequency increases as $\delta^{\frac{1}{2}}$, while the critical Rayleigh number increases as δ.

In contrast, the slow mode is characterized by $\omega^2 = \lambda k^2 + \mathcal{O}(k^4)$, $R = R_0 + \mathcal{O}(k^2)$. This time one finds

$$\omega^2 = \frac{\zeta^2(1-\zeta)}{1+\zeta}\frac{\pi^2 k^2}{\delta} + \mathcal{O}(k^4), \qquad R = \frac{2\zeta^2\pi^2}{1+\zeta} + \mathcal{O}(k^2). \tag{21}$$

It follows that the slow mode is entirely independent of the Prandtl number in the long wavelength limit. Its frequency increases with k, but at fixed k the frequency decreases as $\delta^{-\frac{1}{2}}$. Once again the threshold R_0 is independent of both k and the rotation rate, in contrast to the fast mode (20). In all cases the above predictions are in perfect agreement with the numerical results in Figure 5.

4.2.3. *Regime III′: The Dispersion Relation (10)*

It remains to consider the dispersion relation (10) valid in regime III′. The magnetoconvective mode again takes the form $\omega = \omega_0 + \mathcal{O}(k^2)$, $R = R_0/k^2 + \mathcal{O}(1)$. This time one finds

$$\omega^2 = \hat{\sigma}\zeta(1+\zeta)\pi^2 + \mathcal{O}(k^2), \qquad R = (1+\zeta)\frac{\hat{\sigma}}{\zeta}\frac{\pi^2\delta}{k^2} + \mathcal{O}(1). \tag{22}$$

Here $\hat{\sigma}$ is the scaled Prandtl number. Thus at leading order the frequency of the magnetoconvective mode is independent of δ, while the threshold decreases both with increasing k and at fixed k with decreasing δ.

The fast mode takes the different form $\omega^2 = \lambda/k^2 + \beta + \mathcal{O}(k^2)$, $R = R_0 + \mathcal{O}(k^2)$, where λ, β and R_0 are $\mathcal{O}(1)$ quantities. Straightforward analysis now shows that

$$\omega^2 = \frac{\pi^2\hat{\sigma}^2\delta}{k^2} + 2\hat{\sigma}\zeta(1-\zeta)\pi^2 + \mathcal{O}(k^2), \qquad R = 2\zeta^2\pi^2 + \mathcal{O}(k^2). \tag{23}$$

These relations describe not only the decrease of the frequency of the fast mode with increasing k but indicate in addition that at fixed k the frequency increases as $\delta^{\frac{1}{2}}$, while the critical Rayleigh number is unaffected by changes in the rotation rate and is asymptotically independent of k, in contrast to the result (20) for $p = \frac{1}{2}$. This is a consequence of the fact that the limits $k \to 0$ and $\sigma \to 0$ do not commute.

In contrast, the slow mode is independent of the Prandtl number and hence continues to be described by the asymptotic relations (21). It follows that in the limit $k \to 0$ the slow modes set in only slightly before the fast modes, the difference in thresholds decreasing with decreasing ζ. These predictions are in perfect agreement with the numerical results in Figures 6-8.

5. Discussion

In this note we have explored the properties of the dispersion relation characterizing the onset of convection in a rotating layer in the presence of an imposed vertical magnetic field. Although the basic dispersion relation is valid for both two- and three-dimensional disturbances we have emphasized the properties of two-dimensional roll-like perturbations. For sufficiently small values of the two Prandtl numbers σ and ζ the dispersion relations admits three modes. We have called these the slow and fast MAC (Magneto-Archimedean-Coriolis) modes by analogy with earlier work by Braginskii, with a third mode called the magnetoconvective mode. We have focused on the astrophysically relevant regime of strong rotation and strong magnetic field, and identified three distinct asymptotic regimes, characterized by the relative strength of the rotation and magnetic field. Each mode has a characteristic dispersion curve $\omega(k)$, where k is the wavenumber. As rotation increases the usual overstable convection mode turns into the slow MAC mode. The frequency of this mode is low and increases with increasing k, in contrast to the fast MAC mode which is essentially an inertial wave whose frequency increases with the rotation rate but decreases with k. The magnetoconvective mode requires sufficiently fast rotation for its presence but its frequency is independent of the rotation and almost independent of k, at least until the dissipative regime is reached. The critical Rayleigh numbers of all three modes decrease as k^{-2} reaching a minimum somewhere between $k = \mathcal{O}(Q^{\frac{1}{6}})$ and $\mathcal{O}(Q^{\frac{1}{4}})$, and then increase at higher wavenumbers. In-between the modes undergo a sequence of complicated interactions among themselves and the steady state instability, mediated by the creation and destruction of Takens-Bogdanov points. We have explored this interaction as parameters are varied both within the exact dispersion relation and within several simplified asymptotic regimes, in all of which the perturbations remain in magnetostrophic balance. Somewhat similar transitions occur in other problems described by higher order dispersion relations, as shown already by Pearlstein [10]. We anticipate that these interactions continue into the nonlinear regime, and will describe the corresponding results elsewhere.

Acknowledgements

We acknowledge partial support from NASA grant NASW-99026 (KJ) and NSF grant DMS-0072444 (EK).

References

[1] S. Chandrasekhar, *Hydrodynamic and Hydromagnetic Stability*, Oxford University Press (1961).
[2] P. J. Amado, *Observatory* **118**, 247 (1998).
[3] S. I. Braginskii, *Geomag. Aeron.* **7**, 851 (1967).
[4] S. M. Cox and P. C. Matthews, *Physica* **D 149**, 210 (2001).
[5] K. Julien and E. Knobloch, *Phys. Fluids* **9**, 1906 (1997).
[6] K. Julien, E. Knobloch and S. Tobias, *Physica* **D 128**, 105 (1999).
[7] T. Clune and E. Knobloch, *Phys. Rev.* **E 47**, 2536 (1993).
[8] A. P. Bassom and K. Zhang, *Geophys. Astrophys. Fluid Dyn.* **76**, 223 (1994).
[9] P. C. Matthews, *J. Fluid Mech.* **387**, 397 (1999).
[10] A. Pearlstein, *J. Fluid. Mech.* **103**, 389 (1981).

CHAPTER 6

PATTERN FORMATION ON A SPHERE

Paul C. Matthews

School of Mathematical Sciences, University of Nottingham, Nottingham NG7 2RD, UK

Pattern formation on the surface of a sphere is described by equations involving interactions of spherical harmonics of degree l. When l is even, the bifurcation is transcritical and the leading-order equations are determined uniquely by the symmetry. Existence and stability results are described for the general case of even l and specifically for even values of l up to 28. Using either a variational or eigenvalue criterion, the preferred solution has icosahedral symmetry for $l = 6$, $l = 10$ and $l = 12$. These states are unstable at onset but can become stable at a saddle-node bifurcation if cubic terms are included. The resulting bifurcation diagrams are given for $l = 4$ and $l = 6$.

1. Introduction

This paper is concerned with pattern formation on the surface of a sphere, or, more generally, the question of instability of any system that possesses a spherically symmetric steady state. One of the earliest statements of the problem was in the classic paper of Turing [1] on the problem of morphogenesis – how an embryo, which is initially a spherical ball of cells, starts to develop its shape. Turing posed the problem as follows: "A system which has spherical symmetry, and whose state is changing because of chemical reactions and diffusion, will remain spherically symmetrical for ever. ... It certainly cannot result in an organism such as a horse, which is not spherically symmetrical." Turing answered the question by proposing an instability mechanism that combines chemical reactions with molecular diffusion, initiating the theory of pattern formation in reaction–diffusion equations. Most of Turing's paper was concerned with the case of a circular ring of cells, where the mathematics is relatively straightforward, but he briefly discussed the more complicated case of spheres, where "there are a number of different patterns with the same wave-length, which can be superposed with various amplitude factors". This degeneracy gives the problem great mathematical complexity, which is unlikely to ever be fully resolved, since as the degree l of the spherical harmonics increases, so does the number $2l+1$ of linearly independent eigenfunctions with the same eigenvalue.

Recent numerical simulations of reaction–diffusion equations have shown a variety of interesting and complicated patterns [2–4]. Another biological application of the breaking of spherical symmetry is the formation of non-spherical structure in tumors [2, 5]. Physically motivated applications include the buckling of a sphere under uniform external pressure [6, 7], and the much-studied problem of convection in a spherical shell [8–11], where patterns are formed by the arrangement of rising and sinking plumes of fluid.

Theoretical investigation of symmetry breaking from spherical, or $O(3)$ symmetry has exercised mathematicians for three decades. Significant progress was made by Busse [8], who studied the problem in the context of convection but was clearly aware that the resulting equations did not depend on the details of the problem; this work formulated an important variational principle and identified a number of patterns for $l = 4$, $l = 6$ and $l = 8$. A more abstract approach, employing group theory and representation theory, was pioneered by Sattinger [12], who showed that the quadratic terms in the bifurcation equations are uniquely determined by the spherical symmetry. Sattinger also discussed the variational structure of the equations, and obtained explicit results on the existence and stability of solutions for $l = 4$. Ihrig and Golubitsky [13] made a systematic study of the subgroups of $O(3)$ and computed the dimension of their fixed-point subspaces for all values of l (some of these were given earlier by Michel [14]); using these results they were able to apply the equivariant branching lemma to demonstrate the existence of stationary patterns with several different symmetry groups. Probably the most comprehensive work on the subject is the paper of Chossat, Lauterbach and Melbourne [15], who derived a number of general results on the existence and stability of solution branches, and carried out detailed calculations for the cases $l = 3$, 4 and 5.

The subject continues to generate interest in the 21st century, due to its various applications and mathematical complexity. Further progress on the existence and stability of solutions has been made recently by Matthews [4, 16] and Callahan [17], while numerical simulations of convection in a spherical shell continue to generate patterns that present theoretical challenges [11].

2. Spherical Harmonics and their Properties

The spherical harmonics $Y_l^m(\theta, \varphi)$ form the natural basis for the description of patterns in spherical coordinates, $0 \leq \theta \leq \pi$, $0 \leq \varphi < 2\pi$. They are eigenfunctions of the Laplacian on the surface of a sphere, with eigenvalue $-l(l+1)/R^2$, where R is the radius of the sphere. So if a pattern formation problem has a wavenumber k_0 at which the growth rate is maximized, this selects a particular value of the integer l as the value which is closest to satisfying

$$l(l+1) = R^2 k_0^2. \tag{1}$$

Hence, the value of l increases as the size of the sphere increases, and l may be thought of as a kind of discrete wavenumber. It is natural to scale lengths with the radius R, so that the eigenvalue of the Laplacian is simply $-l(l+1)$.

The function $Y_l^m(\theta, \varphi)$ is proportional to $\exp im\varphi$ and a polynomial of degree l in $\sin\theta$ and $\cos\theta$. For any value of l there are $2l+1$ spherical harmonics, with $-l \leq m \leq l$, all with the same eigenvalue. The spherical harmonics are complex (except for $m = 0$) and obey

$$Y_l^{-m}(\theta, \varphi) = (-1)^m Y_l^{m*}(\theta, \varphi), \tag{2}$$

so we may restrict attention to $m > 0$. The functions are orthonormal on the sphere, so

$$\int_0^{2\pi} \int_0^{\pi} Y_l^m(\theta, \varphi)\, Y_l^{n*}(\theta, \varphi) \sin\theta \, d\theta \, d\varphi = \delta_{mn}. \tag{3}$$

Two important symmetry properties are the reflection symmetry

$$Y_l^m(\theta, -\varphi) = Y_l^{m*}(\theta, \varphi), \tag{4}$$

and the inversion symmetry

$$Y_l^m(\pi - \theta, \pi + \varphi) = (-1)^l Y_l^m(\theta, \varphi), \tag{5}$$

relating two opposite points on the sphere. The symmetry (5) means that for even l, all the spherical harmonics, and hence all the patterns they can generate, have the property that opposite points on the sphere are equivalent.

The spherical harmonics of degree $l = 4$ are shown in Figure 1. Observe that Y_l^0 is axisymmetric, while Y_l^m is invariant under rotation by an angle $2\pi/m$ and changes sign under rotation through π/m. Only the five real parts are shown in the figure, since the remaining four patterns making up

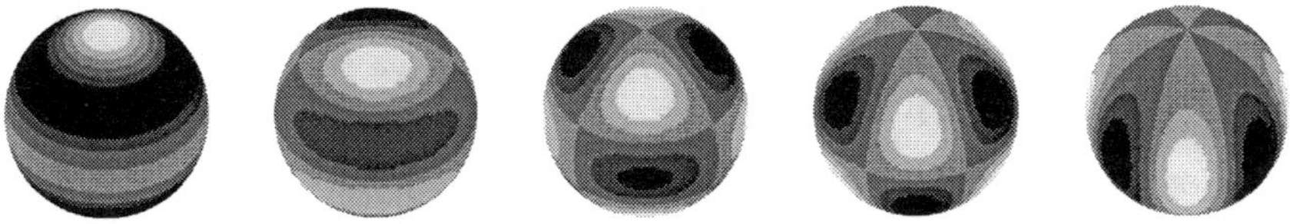

Fig. 1. Spherical harmonics of degree $l = 4$, for $m = 0, 1, 2, 3, 4$. Only the real parts are shown. For $m > 0$, the imaginary parts correspond to the same pattern rotated through an angle $\pi/2m$.

the nine-dimensional space are obtained by rotating the four patterns with $m > 0$ by an angle $\pi/2m$.

Near a bifurcation from a spherically symmetric state, a physical variable w is expanded in terms of spherical harmonics, and this expansion is reduced to a $2l + 1$-dimensional centre manifold,

$$w(\theta, \varphi, t) = \sum_{m=-l}^{l} z_m(t) Y_l^m(\theta, \varphi). \tag{6}$$

Nonlinearity introduces terms with different values of l, but these are slaved to the modes with the particular value of l selected by the linear stability problem.

Nonlinear ordinary differential equations can then be obtained for $z_m(t)$. Truncated to quadratic order, these equations have the form

$$\dot{z}_m = \lambda z_m + \beta \sum_{j=-l}^{l} \sum_{k=-l}^{l} c(j, k, m) z_j z_k, \tag{7}$$

where β is an arbitrary scaling constant. The coefficients $c(j, k, m)$ are found by imposing that the system of equations must be consistent with the spherical symmetry. To do this it is necessary to introduce some of the language of group and representation theory, as discussed below.

3. Groups, Symmetries and Patterns

The symmetry group $O(3)$ is generated by reflections and rotations of the sphere. Any element g of $O(3)$ transforms a linear combination of spherical harmonics of degree l into a different linear combination, and therefore corresponds to a $(2l + 1) \times (2l + 1)$ matrix. This set of matrices is known as a *representation* of the group $O(3)$. For example, the inversion symmetry (5) is represented by the identity matrix I if l is even, or $-I$ if l is odd. A rotation $\varphi \to \varphi+\varphi_0$ is represented by a diagonal matrix in which the (m, m) entry is $\exp im\varphi_0$. This representation is irreducible: this means that there is no proper subspace of the $2l + 1$-dimensional space that is left invariant by all elements of $O(3)$.

The system of equations (7) must be *equivariant* with respect to $O(3)$. This means the equations are of the form $\dot{z} = f(z)$ where

$$f(gz) = gf(z) \quad \text{for all } g \in O(3). \tag{8}$$

The notation gz here really means $M(g)z$, where M is the representation matrix corresponding to the symmetry g, but it is conventional to simplify notation by writing gz.

Imposing the inversion symmetry (5) provides no information if l is even; but for odd l we have $f(-z) = -f(z)$, so f is an odd function of z and there can be no quadratic terms in the equations. Thus the bifurcation from the spherically symmetric state is in general transcritical for even l, but a pitchfork for odd l.

Equivariance with respect to rotations $\varphi \to \varphi + \varphi_0$ is straightforward to impose in view of the simple action of this symmetry on the spherical harmonics given above. For example, for $l = 4$, the only quadratic terms that are permitted in the equation for z_3 are $z_0 z_3$, $z_1 z_2$ and $z_4 z_1^*$, each with an arbitrary numerical coefficient. Imposing the reflection symmetry (4) implies that these coefficients must all be real. Imposing equivariance with respect to other rotations is more complicated, and is usually done using "ladder" operators, corresponding to infinitesimal rotations. The end result [12] is that the quadratic terms are uniquely determined, up to an arbitrary scaling factor (β in (7)). Given this fact, a simple way to derive the quadratic term is to consider the equation $\dot{w} = \lambda w + w^2$ and obtain the equations for $\dot{z}$ using the orthogonality condition (3). This yields (7), where the coefficients are

$$c(j, k, m) = \int_0^\pi \int_0^{2\pi} Y_l^j Y_l^k Y_l^{m*} \sin\theta \, d\varphi \, d\theta. \tag{9}$$

These are known as *Clebsch–Gordan coefficients*; they have many properties [16] that are important in determining the existence and stability of stationary solutions, but these will not be given here.

Ideally, we would like to be able to find all the stationary solutions of (7), but this is impractical except for small values of l, because of the large number of equations and the fact that solutions exist in *group orbits*, related by the spherical symmetry, that are in most cases three-dimensional.

A crucial simplification is to consider *invariant subspaces* of the $2l+1$-dimensional subspace V. A subspace $U \subset V$ is invariant if $f(u) \in U$ for all $u \in U$. Invariant subspaces are generated by the subgroups H of $O(3)$: the *fixed-point subspace* of H is defined by

$$\text{Fix}(H) = \{v \in V : hv = v \text{ for all } h \in H\}. \tag{10}$$

It is clear that Fix(H) is an invariant subspace, since $v \in \text{Fix}(H) \Rightarrow hv = v$ for all $h \in H \Rightarrow f(v) = f(hv) = hf(v) \Rightarrow f(v) \in \text{Fix}(H)$. Since trajectories cannot leave invariant subspaces, any fixed point found in an invariant subspace must also be a fixed point in the full $2l+1$-dimensional space.

Examples of invariant subspaces include the one-dimensional subspace in which $z_m = 0$ for $m \neq 0$, which is Fix($O(2)$), and the $l+1$-dimensional subspace in which z_m is real, which is the fixed-point subspace of the group Z_2 generated by the reflection $\varphi \to -\varphi$.

In order to derive invariant subspaces systematically it is necessary to determine all the subgroups of $O(3)$ and the dimensions of their fixed-point subspaces. In fact the important subgroups are the *isotropy subgroups*. A subgroup H is said to be an isotropy subgroup if there exists a point $v \in V$ for which

$$H = \{g \in O(3) : gv = v\}, \tag{11}$$

and a point satisfying (11) is said to have isotropy H. Not all subgroups are isotropy subgroups, and whether a subgroup is an isotropy subgroup or not depends on the choice of l. For example, for $l = 4$ the isotropy of the point $z = (0,0,0,0,1)$, corresponding to the spherical harmonic Y_4^4, is $D_4 \times Z_2^c$, since this pattern (see Figure 1) has square symmetry plus the inversion symmetry (5). Thus $D_4 \times Z_2^c$ is an isotropy subgroup, and the dimension of its fixed-point subspace is $\text{Dim}(\text{Fix}(D_4 \times Z_2^c)) = 2$, since any point $z = (x,0,0,0,y)$ representing a combination of Y_4^0 and $Re(Y_4^4)$ possesses this symmetry. However, $D_5 \times Z_2^c$ is not an isotropy subgroup for $l = 4$, since $\text{Fix}(D_5 \times Z_2^c) = (x,0,0,0,0)$, and the isotropy subgroup of $(x,0,0,0,0)$ is $O(2) \times Z_2^c$.

The isotropy subgroups H of $O(3)$ and the dimension of their fixed-point subspaces have been derived in earlier work [13–15]. For even l, the symmetry (5) is always present, so the subgroups are all of the form $H \times Z_2^c$, where H is a subgroup of $SO(3)$. A partial lattice of these subgroups [13] is shown in Figure 2. The isotropy subgroups in the case of even l are listed in Table 1. Here, the notation $[x]$ is used to indicate the largest integer less than or equal to x, $\mathcal{D}^l(H)$ is used as an abbreviation for Dim(Fix(H)), and

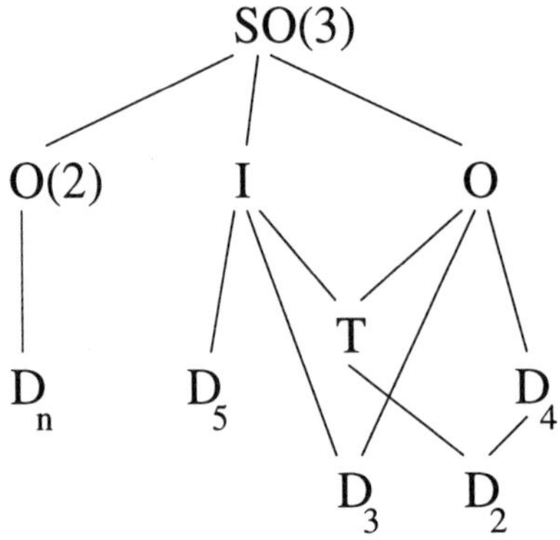

Fig. 2. Partial lattice of subgroups of $SO(3)$.

the additional Z_2^c symmetry is not explicitly mentioned since it is always present. The group I is the symmetry group of rotations of the icosahedron or dodecahedron, and $I \times Z_2^c$ is the symmetry group of the same object including reflections. Similarly, O is the symmetry of rotations of the cube (or octahedron) and $O \times Z_2^c$ includes reflections of the cube. However, $T \times Z_2^c$ is not the symmetry group of a tetrahedron including reflections (which, confusingly, is isomorphic to O!); an example of a pattern with symmetry $T \times Z_2^c$ is shown in Figure 4.

An isotropy subgroup H is said to be *maximal* if $\mathrm{Dim}(\mathrm{Fix}(H))$ is positive and $\mathrm{Dim}(\mathrm{Fix}(K))$ is zero for all K with $H \subset K \subset O(3)$. For even l, only $O(2) \times Z_2^c$, $I \times Z_2^c$ and $O \times Z_2^c$ can be maximal isotropy subgroups. The other isotropy subgroups are said to be *submaximal.*

Another important group-theoretic definition is the *normalizer* $N(H)$ of a subgroup H, defined as the largest subgroup of $O(3)$ in which H is normal. The normalizers are also listed in Table 1. It can easily be shown that $N(H)$ maps $\mathrm{Fix}(H)$ to itself, and hence that the quotient group $N(H)/H$ describes the "residual symmetry" within $\mathrm{Fix}(H)$. Thus if $N(H) = H$, as is the case, for example, for $H = I$ or $H = O$, there is no symmetry in the restriction of the system of equations to the subspace $\mathrm{Fix}(H)$. But for the group T, $N(T) = O$ and so there is a symmetry $O/T \equiv Z_2$ within $\mathrm{Fix}(T)$; the fixed-point subspace of this symmetry is $\mathrm{Fix}(O)$.

A crucial result in the search for bifurcating branches of stationary solutions of (7) is the *equivariant branching lemma*, which states that a unique branch of solutions with isotropy H exists if H is an isotropy subgroup and $\mathrm{Dim}(\mathrm{Fix}(H)) = 1$ (see [13, 18] for a precise statement of the lemma and

Table 1. Isotropy subgroups of $O(3)$ for even l, with normalisers $N(H)$ and dimension of fixed-point subspaces. The additional symmetry Z_2^c is suppressed.

Group H	Symmetry	$\|H\|$	$N(H)$	l	$\mathcal{D}^l(H)$
$O(2)$	Circle	∞	$O(2)$	all l	1
I	Icosahedron	60	I	$l \neq 2, 4, 8, 14$	$1 - l/2 + [l/3] + [l/5]$
O	Cube	24	O	$l \neq 2$	$1 - l/2 + [l/3] + [l/4]$
T	Tetrahedron	12	O	$l \neq 2, 4, 8$	$1 - l/2 + 2[l/3]$
D_n, $n > 2$	Regular n-gon	$2n$	D_{2n}	$l \geq n$	$1 + [l/n]$
D_2	Rectangle	4	O	all l	$1 + l/2$
Z_n	Directed n-gon	n	$O(2)$	$l \geq 2n$	$1 + 2[l/n]$

the necessary conditions). Applying this lemma for the case of even l shows that exactly one solution exists in the following cases: with isotropy $O(2)$ for all even l, with isotropy O for $l = 4, 6, 8, 10, 14$ and with isotropy I for $l = 6, 10, 12, 16, 18, 20, 22, 24, 26, 28, 32, 34, 38, 44$ [13].

4. Solution Branches of (7) for Even l

In this section, results on the existence and stability of solution branches in the quadratic truncation (7) are summarized, for the case when l is even (recall that for odd l, all quadratic terms are zero).

4.1. *Existence of Solutions*

For isotropy subgroups H with $\mathrm{Dim}(\mathrm{Fix}(H)) = 1$, the equivariant branching lemma can be used to deduce the existence of the solutions listed above. In the case $\mathrm{Dim}(\mathrm{Fix}(H)) > 1$, there are two complications. Firstly, there is no general analogue of the equivariant branching lemma in dimensions greater than one, and examples can be found where the number of solution branches may be zero or greater than one. However, it is known that if the dimension is odd, at least one solution exists for maximal isotropy subgroups. Secondly, for submaximal isotropy subgroups it must be checked carefully whether the solution found has in fact greater symmetry — for example, when seeking solutions with isotropy T, solutions with isotropy I and O are also found.

Methods for demonstrating the existence or non-existence of solutions in subspaces with $\mathcal{D}(H) \equiv \mathrm{Dim}(\mathrm{Fix}(H)) > 1$ include solution counting, exploiting general properties of the Clebsch–Gordan coefficients, and explicit construction of the equations and their solutions.

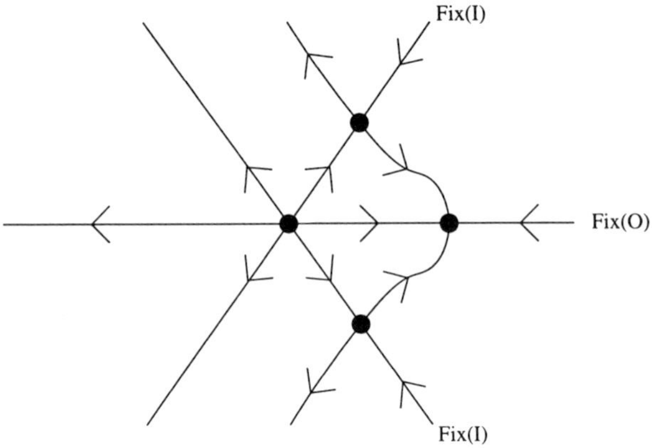

Fig. 3. Phase portrait within Fix(T) for $l = 6$ and $l = 10$ with $\lambda > 0$. The solution with isotropy O is stable in this subspace but is unstable to perturbations outside the subspace.

Although solutions are not guaranteed, in fact solutions do exist generically when $\mathcal{D}(H) > 1$. This is a consequence of Bezout's theorem, which states that the number of complex solutions to p coupled equations of degree q is q^p. In the case of interest here, $q = 2$, and the coefficients are real so that solutions are either real or occur in complex conjugate pairs, so the number of real solutions is even in general. Since one solution is known to exist (the basic state with spherical symmetry, $z = 0$), at least one other solution is expected, and the total number of non-zero solutions is in general odd and bounded above by $2^p - 1$. In fact, two non-degeneracy conditions have to be checked to be certain of the existence of solutions [16,19]. These correspond to the possibility of multiple roots or solutions at infinity. As an example of this solution-counting approach, we expect to find either one or three branches of solutions with isotropy O for $l = 12$, 16, 18, 20, 22, 26, since for these values of l, $\mathcal{D}(O) = 2$. Direct computation reveals that there is one solution for $l = 20$ and three solutions in the other cases. For $l = 24$, $\mathcal{D}(O) = 3$ so we expect 1, 3, 5 or 7 solutions, and direct computation reveals that there are 5.

This approach of counting solutions can also sometimes be used to rule out the existence of solutions with submaximal isotropy. For example, in the cases $l = 6$ and $l = 10$, $\mathcal{D}(T) = 2$ and $\mathcal{D}(O) = \mathcal{D}(I) = 1$. Since $T \subset O$ and $T \subset I$, the two-dimensional subspace Fix(T) includes the one-dimensional subspaces Fix(I) and Fix(O). Furthermore, since $N(T) = O$, there is a symmetry in Fix(T) which acts as a reflection in the line Fix(O). As a consequence of this symmetry, Fix(T) includes two copies of Fix(I). The maximum number of non-zero solutions in Fix(T) is $2^2 - 1 = 3$, and these are all accounted for, since Fix(T) must include the solution in Fix(O) and two copies of the solution in Fix(I), both of which are known to exist by the equivariant branching lemma. Therefore, there can be no solution branch with isotropy T in the quadratic truncation for $l = 6$ or $l = 10$. To clarify the argument, a sketch of the phase portrait in Fix(T) is given in Figure 3.

Existence of solutions with submaximal isotropy D_n in two-dimensional subspaces can be established using the properties of the Clebsch–Gordan coefficients. This approach was first used by Chossat (see [15]) and has been extended by Matthews [16]. From Table 1, Fix(D_n) is two-dimensional if $l/2 < n \leq l$. This subspace is obtained by setting z_n to be a real number x_n and setting z_m to zero for all m except $m = n$ and $m = 0$. The equations in Fix(D_n) are then

$$\dot{z}_0 = \lambda z_0 + \beta c(0,0,0)z_0^2 + 2\beta c(n,-n,0)x_n x_{-n}, \tag{12}$$

$$\dot{x}_n = \lambda x_n + 2\beta c(0,n,n)z_0 x_n, \tag{13}$$

which can be simplified, using the properties of the coefficients, to

$$\dot{z}_0 = \lambda z_0 + \beta c(0,0,0)z_0^2 + 2\beta c(0,n,n)x_n^2, \tag{14}$$

$$\dot{x}_n = \lambda x_n + 2\beta c(0,n,n)z_0 x_n. \tag{15}$$

These equations have a reflection symmetry $x_n \to -x_n$ arising from the normalizer. In addition to the solution $x_n = 0$, $z_0 = -\lambda/\beta c(0,0,0)$, with isotropy $O(2)$, a further pair of solutions exists if

$$c(0,0,0)/2c(0,n,n) < 1. \tag{16}$$

These two solutions should be regarded as equivalent — in fact one is a rotated form of the other. Hence the existence of solution branches with isotropy D_n depends on the ratio of certain of the Clebsch–Gordan coefficients. In some cases it can be shown that the ratio in (16) is negative, so that a solution does exist. This is the case, for example, for $n = l$ if $l = 2$ mod 4, $l \geq 6$; so a solution with isotropy D_l exists for $l = 6, 10, 14, \ldots$ [15]. In other cases it can be shown that (16) is not satisfied, for example, for $n = l$ if $l = 0$ mod 4, $l > 8$ [16], so for $l = 8, 12, 16, \ldots$ there is no solution with isotropy D_l.

The results regarding the existence of solutions with isotropy D_n are summarized in Table 2. From the table it appears that for any fixed, large value of l there is an alternation between existence and non-existence of

Table 2. Conditions for existence and non-existence of solution branches with isotropy D_n in the quadratic truncation for even l.

Isotropy	Exist/not	Value of l mod 4	Condition on l
D_l	Exist	$l = 2 \bmod 4$	$l \geq 6$
D_l	Not exist	$l = 0 \bmod 4$	$l \geq 8$
D_{l-1}	Exist	$l = 0 \bmod 4$	$l \geq 8$
D_{l-1}	Not exist	$l = 2 \bmod 4$	$l \geq 18$
D_{l-2}	Exist	$l = 2 \bmod 4$	$l \geq 10$
D_{l-2}	Not exist	$l = 0 \bmod 4$	$l \geq 28$
D_{l-3}	Exist	$l = 0 \bmod 4$	$l \geq 12$
D_{l-3}	Not exist	$l = 2 \bmod 4$	$l \geq 34$
D_{l-4}	Exist	$l = 2 \bmod 4$	$l \geq 18$

solutions with isotropy D_{l-k} as k increases. The precise statement is the following lemma from [16].

Lemma 1: *For any given value of k, there exists an integer l_0 such that for $l \geq l_0$ and l even, a solution of (7) with isotropy D_{l-k} exists if and only if $k + l/2$ is odd.*

Finally in this section, we consider the explicit construction of the equations (7) and their stationary solutions, for particular even values of l. This approach was followed in [6, 8, 10, 12, 15–17]. This becomes increasingly difficult as l increases, since the size of the invariant subspaces increases with l. A useful result [15] is that generically there are no stationary solutions with isotropy Z_n or with no symmetry. Loosely speaking, such states do not have enough symmetry to be stationary. Thus we only need to consider the subgroups D_n, I, O, T and $O(2)$.

For small values of l, solutions were obtained by Busse [8] in the 1970s. For $l = 2$, there is only one branch of solutions, with isotropy $O(2)$. For $l = 4$, Busse found two branches of solutions, with isotropy $O(2)$ and O, and more recent work has confirmed that these are the only solutions [15]. For $l = 6$, the equivariant branching lemma guarantees the existence of solutions with isotropy $O(2)$, O and I, and the result given in the first row of Table 2 shows that there is also a solution with submaximal isotropy D_6. A systematic investigation of all the invariant subspaces down to $\mathrm{Fix}(D_2)$, combined with the known bounds on the number of possible solutions, can be used to confirm that there are no further solution branches [16]. For

$l = 8$, there is a degeneracy in the coefficients, since $c(0, 5, 5) = c(0, 6, 6)$, first noted by Busse [8]. Solutions with isotropy $O(2)$, O, D_3, D_4, D_5, D_6, D_7 exist, but as a result of the degeneracy there is also a (non-generic) family of solutions involving z_0, z_5 and z_6 that possess no symmetry (except for the inversion symmetry (5) that is always present).

Detailed results on the existence of solution branches lying in invariant subspaces with dimension up to 3 were given by Matthews [16] for values of l up to $l = 18$. These results are summarized in Table 3, together with further results up to $l = 28$. Note that the lowest value of l for which a solution with isotropy T exists is $l = 12$. A solution with isotropy T also exists for $l = 16$, but not for $l = 14$ or $l = 20$ (for $l = 18$ and for $l \geq 22$, $\mathcal{D}^l(T) > 3$, so the existence of solutions with isotropy T is not known). For isotropy O, we expect either one or three solutions if $\mathcal{D}^l(O) = 2$ and 1, 3, 5 or 7 solutions if $\mathcal{D}^l(T) = 3$, as discussed above. For D_n with $l/2 < n \leq l$, $\mathcal{D}^l(D_n) = 2$ and there may or may not be a solution. If $l/3 < n \leq l/2$, $\mathcal{D}^l(D_n) = 3$ and this three-dimensional subspace includes the two-dimensional subspace $\mathrm{Fix}(D_{2n})$, which may contain three solutions. Because of the normalizer symmetry, there is a reflection symmetry in $\mathrm{Fix}(D_n)$, so there can be at most two pairs of equivalent solutions with isotropy D_n since the total number of non-zero solutions cannot exceed $2^3 - 1 = 7$. Hence there are either zero, one or two branches of solutions with isotropy D_n when $\mathcal{D}^l(D_n) = 3$.

Four of the patterns corresponding to stationary solutions in the case $l = 12$ are shown in Figure 4. Note that the pattern with isotropy T possesses some of the symmetries of the cube (reflection in three orthogonal planes and a rotation through $2\pi/3$) but lacks some of the reflection symmetries. Further illustrations of the patterns can be found in [4, 8, 16].

4.2. *Stability of Solutions*

Having obtained the stationary solution branches that exist for even l in the quadratic truncation (7), the next step is to consider their stability. Stability is determined by the eigenvalues of the Jacobian matrix J_{ij} of (7), which is

$$J_{ij} = \lambda\delta_{ij} + 2\beta \sum_{k=-l}^{l} c(j, k, i) z_k. \tag{17}$$

A number of important results follow directly from (17) [13, 15]. Since J_{ij} is symmetric, the eigenvalues are real and the eigenvectors are orthogonal.

Table 3. Solution branches known to exist in the quadratic truncation for even l. All solutions with isotropy H where $\mathcal{D}^l(H) \leq 3$ are given, but for $l \geq 8$ further solutions may exist in higher-dimensional subspaces. In cases where there is more than one solution with the same isotropy, the number of solutions is given in square brackets.

l	$\mathcal{D}^l(H) = 1$	$\mathcal{D}^l(H) = 2$	$\mathcal{D}^l(H) = 3$
2	$O(2)$		
4	$O(2)$, O		
6	$O(2)$, O, I	D_6	
8	$O(2)$, O	D_7,D_6,D_5	D_4, D_3
10	$O(2)$, O, I	D_{10}, D_9, D_8, D_7, D_6	D_4
12	$O(2)$, I	$O[3]$, D_{11}, D_{10}, D_9, D_8	T, D_6, D_5
14	$O(2)$, O	D_{14}, D_{13}, D_{12}, D_{11}, D_{10}, D_9, D_8	$D_7[2]$, $D_6[2]$, $D_5[2]$
16	$O(2)$, I	$O[3]$, D_{15}, D_{14}, D_{13}, D_{12}, D_{11}, D_9	T, $D_8[2]$, D_7, D_6
18	$O(2)$, I	$O[3]$, D_{18}, D_{16}, D_{15}, D_{14}, D_{13}, D_{11}	$D_9[2]$, $D_8[2]$, $D_7[2]$
20	$O(2)$, I	O, D_{19}, D_{18}, D_{17}, D_{16}, D_{15}, D_{14}, D_{13}, D_{12}, D_{11}	$D_{10}[2]$, $D_9[2]$, $D_8[2]$, $D_7[2]$
22	$O(2)$, I	$O[3]$, D_{22}, D_{20}, D_{19}, D_{18}, D_{17}, D_{16}, D_{14}, D_{13}, D_{12}	D_{11}, $D_{10}[2]$, $D_9[2]$, $D_8[2]$
24	$O(2)$, I	D_{23}, D_{22}, D_{21}, D_{20}, D_{19}, D_{18}, D_{17}, D_{16}, D_{15}, D_{14}, D_{13}	$O[5]$, D_{12}, D_{11}, $D_9[2]$
26	$O(2)$, I	$O[3]$, D_{26}, D_{24}, D_{23}, D_{22}, D_{21}, D_{20}, D_{19}, D_{18}, D_{17}, D_{16}, D_{15}, D_{14}	$D_{13}[2]$ D_{12}, $D_{11}[2]$, $D_9[2]$
28	$O(2)$, I	D_{27}, D_{25}, D_{24}, D_{23}, D_{22}, D_{21}, D_{20}, D_{19}, D_{18}, D_{17}, D_{16}, D_{15}	$O[5]$, $D_{14}[2]$, $D_{13}[2]$, $D_{12}[2]$, D_{11}, $D_{10}[2]$

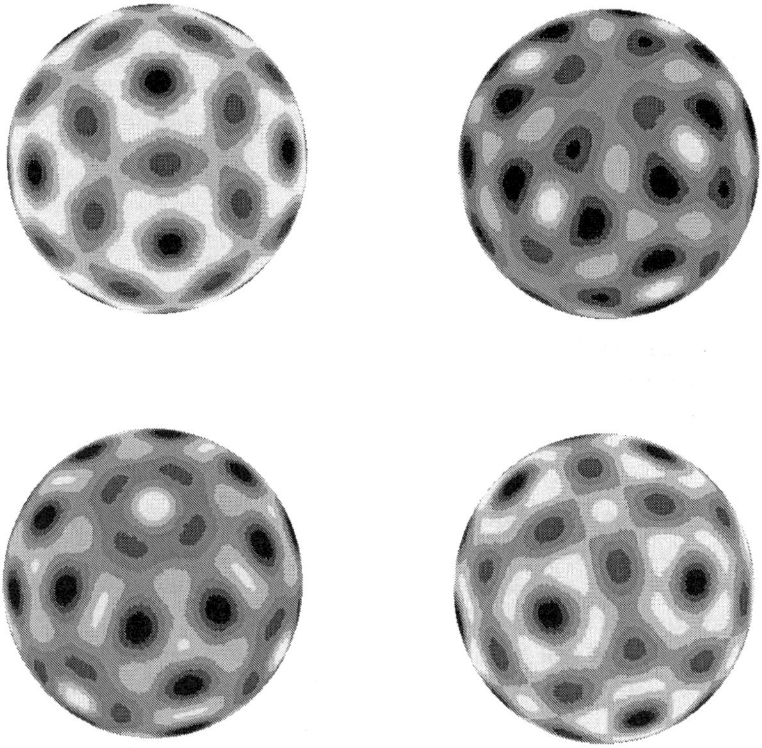

Fig. 4. Four of the stationary solutions for $l = 12$. Top left: isotropy I; top right: isotropy T; bottom left and right: two of the three distinct solutions with isotropy O.

If z_j is a stationary solution, then $J_{ij}z_j = -\lambda z_i$, so z_j is also an eigenvector of J_{ij} with eigenvalue $-\lambda$. These are sometimes referred to as the "radial" eigenvalue and eigenvector, and all other eigenvectors must be transverse to z_j. The eigenvalue $-\lambda$ means that all solution branches are unstable for $\lambda < 0$. It also follows from (17) that the trace of J_{ij}, which is the sum of the eigenvalues, is $(2l+1)\lambda$, and hence any solution branch must be unstable for $\lambda > 0$. This result, that all solution branches are unstable in (7), is in fact a general result for multiple transcritical bifurcations with symmetry [18].

Because of the spherical symmetry, most solutions have three zero eigenvalues, which can be thought of as corresponding to infinitesimal rotations about three perpendicular axes. The exception is the axisymmetric solution, which only has two zero eigenvalues because it is unchanged by one of these rotations.

Although all solution branches are unstable, there are two ways to distinguish a "preferred" solution among the many solution branches. Firstly, the system (7) is variational, *i.e.* it can be written as the gradient of a scalar, so a solution can be identified that maximizes this scalar quantity [8, 12]. In fact this condition is simply that the quantity λI_2 should be maximized, where

$$I_2 = \sum_{j=-l}^{l} |z_j|^2 \tag{18}$$

is the unique second-order quantity that is $O(3)$-invariant.

Alternatively, we can anticipate that solution branches may turn round at a saddle-node bifurcation for negative λ, in which case the radial eigenvalue changes sign (this is discussed further in Section 5 below). If this happens, a solution branch for which the only positive eigenvalue when $\lambda < 0$ is the radial eigenvalue $-\lambda$ can become stable. A solution with this property is said to be *transversely stable.* These two criteria for identifying a preferred solution are closely related. Sattinger [12] showed that the preferred solution according to the variational criterion is transversely stable. The converse is not true, as there may be more than one transversely stable solution.

Transversely stable solutions can be found by explicitly computing the eigenvalues of each solution branch. The results are summarized in Table 4. In each case, the value of I_2 was also computed as a check. For $l = 8$, there is a degeneracy in the coefficients, as discussed above; as a result of this, five solutions (with isotropy D_3, D_4, D_5, D_6, D_7) are transversely stable with the same value of I_2 and seven zero eigenvalues. For $l = 18$ there are two transversely stable solutions, with isotropy D_{15} and I, and the one with isotropy D_{15} is selected by the variational condition. For $l = 20$, 22, 26, none of the solutions listed in Table 3 are transversely stable. Since we know that at least one transversely stable solution must exist, it must lie in a subspace of dimension at least four.

A notable feature of Table 4 is that except for $l = 2$, the axisymmetric solution with isotropy $O(2)$ does not appear to be transversely stable. In fact the following general result can be proved [16].

Lemma 2: *For any even $l > 2$, the solution branch with isotropy $O(2)$ is not transversely stable, and hence cannot satisfy the variational criterion*

Table 4. Transversely stable solution branches for even l. For $l = 8$ there is a degeneracy in the coefficients so results are not given. For $l = 18$, there are two transversely stable solutions. No transversely stable solution exists in subspaces of dimension up to three for $l = 20, 22, 26$.

2	4	6	8	10	12	14	16	18	20	22	24	26	28
$O(2)$	O	I	-	I	I	D_6	T	D_{15}, I	?	?	D_{20}	?	D_{12}

of Busse [8] and Sattinger [12].

This is proved by writing down the Jacobian for the axisymmetric solution, which is diagonal, and hence obtaining the first few eigenvalues corresponding to perturbations of the form $Y_l^m(\theta, \varphi)$ in terms of the Clebsch–Gordan coefficients. Using recurrence relations involving the coefficients, these eigenvalues can be written explicitly in terms of l and λ, and it can be deduced that there are always two double eigenvalues of opposite sign (except when $l = 2$), in addition to the eigenvalue $-\lambda$. This shows that there are always at least two positive eigenvalues, so the axisymmetric state cannot be transversely stable. A similar approach has been used for the case of odd l, where the bifurcation is a pitchfork, to show that the axisymmetric state can never be stable in that case [15].

5. Solution Branches for Even l including Cubic Terms

Since solution branches in the quadratic truncation (7) can never be stable, it is natural to consider the effects of cubic terms, to see whether any solutions can be stabilized at a saddle-node bifurcation as suggested above. This is self-consistent if the quadratic term is small, corresponding to an unfolding of the codimension-two point where an eigenvalue of the basic spherical state is zero and the quadratic terms are zero, *i.e.* the coefficient β in (7) is zero. This approach is commonly followed in planar pattern formation problems, where hexagons generically bifurcate transcritically and are unstable near the bifurcation. Of course, if β is not small, finite amplitude states occur that cannot be analyzed in a consistent manner.

For the case $l = 2$, the inclusion of cubic terms is straightforward [5,18]. There is a unique equivariant cubic term to be added to (7), which is simply $I_2 z_m$, where I_2 is the invariant given in (18). The order of the system can be reduced from five to two, and the resulting second-order system can be written in terms of a complex variable Z as

$$\dot{Z} = \lambda Z + \beta Z^{*2} + \gamma |Z|^2 Z, \tag{19}$$

which is the normal form for a stationary bifurcation with D_3 symmetry. There is only one distinct branch of solutions, with isotropy $O(2)$, which is unstable for small $|Z|$. For $\gamma < 0$, there is a saddle-node bifurcation at $\lambda = \beta^2/4\gamma$ at which the branch becomes stable.

For larger values of l, the inclusion of cubic terms is much more complicated. In addition to the algebraic complexity, it is no longer the case that the equations are uniquely determined from the spherical symmetry. In fact there are $1 + [l/3]$ independent equivariant cubic terms [15]. Thus there are two cubic terms for $l = 4$ and three for $l = 6$.

The case $l = 4$ was investigated by Geiger *et al.* [10], who showed that, depending on the ratio of the two independent cubic terms, there are two possible bifurcation diagrams in which solution branches can become stable. In one of these cases, the branch of solutions with isotropy O becomes stable at a saddle-node bifurcation but loses stability as λ is increased. There is an interval in which the $O(2)$ solution is stable, but at large λ a D_4 branch is stable. In the other case, the O branch becomes stable at a saddle-node bifurcation and then remains stable. The bifurcation diagram for this second case is shown in Figure 5. Note that this picture is in agreement with the argument of the previous section, since the O solution is transversely stable (Table 4).

In fact, several features of this bifurcation diagram can be deduced without any explicit computation involving the cubic terms. In the quadratic truncation, all eigenvalues on the two primary branches are proportional to λ, so they all change sign at $\lambda = 0$, except for the zero eigenvalues. But for large λ, the cubic terms are dominant, so there must be two almost-equivalent branches, with the same eigenvalues. Hence, all the eigenvalues must change sign at secondary bifurcations somewhere along the branch. For the branch with isotropy $O(2)$, there are three pairs of non-zero eigenvalues corresponding to perturbations involving Y_4^2, Y_4^3 and Y_4^4, so there must be secondary bifurcations to branches with isotropy D_2, D_3 and D_4. For the O branch, in the quadratic truncation there is a single eigenvalue $-\lambda$, a double eigenvalue and a triple eigenvalue [12]. The single eigenvalue must change sign at a saddle-node bifurcation, so there must be secondary bifurcations involving two and three zero eigenvalues on this branch, as shown in Figure 5. Furthermore, the general theory of bifurcation from a state with symmetry O [20, 21] can be used to determine what happens at these bifurcations: when two eigenvalues change sign, there must be a transcritical bifurcation to a branch with isotropy D_4, and when three eigenvalues change sign, there is a pitchfork bifurcation to a D_2 branch and

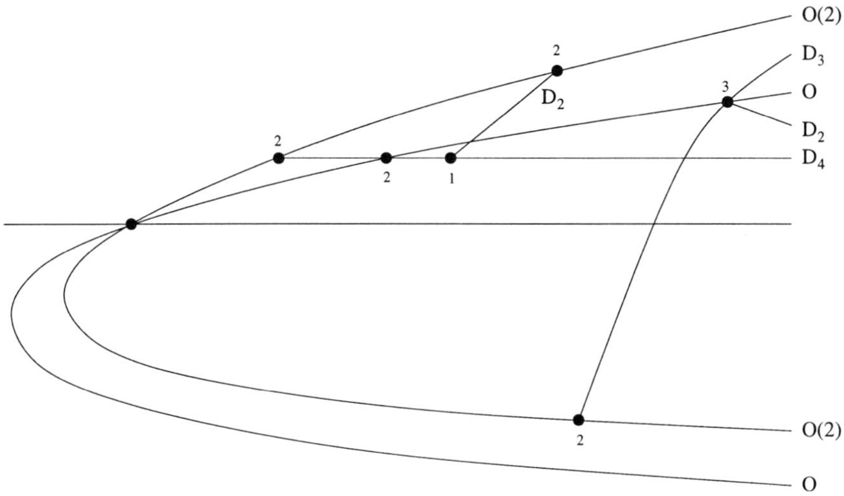

Fig. 5. Bifurcation diagram for $l = 4$. Bifurcations are marked with filled circles, except for the saddle-node bifurcations. Other crossings are not bifurcations. The number beside each bifurcation indicates the number of eigenvalues that change sign. The only stable solutions lie on the branches with isotropy O, on the lower branch below the saddle-node bifurcation and on the upper branch to the right of all the bifurcations.

a transcritical bifurcation to a D_3 branch. Thus, almost all the qualitative features of Figure 5 can be deduced without any calculations.

It may seem that there is little point in considering the effect of the cubic terms for larger l, in view of the increasing number of independent cubic equivariants. But there are two reasons why this is worthwhile. Firstly, as shown above, many secondary bifurcations can be deduced without any calculation. Secondly, it has been shown by Callahan [17] that for reaction–diffusion equations of the form

$$u_t = \nabla^2 u + f(u, v), \tag{20}$$

$$v_t = d\nabla^2 v + g(u, v), \tag{21}$$

as originally introduced by Turing [1], the cubic coefficients appear in fixed, model-independent ratios when the parameters are tuned so that the quadratic terms are small. Thus there is a large class of systems (including, for example, the much-studied Brusselator and Schnakenberg systems) in which the dynamics is universal. In fact, this universality extends beyond these reaction–diffusion systems. The equations for $l = 4$ were computed for the Swift–Hohenberg equation [22],

$$w_t = \lambda w - (1 + \nabla^2)^2 w - \varepsilon w^2 - w^3, \tag{22}$$

using Maple to do the computation, and the resulting system exhibited the same ratio of the two cubic terms given by Callahan. This may be because reaction–diffusion equations can be reduced to (22) near the onset of pattern formation. The computation was repeated using the cubic term $w|\nabla w|^2$, with the same result. Using these equations, the branches of solutions and their eigenvalues were computed, and the resulting bifurcation diagram is shown in Figure 5. The scaling is arbitrary, and the vertical axis has no real meaning, but the bifurcations are shown in the correct order. The diagram is equivalent to that shown by Geiger *et al.* [10], who also gave the signs of the eigenvalues of each solution.

The computation of the equations (7) including cubic terms was repeated for (22) for $l = 6$. There are three equivariant cubic terms, but as for $l = 4$, they appear with the same fixed ratios as given by Callahan [17] for reaction–diffusion systems. The existence and stability of some, but not all, solution branches was computed and the resulting partial bifurcation diagram is shown in Figure 6. Solutions with isotropy D_3 and D_2 were not computed, but some of these are indicated where they are known to bifurcate from other branches.

Again, some bifurcations in Figure 6 can be deduced using the fact that all eigenvalues must change sign at some point on the branch. For the $O(2)$ branch, we can deduce bifurcations to branches with isotropy D_2, D_3, D_4, D_5 and D_6. For the O branch, in the quadratic truncation there are two single eigenvalues, one double eigenvalue and two triple eigenvalues [16], so from the general theory [20, 21] we know that there is a saddle-node bifurcation, a pitchfork bifurcation to a solution with isotropy T, a transcritical bifurcation to a D_4 branch and two bifurcations to D_3 and D_2 branches (as for $l = 4$). For the I branch we need to make use of the theory for bifurcation from icosahedral symmetry, that has only recently been established [21, 23]. The group I has irreducible representations of dimension 1, 3, 4 and 5, and the eigenvalues in the quadratic truncation have multiplicity 1, 4 and 5, so apart from the saddle-node bifurcation there must be one bifurcation in the 4-dimensional representation and one in the 5-dimensional representation. These bifurcations are known to be transcritical, with branches with isotropy T and D_3 in the 4-dimensional case and D_5 and D_3 in the 5-dimensional case [21, 23]. However, this argument does not hold for the D_6 branch, because there is another D_6 branch at large λ (bifurcating from

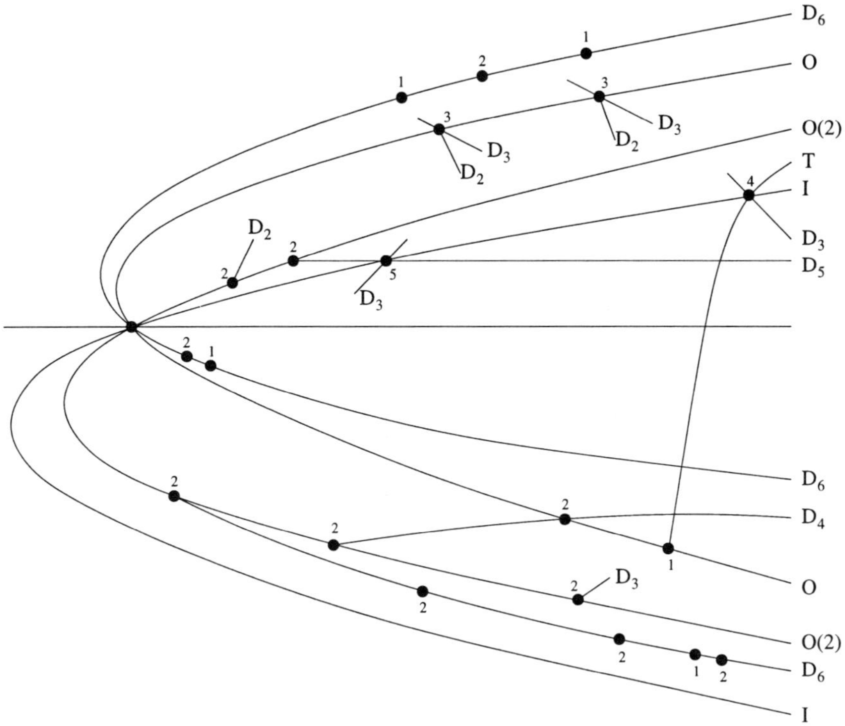

Fig. 6. Partial bifurcation diagram for $l = 6$, with notation as in Figure 5. All bifurcations on the primary branches, with isotropy O, D_6, $O(2)$ and I are shown. Branches with isotropy D_3 and D_2 were not explicitly computed, but these are marked as short lines where they are known to exist. Secondary branches from the D_6 branch are not marked, but bifurcation labelled 1 are pitchforks creating a D_3 branch and those marked 2 are transcritical involving a D_2 branch. The solution with isotropy I is stable on the lower branch below the saddle-node bifurcation and on the upper branch to the right of all the other bifurcations; all other solution branches are unstable.

the $O(2)$ branch).

A further feature of Figure 6 that can be deduced is that there can only be one solution with isotropy T. This follows from the solution-counting argument of Section 4.1. Since $\mathrm{Dim}(\mathrm{Fix}(T)) = 2$, there can be no more than $3^2 - 1 = 8$ non-zero solutions in $\mathrm{Fix}(T)$. These include the two O solutions, two copies of the two I solutions (in view of the symmetry in $\mathrm{Fix}(T)$, see Figure 3) and two copies of any T solution, so there can only be one solution with isotropy T.

Only the solution with isotropy I in Figure 6 can be stable. It is stable on the lower branch, below the saddle-node bifurcation, and on the upper branch, to the right of the two secondary bifurcations. For large λ, the $O(2)$ solution has four pairs of positive eigenvalues and the O solution has one single and one triple positive eigenvalue. The uppermost and lowermost D_6 branches have only one positive eigenvalue, but the middle of the three branches (which has very small amplitude at large λ) has nine positive eigenvalues. Some of these stability results were given by Callahan [17], and the results are in agreement in all cases.

6. Discussion

In this paper we have considered existence and stability of stationary patterns on a sphere, focussing on the case where the degree l of the spherical harmonics is even. In the quadratic truncation, the nonlinear terms are uniquely determined, and results are given up to $l = 28$; all solutions are unstable but the variational criterion [12] favors icosahedral patterns for $l = 6$, $l = 10$ and $l = 12$. In the more complicated cubic truncation, only $l = 2$, $l = 4$ and $l = 6$ are considered, but these results confirm that the transversely stable branch in the quadratic truncation can become stable at a saddle-node bifurcation.

A natural question is whether the cubic truncation is sufficient to capture the "essential dynamics" of the system. In fact, in some ways the cubic truncation is not sufficient. For example, for $l = 2$, the only steady states have isotropy $O(2)$ in the cubic truncation, whereas we might expect a secondary bifurcation to a state with isotropy D_2; this state can only be found if fifth-order terms are included [18]. A second reason why the cubic truncation is insufficient is that it has been shown by Michel (see [15]) that all the cubic equivariant terms are variational, and so can only exhibit simple dynamics.

Returning to the applications of the theory, it is of interest to consider the most relevant values of l. For tumor growth models (and also models of embryo development) it is generally the case that the first mode to become unstable as the size of the sphere increases is $l = 2$ [5], which will lead to the formation of an oblate or prolate spheroid. In convection problems, the relevant value of l depends on the thickness of the spherical shell. For relatively thick layers, such as the Earth's mantle, small values of l are appropriate, and numerical simulations have found the cubic solution for $l = 4$ predicted by the theory above [9]. But for thinner shells, simulations motivated by Mercury's mantle have $l \approx 18$ and exhibit complicated patterns, including large spirals [11]. Finally, an intriguing but rather speculative potential application is in the structure of "spherical" viruses, many of which (such as that responsible for the common cold) are known to have an icosahedral structure [24, 25].

References

[1] A.M. Turing, *Phil. Trans. Roy. Soc.* ***B*** **237**, 37 (1952).
[2] M.A.J. Chaplain, M. Ganesh and I.G. Graham, *J. Math. Biol.* **42**, 387 (2001).
[3] C. Varea, J.L. Aragon and R.A. Barrio, *Phys. Rev.* ***E*** **60**, 4588 (1999).
[4] P.C. Matthews, *Phys. Rev.* ***E*** **67**, 036206 (2003).
[5] H. Byrne and P. Matthews, *IMA J. Math. Appl. Med. Biol.* **19**, 1 (2002).
[6] G.H. Knightly and D. Sather, *Arch. Rat. Mech. Anal.* **72**, 315 (1980).
[7] H. Pleiner, *Phys. Rev.* ***A*** **42**, 6060 (1990).
[8] F.H. Busse, *J. Fluid Mech.* **72**, 67 (1975).
[9] G. Schubert, G.A. Glatzmeier and B. Travis, *Phys. Fluids* ***A*** **5**, 1928 (1993).
[10] C. Geiger, G. Dangelmayr, J.D. Rodriguez and W. Guttinger, *Fields Inst. Comm.* **5**, 225 (1996).
[11] P. Zhang, X. Liao and K. Zhang, *Phys. Rev.* ***E*** **66**, 055203 (2002).
[12] D.H. Sattinger, *J. Math. Phys.* **19**, 1720 (1978).
[13] E. Ihrig and M. Golubitsky, *Physica* ***D*** **13**, 1 (1984).
[14] M. Michel, *Rev. Mod. Phys.* **52**, 617 (1980).
[15] P. Chossat, R. Lauterbach and I. Melbourne, *Arch. Rat. Mech. Anal.* **113**, 313 (1990).
[16] P.C. Matthews, *Nonlinearity* **16**, 1449 (2003).
[17] T.K. Callahan, *Physica* ***D*** **188**, 65 (2004).
[18] M. Golubitsky, I. Stewart and D.G. Schaeffer, *Singularities and Groups in Bifurcation Theory* (Springer, 1988).
[19] A. Lari-Lavassani, W.F. Langford and K. Huseyin, *Dynamics and Stability of Systems* **9**, 345 (1994).
[20] I. Melbourne, *Dynamics and Stability of Systems* **1**, 293 (1986).
[21] P.C. Matthews, *LMS J. Comput. Math.* **7**, 101 (2004).
[22] J.B. Swift and P.C. Hohenberg, *Phys. Rev.* ***A*** **15**, 319 (1977).
[23] R.B. Hoyle, *Physica* ***D*** **191**, 261 (2004).
[24] F.H.C. Crick and J.D. Watson, *Nature* **177**, 473 (1956).
[25] N.J. Dimmock, A.J. Easton and K.N. Leppard, *Introduction to Modern Virology* (Blackwell, Oxford, 2001).

CHAPTER 7

CONVERGENCE PROPERTIES OF FOURIER MODE REPRESENTATIONS OF QUASIPATTERNS

Alastair M. Rucklidge

Department of Applied Mathematics, University of Leeds, Leeds LS2 9JT, UK

Spatial Fourier transforms of quasipatterns observed in Faraday wave experiments suggest that the patterns are well represented by the sum of 8, 10 or 12 Fourier modes with wavevectors equally spaced around a circle. We show that nonlinear interactions of n such Fourier modes generate new modes with wavevectors that approach the original circle no faster than a constant times n^{-2}. These close approaches lead to small divisors in the standard perturbation theory used to compute properties of these patterns, and we show that the convergence of the standard method is questionable in spite of the bound on the small divisors.

1. Introduction

One well studied example of a pattern-forming instability is the Faraday wave problem of the formation of waves on the surface of a layer of fluid as it is driven by vertical vibrations. This system has been subjected to intensive scrutiny in laboratory experiments and has come to be regarded as an archetypal pattern forming system. Clear examples of pattern formation occur in a wide range of other systems, including Rayleigh–Bénard convection, liquid crystals in externally imposed electric fields, nonlinear optics, directional solidification, vibrated granular media, chemical reactions and catalytic oxidation.

The simplest patterns, stripes, squares and hexagons, have reflection, rotation and translation symmetries. A comprehensive and very successful theory has been developed to analyze the creation of these patterns from an initial featureless state. This theory, which is based on computing the amplitudes of the various waves (or modes) that make up the pattern, is known as equivariant bifurcation theory, and is expounded in detail in a series of texts (see, for example, [8]).

In order to apply rigorous mathematical theories to explain experimental results and other occurrences of pattern formation in the natural world, there are naturally a series of idealizations and approximations that must be made. One supposes that in the absence of any driving force, the system will remain featureless, and that if the forcing is turned up, it must reach a critical level before it can overcome any inherent dissipation in the system. If the level of forcing (which is a parameter under the control of the experimentalist) exceeds this critical value, the featureless state will be unstable, and any small disturbances will grow. These cannot grow for ever, and one possible outcome is that the system will settle down to a steady state with some degree of spatial structure: a pattern.

Two further idealizations are often made when computing the mathematical properties of patterns. First, the experimental boundaries are ignored, and so in effect the experiment is supposed to be taking place in a container of infinite size; and second, the observed pattern is supposed to have perfect spatial periodicity. By only considering patterns that are periodic in space, rigorous theory can be applied to prove the existence of stripe, square and hexagon (and other) solutions of the nonlinear partial differential equations (PDEs) that model the experimental situation. Given that in some highly controlled experiments the idealization of spatial periodicity appears to hold over dozens of repeats of the pattern, these assumptions are perfectly reasonable when the objective is to understand the nature of these periodic patterns.

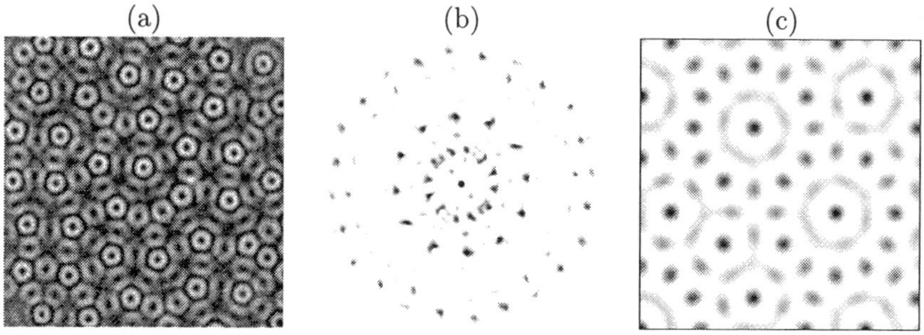

Fig. 1. Quasipatterns: (a) 12-fold quasipattern observed in a two-frequency forced Faraday wave experiment; (b) spatial Fourier transform, showing the 12-fold rotational order in spite of the absence of any translation symmetry (both from [1], with permission). (c) Synthetic quasipattern, constructed from the sum of 12 modes with wavevectors spaced equally around a circle; see equation (8).

However, experiments that are carried out in large domains are quite capable of producing patterns that cannot be analyzed in this way. A notable example of this is quasipatterns, which are most readily found in Faraday wave experiments in which a tray of liquid is subjected to vertical vibrations with two commensurate forcing frequencies [6]. A recent survey of experimental results can be found in [1], and one experimental example of a quasipattern is shown in Figure 1a. This pattern is quasiperiodic in any horizontal direction, that is, the amplitude of the pattern (taken along any direction in the plane) can be regarded as a sum of modes with incommensurate spatial frequencies. In general, quasipatterns exhibit long range rotational order, most evident in their spatial Fourier transform (Figure 1b), but they lack spatial periodicity. In this respect, there are obvious similarities with quasicrystals, which were discovered a decade earlier [11].

Models of quasipatterns have been developed by several researchers without the theoretical background required to justify their use (see below). These models are derived using a perturbation theory approach that is successful for periodic patterns; however, when the method is applied to the case of quasipatterns, a difficulty known as the problem of small divisors arises. This problem appears whenever quasiperiodic behavior is found in a nonlinear set of differential equations and attempts are made to use perturbation theory to compute the quasiperiodic solution by a series of approximations. In many cases, including the case of spatially periodic patterns, it can be proved that this process, if carried to the limit, will indeed converge to a true solution. However, in the case of quasiperiodic behavior, the corrections turn out not to be uniformly small, owing to the appearance of small numbers in the denominators, and convergence is called into question.

This difficulty was faced first by Poincaré in the context of celestial mechanics in the late 19th century. In the absence of any gravitational interaction between planets, each planet in the solar system orbits the Sun with its own period, and the system as a whole is quasiperiodic in time. Poincaré considered the question of whether or not the solar system is quasiperiodic given the presence of weak interactions between the planets. Formally, the problem could be solved by perturbation theory, but Poincaré realized that small divisors called convergence of the perturbation series into question.

The small divisor issue was resolved for this type of problem by Kolmogorov, Arnol'd and Moser (KAM) in the 1950's and 60's, who showed under what circumstances quasiperiodic behavior would be found (see, for

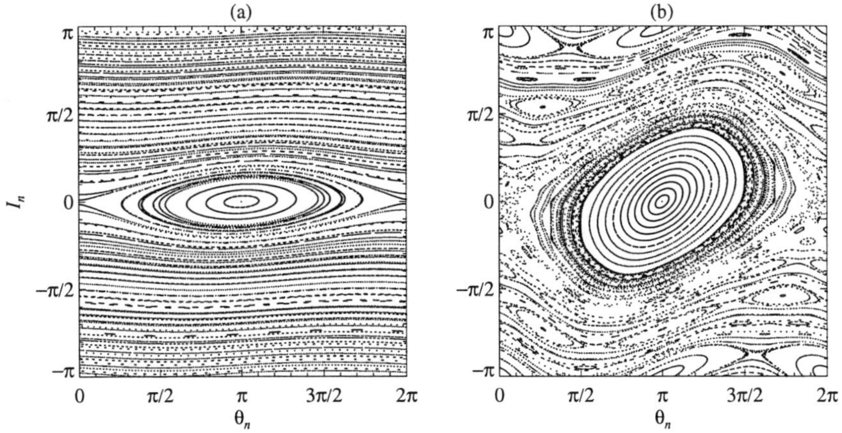

Fig. 2. Trajectories in the standard map (1): (a) $\varepsilon = 0.1$; (b) $\varepsilon = 0.8$. For small ε, several features are apparent: there are fixed points (at $(\theta, I) = (0,0)$ and $(\pi, 0)$), periodic orbits and two types of quasiperiodic orbit: those that have a bounded range of θ (in an island centered on $(\pi, 0)$), and those for which θ increases or decreases monotonically. For larger ε, more islands are visible, as well as chaotic dynamics between the islands, and yet some quasiperiodic trajectories persist.

example, [15]). To take an example, consider the so-called standard map:

$$I_{n+1} = I_n + \varepsilon \sin(\theta_n), \qquad \theta_{n+1} = \theta_n + I_{n+1} \mod 2\pi, \tag{1}$$

which models a freely rotating pendulum in the absence of gravity, subjected to periodic impulsive forces. When $\varepsilon = 0$, all trajectories are of the form $(\theta_n, I_n) = (\theta_0 + nI_0, I_0) \mod 2\pi$, and are periodic with period q if $I_0/2\pi = p/q$ is rational (with p and q integers), and quasiperiodic otherwise. Both periodic and quasiperiodic orbits lie on horizontal lines (invariant curves) in the (θ, I) plane, but the lines are made up of individual periodic points in the first case, while a quasiperiodic orbit will eventually visit a neighbourhood of each point on the line. When ε is perturbed away from zero (see Figure 2a), the question is which of these families of trajectories will persist as invariant curves of the map? The essential content of the KAM theorem is that, for small enough perturbations, and for almost every irrational value of $I_0/2\pi$, there will be an invariant curve close to the unperturbed invariant curve, and the corresponding quasiperiodic trajectory survives the perturbation. The curves that persist are those that satisfy a Diophantine condition, that is, for which there are constants $K > 0$ and

$\delta > 0$ such that $I_0/2\pi$ satisfies

$$\left| p - \frac{I_0}{2\pi} q \right| \geq \frac{K}{(|p| + |q|)^\delta} \tag{2}$$

for every pair of integers p and q, apart from $(0,0)$. The exponent δ is an indication of the 'irrationality' of $I_0/2\pi$, so, for example, $(\sqrt{5}-1)/2$ satisfies (2) with $\delta = 1$. In general, curves with smaller values of δ persist to larger values of the perturbation ε. Invariant curves with rational values of $I_0/2\pi$ are immediately broken up into elliptic and hyperbolic periodic points, with a web of chaotic trajectories near the hyperbolic equilibria (see Figure 2b).

KAM theory has been applied successfully to a variety of problems in which small divisors arise, for instance quasiperiodicity in the solar system and in the dynamics of charged particles in tokamak magnetic fields. However, the methods of KAM (based around canonical coordinate transformations) were developed for problems in which quasiperiodicity occurs in only *one* direction (time), whereas quasipatterns are quasiperiodic in *two* spatial directions. For this reason, KAM theory is not applicable to quasipatterns, at least not directly, and either the theory must be extended to cover this case, or alternative methods must be developed. In principle, similar issues arise in solid-state quasicrystals, though the main theoretical approaches for these are developed around aperiodic Penrose tilings of the plane or three dimensional space, and around projecting higher dimensional periodic lattices down to three dimensions [10], whereas a wave-based approach is more natural for the fluid dynamical quasipatterns.

The purpose of this paper is to draw attention to some of the theoretical difficulties that are preventing progress in the development of a mathematical understanding of two-dimensional quasipatterns. We review progress that has recently been made in coming to terms with the small divisor problem [17]. Section 2 introduces a particularly simple pattern-forming PDE (the Swift–Hohenberg equation) and indicates how the small divisors arise. Limits on the magnitude of these small divisors are calculated in Section 3, and the perturbation theory for the quasipattern solution of the Swift–Hohenberg equation is concluded in Section 4, with an indication that the problem of small divisors does indeed cause the perturbation theory to fail. We conclude with general remarks in the last section.

2. Model Equations

One of the key mathematical questions concerning quasipatterns is one of *existence*: do PDEs that model pattern-forming problems have solutions that are quasiperiodic in space, along the lines of the experimentally observed pattern in Figure 1a? Rather than try to answer this question in the context of a PDE that specifically models the Faraday wave problem, it seems sensible to start with the simplest possible pattern forming PDE: the Swift–Hohenberg equation [19]. In fact, considering the Swift–Hohenberg equation is not such a simplification, since many pattern-forming problems can be cast into this form, or variations [14]. The simplest variant is:

$$\frac{\partial U}{\partial t} = \mu U - (1 + \nabla^2)^2 U - U^3. \tag{3}$$

The equation is posed on the plane, with $\boldsymbol{x} = (x, y) \in \mathbb{R}^2$, and $U(x, y, t) \in \mathbb{R}$ supposed to be bounded as $(x, y) \to \infty$. The parameter μ represents the force that will drive the pattern formation.

In fact, stable quasipatterns are not observed in the Swift–Hohenberg equation with standard cubic nonlinearities, although they have been seen in numerical simulations with a modified linear term to allow marginally stable modes at two wavenumbers [12], in model equations that are essentially the Laplacian of the Swift–Hohenberg equation [3,4], and in the Zhang–Viñals model of the Faraday wave experiment [20]. However, the issue here is one of existence of quasipatterns rather than their stability, so we focus on (3) as a model problem.

This PDE has a spatially uniform trivial solution $U(x, y, t) = 0$, and the stability of this solution can be investigated by linearizing (3). The linearized equation has wave-like solutions: $U = e^{st} e^{i\boldsymbol{k}\cdot\boldsymbol{x}}$, with growth rate s and wavevector $\boldsymbol{k}$, with the growth rate related to μ and $|\boldsymbol{k}|$ by $s = \mu - (1 - |\boldsymbol{k}|^2)^2$. This relation is plotted in Figure 3a in the case $\mu = 0$: with this value of μ, all modes are damped (have negative growth rate) apart from those with wavenumber $|\boldsymbol{k}|$ equal to 1. With μ just above zero, modes with $|\boldsymbol{k}|$ close to 1 will grow, until the nonlinear term in (3) causes the amplitudes of these modes to saturate at a level related to the value of μ.

In many pattern forming problems, standard perturbation theory can be used to compute how the amplitude saturates, with the assumption that the parameter μ and the amplitude of the pattern are both very small. This degree of smallness is explicitly introduced as a small parameter $\varepsilon \ll 1$, and U is written in the form:

$$U = \varepsilon U_1 + \varepsilon^3 U_3 + \varepsilon^5 U_5 + \ldots \tag{4}$$

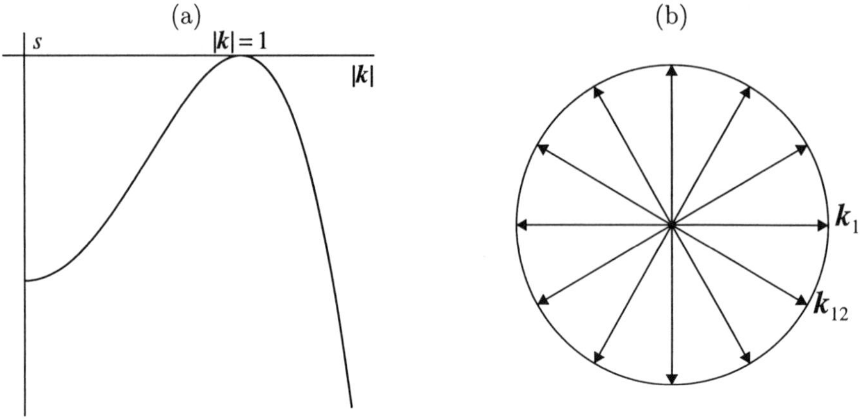

Fig. 3. (a) Schematic growth (decay) rate s of a mode $e^{i\boldsymbol{k}\cdot\boldsymbol{x}}$, as a function of $|\boldsymbol{k}|$ at $\mu = 0$. Modes with $|\boldsymbol{k}| = 1$ are marginally stable. (b) 12 wavevectors on the circle $|\boldsymbol{k}| = 1$. Adding equal amounts of 12 modes with these wavevectors (numbered $\boldsymbol{k}_1$ to $\boldsymbol{k}_{12}$) results in the synthetic pattern in Figure 1c.

The absence of even terms ($\varepsilon^2 U_2$) is because of the symmetry $U \to -U$ in equation (3). The connection between the small forcing μ and the small parameter ε is made explicit by setting $\mu = \varepsilon^2$. The expansion (4) is inserted into the Swift–Hohenberg equation (3) and like powers of ε are collected together:

$$0 = \varepsilon\mathcal{L}(U_1) + \varepsilon^3\left(U_1 + \mathcal{L}(U_3) - U_1^3\right) + \varepsilon^5\left(U_3 + \mathcal{L}(U_5) - 3U_1^2U_3\right) + \ldots,$$

where, to make the presentation simpler, only steady patterns are considered. The linear differential operator $\mathcal{L}(U)$ is $-(1+\nabla^2)^2 U$.

In order for this equation to be satisfied for all parameter values, the coefficient of each power of ε must separately be zero, and so the equation can be solved formally by considering each power of ε in turn. The leading order equation is

$$\mathcal{L}(U_1) = 0. \tag{5}$$

The operator $\mathcal{L}$ acting on a mode $e^{i\boldsymbol{k}\cdot\boldsymbol{x}}$ yields $-(1-|\boldsymbol{k}|^2)^2 e^{i\boldsymbol{k}\cdot\boldsymbol{x}}$, which is zero only when $|\boldsymbol{k}| = 1$, so equation (5) has non-trivial solutions that are made up of linear combinations of modes with wavevectors $\boldsymbol{k}$ on the unit circle. Any set of such wavevectors is possible at this level, but a natural choice to make when studying quasipatterns is

$$U_1(x,y) = \sum_{j=1}^{12} A_j e^{i\boldsymbol{k}_j\cdot\boldsymbol{x}},$$

where the 12 vectors $\boldsymbol{k}_1$ to $\boldsymbol{k}_{12}$ are equally spaced around the circle (Figure 3b). This choice of modes is inspired by the evidence in the Fourier transforms of experimentally observed quasipatterns (as in Figure 1b). In order for U to be real, the amplitudes must satisfy $A_{j+6} = \bar{A}_j$. Setting each A_j to the same real value results in a quasipattern of the form depicted in Figure 1c.

At third order in ε, the equation to solve is:

$$\mathcal{L}(U_3) = -U_1 + U_1^3 = -\sum_{j=1}^{12} A_j e^{i\boldsymbol{k}_j\cdot\boldsymbol{x}} + \sum_{j=1}^{12}\sum_{k=1}^{12}\sum_{l=1}^{12} A_j A_k A_l e^{i(\boldsymbol{k}_j+\boldsymbol{k}_k+\boldsymbol{k}_l)\cdot\boldsymbol{x}}. \tag{6}$$

Notice that U_1^3 contains cubic interactions between the modes in U_1, which take the form of modes with all possible combinations of three of the 12 original wavevectors (allowing repeats). Some combinations (for example, $\boldsymbol{k}_1 + \boldsymbol{k}_1 + \boldsymbol{k}_7 = \boldsymbol{k}_1$) lie on the unit circle, but most ($\boldsymbol{k}_1 + \boldsymbol{k}_2 + \boldsymbol{k}_3$) do not. Modes with different wavevectors are orthogonal, so the coefficients of each mode on the left and the right of equation (6) must be equal. In particular, the coefficient of modes with wavevectors on the unit circle is zero on the left, since $\mathcal{L}$ acting on such a mode is zero. Setting the coefficient of (for example) $e^{i\boldsymbol{k}_1\cdot\boldsymbol{x}}$ to zero on the right results in an equation relating the amplitudes of the modes:

$$0 = A_1 - 3(|A_1|^2 + 2|A_2|^2 + 2|A_3|^2 + 2|A_4|^2 + 2|A_5|^2 + 2|A_6|^2)A_1, \tag{7}$$

with similar equations resulting from the other modes. One solution of the amplitude equations is for all the amplitudes to be zero (the trivial solution); setting all amplitudes to have the same non-zero modulus results in a quasipattern. One particular solution is $A_1 = \ldots = A_{12} = 1/\sqrt{33}$, and so, in terms of the original variables, the pattern is:

$$U(x,y) = \sqrt{\frac{\mu}{33}} \sum_{j=1}^{12} e^{i\boldsymbol{k}_j\cdot\boldsymbol{x}} + \ldots \tag{8}$$

This result suggests that the quasipattern solution is created when μ increases through zero, with an amplitude proportional to $\sqrt{\mu}$.

This might appear to be the end of the story: the amplitude of the quasipattern has been computed as a function of the driving force, and a little more effort leads to an estimate of the stability of the pattern. This kind of calculation has been carried out in a variety of situations, starting either from equations describing the Faraday wave experiment or other experiments, or just using considerations of the symmetry of the quasipattern [2,7,13,16]. All these calculations result in amplitude equations similar to (7), and all suffer from two severe drawbacks.

The first drawback is that equation (7) determines only the amplitudes of the complex numbers A_j, and not their phase. In all, there are six free phases: two of these are fixed by considering resonances that occur at fifth order; two are genuinely free, and are associated with translating (but not changing) the pattern; and two phases (called phason modes) are not determined even by high-order resonances. In this context, the phason modes describe relative translations of two hexagonal sublattices generated by $\boldsymbol{k}_1$, $\boldsymbol{k}_3$, $\boldsymbol{k}_5$ and $\boldsymbol{k}_2$, $\boldsymbol{k}_4$, $\boldsymbol{k}_6$, and may play a role in long-wave instabilities of the quasipattern [5]. However, as they have a marked effect on the appearance of the pattern, they ought to be determined in a satisfactory theory without long-wave considerations.

The second drawback becomes apparent only when an attempt is made to compute higher order corrections to the pattern. Returning to equation (6), all modes with wavevectors on the unit circle have already been taken into account by solving (7). The remaining modes all have wavevectors off the unit circle ($|\boldsymbol{k}| \neq 1$), and so the linear operator $\mathcal{L}$ can be inverted to find U_3:

$$U_3 = - \sum_{|\boldsymbol{k}_j+\boldsymbol{k}_k+\boldsymbol{k}_l|\neq 1} \frac{A_j A_k A_l}{(1 - |\boldsymbol{k}_j + \boldsymbol{k}_k + \boldsymbol{k}_l|^2)^2} e^{i(\boldsymbol{k}_j+\boldsymbol{k}_k+\boldsymbol{k}_l)\cdot\boldsymbol{x}},$$

since the operator $\mathcal{L}^{-1}$ acting on a mode $e^{i\boldsymbol{k}\cdot\boldsymbol{x}}$ yields $-e^{i\boldsymbol{k}\cdot\boldsymbol{x}}/(1-|\boldsymbol{k}|^2)^2$, defined as long as $|\boldsymbol{k}| \neq 1$.

However, if $|\boldsymbol{k}|$ is close to one, $\mathcal{L}^{-1}(e^{i\boldsymbol{k}\cdot\boldsymbol{x}})$ can be arbitrarily large. This does not pose difficulties for computing U_3, but continuing the calculation to higher order results in combinations of vectors that can come arbitrarily close to the unit circle. Specifically, U_3 involves sums of three of the original 12 vectors, and U_N will involve integer combinations of up to N of the 12 vectors $\boldsymbol{k}_1$ to $\boldsymbol{k}_{12}$. If the original choice of vectors had been two, four or six, in an attempt to describe striped, square or hexagonal patterns, the integer combinations of vectors arising at high order would not have come close to the unit circle, instead forming a lattice. Choosing 12 evenly spaced vectors leads to integer combinations of vectors that come arbitrarily close to the unit circle. Small divisors arise when the operator $\mathcal{L}$ is inverted, which raises doubts as to whether or not the power series (4) for U will converge.

3. Small Divisors

Does the smallness of the small divisors arising from inverting $\mathcal{L}$ cause the sum (4) for $U(x, y, t)$ to diverge? To answer this question, the first stage is to derive a Diophantine-like condition for integer combinations of up to N of the 12 original vectors on the unit circle (such combinations arising at order N in the power series for U). It turns out that, for a given N, the smallest nonzero distance from the unit circle of a combination of N vectors is bounded above and below by a constant times N^{-2}.

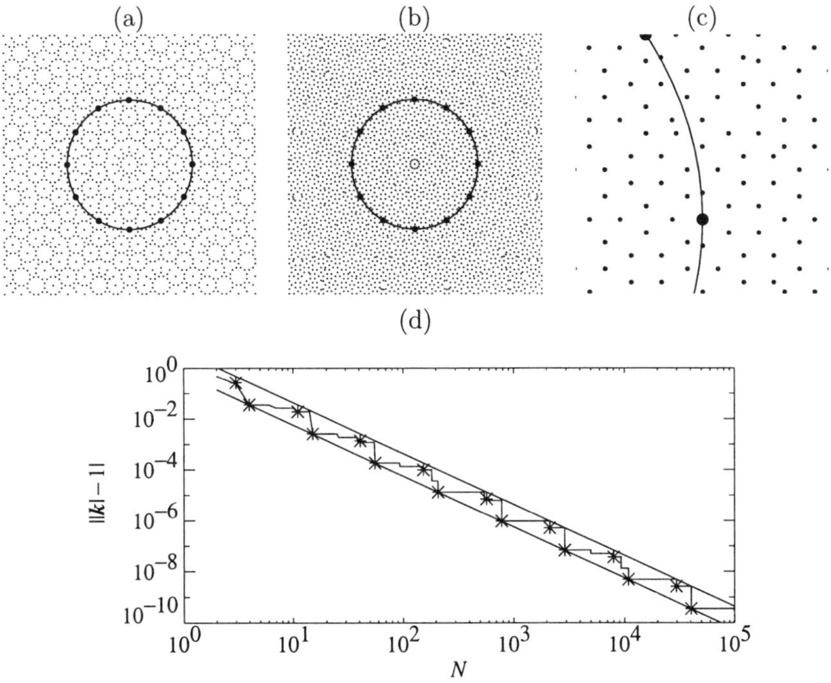

Fig. 4. Positions of combinations of up to N of the original 12 vectors on the unit circle, with (a) $N = 11$, (b) $N = 15$; (c) detail of (b). The circle indicates the unit circle, $|\boldsymbol{k}| = 1$, the large dots are the original 12 wavevectors, and the small dots are integer combinations of these. Note how the density of points increases with N, and the proximity of points to the unit circle decreases with N. (d) Smallest nonzero distances from the unit circle $||\boldsymbol{k}_m| - 1|$ as a function of the total number of modes $|\boldsymbol{m}| = N$. Stars mark distances calculated from equation (10), and straight lines indicate the scaling N^{-2}. After [17].

An explanation of how this is derived begins with Figure 4a–c, illustrating the locations of combinations of up to $N = 11$ and 15 wavevectors. Note how the density of points increases with N, and how the minimum distance between points and the unit circle goes down with N. Figure 4d shows results for the smallest nonzero distance from the unit circle as a function of the total number of vectors. The solid lines in Figure 4d confirm numerically that the scaling for the distance to the unit circle is order N^{-2}, and the stars represent explicit combinations of wavevectors close to the unit circle, which were found as follows.

The vectors $\boldsymbol{k}_1$, $\boldsymbol{k}_2$, ..., $\boldsymbol{k}_{12}$ are labelled anticlockwise around the circle starting with $\boldsymbol{k}_1 = (1, 0)$, with $\boldsymbol{k}_{j+6} = -\boldsymbol{k}_j$ (Figure 3b). Integer combinations of N of these vectors can be written as $\boldsymbol{k}_{\boldsymbol{m}} = \sum_{j=1}^{12} m_j \boldsymbol{k}_j$, with $|\boldsymbol{m}| = \sum_j |m_j| = N$. Including equal and opposite vectors $\boldsymbol{k}_j$ and $\boldsymbol{k}_{j+6}$ will only increase N without coming any closer to the unit circle, so only m_1, ..., m_6 are considered, but these are allowed to be negative. With this restriction, the squared length of a vector $\boldsymbol{k}_{\boldsymbol{m}}$ is:

$$
\begin{aligned}
|\boldsymbol{k}_{\boldsymbol{m}}|^2 = m_1^2 + m_2^2 + m_3^2 + m_4^2 + m_5^2 + m_6^2 \\
&+ m_1m_3 + m_2m_4 + m_3m_5 + m_4m_6 - m_5m_1 - m_6m_2 \\
&+ \sqrt{3}(m_1m_2 + m_2m_3 + m_3m_4 + m_4m_5 + m_5m_6 - m_6m_1).
\end{aligned}
$$

This is of the form $|\boldsymbol{k}_{\boldsymbol{m}}|^2 = 1 + p - rq$, where $r = \sqrt{3}$ is irrational and p and q are integers. If $p - rq$ is close to zero (that is, if r is well approximated by the rational p/q), then $|\boldsymbol{k}_{\boldsymbol{m}}|^2$ can come close to 1 (but can only be exactly 1 if $p = q = 0$).

It is clear that the theory of continued fraction approximations of irrationals will be useful here. The continued fraction expression for $r = \sqrt{3}$ is:

$$
r = \sqrt{3} = 1 + \cfrac{1}{1 + \cfrac{1}{2 + \cfrac{1}{1 + \cfrac{1}{2 + \cdots}}}}.
$$

Since this irrational satisfies a quadratic equation with integer coefficients, $\sqrt{3}$ is called a quadratic irrational.

If the fraction is truncated after l terms, the successive fractions p_l/q_l that approximate $r = \sqrt{3}$ are given in Table 1. The theory of continued

Table 1. Continued fraction approximations to $r = \sqrt{3}$, as a function of the order l of the truncation.

	$l=0$	1	2	3	4	5	6	7	8	9	10
$r=\sqrt{3}$	$\frac{p_l}{q_l} = \frac{1}{1}$	$\frac{2}{1}$	$\frac{5}{3}$	$\frac{7}{4}$	$\frac{19}{11}$	$\frac{26}{15}$	$\frac{71}{41}$	$\frac{97}{56}$	$\frac{265}{153}$	$\frac{362}{209}$	$\frac{989}{571}$

fractions for quadratic irrationals [9] shows that

$$\frac{K_1}{q_l^2} < \left|\frac{p_l}{q_l} - r\right| < \frac{K_2}{q_l^2} \qquad \text{and} \qquad \left|\frac{p_l}{q_l} - r\right| < \left|\frac{p}{q} - r\right|, \tag{9}$$

where K_1, K_2 are constants, q an integer satisfying $0 < q < q_l$. These inequalities mean that the truncated continued fraction expansions p_l/q_l approximate r well, but not too well, as l becomes large, and that if p_l/q_l is the truncation of the continued fraction approximation of an irrational r, no other fraction with a smaller denominator comes closer to r.

Apart from those vectors $\boldsymbol{k_m}$ that fall exactly on the unit circle (which would have $p = q = 0$), the relations in (9) can be used to show that $|\boldsymbol{k_m}|^2$ can approach 1 no faster than order N^{-2}:

$$\big||\boldsymbol{k_m}|^2 - 1\big| \geq \frac{K}{N^2},$$

where $|\boldsymbol{m}| = N$ and K is a constant – this lower limit is shown as a straight line in Figure 4d. See [17] for more details.

The order N^{-2} rate of approach is indeed achieved by special combinations of vectors, which were found after a prolonged examination of the distances plotted in Figure 4d. Choosing

$$\boldsymbol{k_m} = p_l\boldsymbol{k}_4 + (q_l - 1)\boldsymbol{k}_9 + (q_l + 1)\boldsymbol{k}_{11} = (1, p_l - \sqrt{3}q_l), \tag{10}$$

with $|\boldsymbol{m}| = N = p_l + 2q_l$ and $|\boldsymbol{k_m}|^2 - 1 = (p_l - \sqrt{3}q_l)^2$. As N (or equivalently, l or q_l) increases, p_l and q_l are related by $p_l \sim \sqrt{3}q_l + \mathcal{O}(1/q_l)$, so $q_l = \mathcal{O}(N)$, and $|\boldsymbol{k_m}|^2 - 1 = \mathcal{O}(N^{-2})$. These particular choices of $\boldsymbol{k_m}$ are plotted on the graphs in Figure 4d as stars.

In summary, given an integer N, the vector $\boldsymbol{k_m}$ with $|\boldsymbol{m}| = N$ that comes closest to the unit circle (without being on the unit circle) satisfies

$$\frac{K}{N^2} \leq \big||\boldsymbol{k_m}|^2 - 1\big| \leq \frac{K'}{N^2},$$

for constants K and K', for 12 equally spaced original vectors. The numerical evidence in Figure 4d suggests values $K = 0.56$ and $K' = 4.34$.

4. The Question of Convergence

The results of the previous two sections imply that when $\boldsymbol{k_m}$ is close to the unit circle, $\mathcal{L}^{-1}(e^{i\boldsymbol{k_m}\cdot\boldsymbol{x}})$ can be as large as a constant times $N^4 e^{i\boldsymbol{k_m}\cdot\boldsymbol{x}}$, with $N = |\boldsymbol{m}|$. This is so large that it clearly could lead to divergence of the power series (4) for U, particularly when nonlinear interactions of these large contributions are taken into account. This problem of small divisors is not just a feature of the particular Swift–Hohenberg equation (3) used for illustration here, but arises in any calculation of the properties of quasipatterns based on perturbation theory.

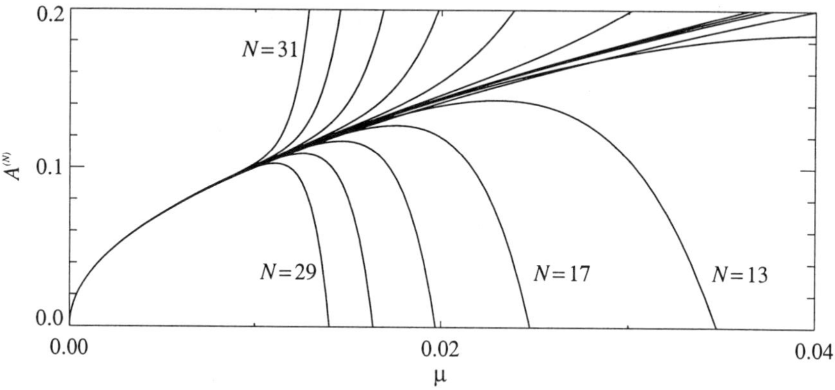

Fig. 5. Amplitude $A^{(N)}$ as a function of μ, for different levels of truncation $N = 1, \ldots, 31$. Increasing the order of truncation leads to graphs of $A^{(N)}$ that diverge for μ closer and closer to zero as N becomes larger. The amplitude has been scaled to remove a factor of $1/\sqrt{33}$. From [17].

This failure of convergence can be illustrated dramatically in the particular Swift–Hohenberg example by carrying out the perturbation theory calculation to high order (33rd order in this case). If the series (4) is truncated to include powers of ε up to and including $N+2$, the resulting expression for $U^{(N)}$ is of the form

$$U^{(N)} = A^{(N)} \sum_{j=1}^{12} e^{i\boldsymbol{k}_j\cdot\boldsymbol{x}} + \text{other modes},$$

so $A^{(1)} = \sqrt{\mu/33}$, from (8). The amplitude $A^{(N)}$ of the basic quasipattern is shown as a function of μ in Figure 5, for $N = 1, \ldots, 31$. In this calculation, only modes with wavenumbers up to $\sqrt{5}$ were kept, to keep the total number of modes within manageable limits. Even so, there were more than 15000 modes generated at the highest order – without this truncation, there would have been almost 2 million. Since the modes that were dropped from the calculation were the most heavily damped, their contribution to the total amplitude was quite small (of the order of 1%), and restricting the number of modes in this way had no effect on how close combinations of wavevectors could get to the unit circle.

It is clear in Figure 5 that, at each level of truncation N, the graph of $A^{(N)}$ against μ diverges at a value of μ that decreases as N becomes larger. The value of μ at which the sum up to order N diverges is related to the smallest distance from the unit circle achieved by combinations of N of the 12 original wavevectors. Since this distance goes to zero as N increases, the sum $A^{(N)}$ will continue to diverge closer and closer to $\mu = 0$. In contrast, the equivalent calculation for spatially periodic patterns has a non-zero radius of convergence [17].

5. Discussion and Speculation

The main conclusion of the calculation is that even if perturbation theory does generate a convergent series approximation to the quasipattern for small enough μ, the series certainly diverges if the parameter μ is bigger than about 0.01. It might be possible that the series does converge for smaller μ, though there is a strong argument that this is not the case. However, even if the series does diverge for all nonzero μ, a low-order truncation may still give a useful asymptotic approximation of the quasipattern, assuming that the equations do have a quasipattern solution. It is on this basis that other researchers have proceeded.

There are two related issues at stake. First, existence: do pattern forming PDEs (like the two-dimensional Swift–Hohenberg equation) have quasipattern solutions? A more general formulation of this question, using the Swift–Hohenberg equation as an example, becomes apparent by setting $\mu = \varepsilon^2$ in (3), scaling U by ε and seeking a steady solution. The resulting equation can be written as

$$\mathcal{L}(U) = \varepsilon^2(-U + U^3),$$

which incidentally demonstrates that this is not a singularly perturbed problem. When $\varepsilon = 0$, *any* linear combination of waves with wavevectors on the unit circle solves this equation. The question is, which of these solutions persist to small but positive ε? Current theory can so far only answer this question for those solutions that are spatially periodic. The limits on the rate of approach of wavevectors to the unit circle will play a central role in an eventual existence theory for quasipatterns.

The second issue is, given the small divisor problem, are there methods that yield useful approximations to quasipattern solutions? Standard perturbation theory does not converge sufficiently rapidly (or slowly) to provide an answer unequivocally one way or the other. However, if quasipattern solutions exist, then the series ought to provide an asymptotic approximation to those solutions. Nonetheless, this approach will be left with difficulties, such as the undetermined phason modes, and so should not be regarded as a reliable way of computing properties of quasipatterns.

What is needed is a method that converges more rapidly. Each order in the standard theory gains a factor of ε^2 as well as large factors from any small divisors that arise. There are other methods, developed for proofs of KAM theory, that converge more rapidly, and these may be required for a rigorous treatment of quasipatterns as well. The difference between the KAM situation and that of quasipatterns is that in the KAM case, the solutions of interest are quasiperiodic in only one dimension (time), while in the second, quasipatterns are quasiperiodic in two space directions.

There are alternative approaches to analysing quasipatterns, for instance based on successive approximation of a quasipattern by a periodic pattern with increasingly large periodicity, defined on square or hexagonal lattices. For example, approximate 12-fold quasipatterns can be constructed using modes with wavevectors $(1,0)$, $(2p_lq_l/(p_l^2+q_l^2),(p_l^2-q_l^2)/(p_l^2+q_l^2))$ and so on, where $\frac{p_l}{q_l}$ is a truncated continued fraction approximation to $\sqrt{3}$. These generate patterns that are periodic on domains of size $p_l^2+q_l^2$ times the original wavelength: 5, 34, 65, …, for $l = 1, 2, 3, \ldots$, and have angles between their wavevectors of 36.9°, 28.1°, 30.5°, …. The wavevectors all have unit wavenumber, since $(p_l^2-q_l^2, 2p_lq_l, p_l^2+q_l^2)$ form Pythagorean triplets – see [4] for more details. Similarly, 12-dimensional representations of the group $D_6 \ltimes T^2$ can be chosen so that the modes are nearly equally spaced and yet they generate a hexagonal lattice ([18], and by allowing the wavevectors to have slightly different lengths, there are even more possibilities. The drawback with approximating quasipatterns by periodic patterns in these ways is that the range of validity of the normal forms derived shrinks to zero as the approximation improves.

Acknowledgments

I am grateful to many people who have helped shape these ideas, in one way or another, over a period of several years. This research is supported by the Engineering and Physical Sciences Research Council.

References

[1] Arbell, H. & Fineberg, J. 2002 . *Phys. Rev. E* **65**, 036224.
[2] Chen, P. & J.Viñals 1999 . *Phys. Rev. E* **60**, 559–570.
[3] Cox, S.M. & Matthews, P.C. 2001 . *Physica* **149D**, 210–229.
[4] Dawes, J.H.P., Matthews, P.C. & Rucklidge, A,M. 2003 . *Nonlinearity* **16**, 615–645.
[5] Echebarria, B. & Riecke, H. 2001 . *Physica* **158D**, 45–68.
[6] Edwards, W.S. & Fauve, S. 1994 . *J. Fluid Mech.* **278**, 123–148.
[7] Golovin, A.A., Nepomnyashchy, A.A. & Pismen, L.M. 1995 . *Physica* **81D**, 117–147.
[8] Golubitsky, M. & Stewart, I. 2002 *The Symmetry Perspective: From Equilibrium to Chaos in Phase Space and Physical Space.* Basel: Birkhäuser.
[9] Hardy, G.H. & Wright, E.M. 1960 *An Introduction to the Theory of Numbers, 4th edition.* Oxford: Clarendon Press.
[10] Janot, C. 1994 *Quasicrystals: a Primer, 2nd edition.* Oxford: Clarendon Press.
[11] Levine, D. & Steinhardt, P.J. 1984 . *Phys. Rev. Lett.* **53**, 2477–2480.
[12] Lifshitz, R. & Petrich, D.M. 1997 . *Phys. Rev. Lett.* **79**, 1261–1264.
[13] Lyngshansen, P. & Alstrom, P. 1997 . *J. Fluid Mech.* **351**, 301–344.
[14] Melbourne, I. 1999 . *Trans. Am. Math. Soc.* **351**, 1575–1603.
[15] Moser, J. 1973 *Stable and Random Motions in Dynamical Systems.* Princeton: Princeton University Press.
[16] Pismen, L.M. 1981 . *Phys. Rev. A* **23**, 334–344.
[17] Rucklidge, A.M. & Rucklidge, W.J. 2003 . *Physica* **178D**, 62–82.
[18] Silber, M., Topaz, C.M. & Skeldon, A.C. 2000 . *Physica* **143D**, 205–225.
[19] Swift, J. & Hohenberg, P.C. 1977 . *Phys. Rev. A* **15**, 319–328.
[20] Zhang, W. & J.Viñals 1996 . *Phys. Rev. E* **53**, R4283–R4286.

PART II

Localized Patterns, Waves, and Weak Turbulence

CHAPTER 8

PHASE DIFFUSION AND WEAK TURBULENCE

Joceline Lega

Department of Mathematics, University of Arizona,
617 N. Santa Rita, P.O. Box 210089,
Tucson, AZ 85721-0089, USA

This chapter gives a brief overview of phase diffusion, phase instabilities, and weak turbulence as described by classical envelope equations. In particular, phase turbulence, hole-mediated turbulence, and defect-mediated turbulence in the complex Ginzburg-Landau equation are discussed in detail.

1. Introduction

Spatially extended systems driven far from equilibrium often respond to an external forcing by forming a pattern, which is a periodic structure that breaks some of the original symmetries of the system [1–5]. Near this symmetry-breaking bifurcation, the nonlinear dynamics of the system may be described in terms of envelope equations, which are partial differential equations ruling the slow-time evolution of the (often complex) envelope of the pattern [6–8]. Such equations are typically formally derived by means of multiple-scales analysis, but recent analytical results have confirmed their near-threshold validity in some cases [9–16]. Most patterns break the time or space translation symmetries, or both. As a consequence, the corresponding envelope equations are gauge-invariant, which means that the phase of the complex order parameter describing the pattern may be arbitrarily shifted by a constant term. This in turns allows for a slow diffusive behavior of the phase [17–19], which may lead to instabilities [20–22] if the parameters of the system are such that the corresponding diffusion coefficient is negative. The goal of this chapter is to give an elementary description of the various phase instabilities commonly observed in envelope equations, and to discuss the weakly turbulent regimes that ensue.

2. Phase Instabilities, Phase Equations and Phase Turbulence

In this section, we give a brief overview of the various phase (or modulational) instabilities that occur in pattern-forming systems, as described by envelope equations. We do not discuss phase-diffusion equations for patterns far from threshold [23, 24].

2.1. *The Nature of Phase Instabilities*

Because most patterns break translational invariance, either in time or in space, the corresponding envelope equations are invariant under the gauge transformation $A \to A\exp(i\varphi)$, where φ is a real constant and A is the complex amplitude of the pattern. A classical example is that of the (scaled) complex Ginzburg-Landau (CGL) equation, which describes the near-threshold dynamics of a homogeneous structure undergoing a Hopf bifurcation [6], and thus breaking time translational invariance. The dimensionless form of this equation is

$$\frac{\partial A}{\partial t} = \mu A + (1 + i\alpha)\Delta A - (1 + i\beta)|A|^2 A, \tag{1}$$

where μ, α and β are real parameters which respectively represent the distance from the bifurcation threshold, dispersion, and nonlinear renormalization of temporal frequency. The complex order parameter A is the amplitude or envelope of the bifurcated structure, which is described as

$$u(x, y, t) = u_0 + A(X, Y, T)\exp(i\omega_0 t) + c.c. + \text{h.o.t.}, \tag{2}$$

where $u(x, y, t)$ corresponds to the oscillatory pattern, u_0 is the constant solution which undergoes a Hopf bifurcation at threshold (i.e. when $\mu = 0$), X, Y and T are slow spatial and temporal variables (typically $X = \varepsilon x$, $Y = \varepsilon y$, $T = \varepsilon^2 t$ and $\mu \propto \varepsilon^2$), and h.o.t. stands for terms of higher order in powers of A and of its complex conjugate $\bar{A}$. The gauge invariance of CGL stems from the fact that since u_0 is time-independent, one can freely choose the origin of the time variable t. Indeed, if $u(x, y, t)$ is a solution, so is $u(x, y, t+t_0)$, where t_0 is an arbitrary real number. Given the definition of A in Eq. (2), this in turn implies that if A solves CGL, so does $A\exp(i\omega_0 t_0)$. As a consequence, CGL must be gauge invariant.

Similarly, a bifurcation from an homogeneous state toward a stripe pattern in an isotropic system breaks the space translational invariance. The corresponding (dimensionless) envelope equation is the Newell-Whitehead-

Segel (NWS) equation [7,8], which reads

$$\frac{\partial A}{\partial t} = \mu A + \left(\frac{\partial}{\partial x} - \frac{i}{2k_0}\frac{\partial^2}{\partial y^2}\right)^2 A - |A|^2 A. \qquad (3)$$

Here, k_0 is the critical wavenumber of the pattern, which is described by

$$u(x,y,t) = u_0 + A(X,Y,T)\exp(ik_0 x) + c.c. + \text{h.o.t.},$$

with $X = \varepsilon x$, $Y = \sqrt{\varepsilon} y$ and $T = \varepsilon^2 t$. Arbitrariness in the choice of the origin of the x variable leads to the invariance of NWS under the transformation $A \to A\exp(i\varphi)$.

An important consequence of the gauge invariance of envelope equations is that the linearization about any solution A_0 has a marginal mode (neglecting boundary conditions), iA_0. In particular, if A_0 corresponds to a plane wave, then zero is in the continuous spectrum of the linearization about this solution. The continuous part of the spectrum typically has two branches, one anchored at zero and one bounded away from zero. They are respectively called the *phase* and *amplitude* branches of the spectrum. A *phase instability* occurs when the phase branch of the continuous spectrum crosses the imaginary axis. This is illustrated in Figure 1, in the case of CGL.

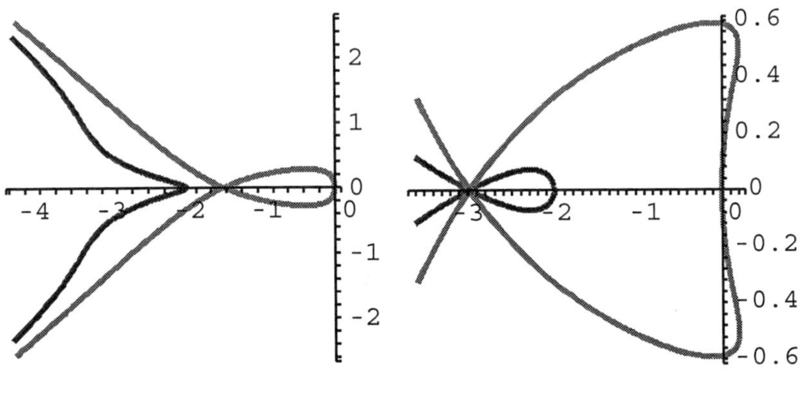

Fig. 1. Continuous spectrum of the linearization of 1d-CGL about a plane wave solution. Left: the plane wave is spectrally stable; Right: the plane wave is phase unstable.

2.2. *The Eckhaus, zig-zag and Benjamin-Feir Instabilities*

Typical phase instabilities are the Eckhaus instability [20, 22, 25], the zig-zag instability [26], and the Benjamin-Feir instability [21, 22, 27]. A characterization of these instabilities is easily obtained by linearizing NWS or CGL about a stationary pattern $A = R\exp(ikx)$ or a plane wave solution $A = R\exp[i(kx + \omega t)]$ respectively. It is useful to write the perturbation in the form $w\exp(ikx)$ – respectively $w\exp[i(kx+\omega t)]$ – with w complex. The corresponding linear system then decouples in Fourier space as an infinite number of 2×2 systems of ordinary differential equations, parametrized by the wavevector (q_x, q_y) of the perturbation in Fourier space. One can then look for an expansion of the eigenvalues of this system in powers of q_x and q_y. A phase instability occurs when the real part of the "phase" eigenvalue, $\mathrm{Re}\, e(\lambda) = -D_x(k)q_x^2 - D_y(k)q_y^2 + o(q_x^2, q_y^2)$, of this 2×2 system is positive for q_x or q_y small. This happens when $D_x(k) < 0$ or $D_y(k) < 0$. For instance, linearization of the Newell-Whitehead-Segel equation about the solution $A = \sqrt{\mu - k^2}\exp(ikx)$ leads to the following criteria:

- An Eckhaus instability $(D_x(k) < 0)$ occurs if $\mu - 3k^2 < 0$,
- A zig-zag instability $(D_y(k) < 0)$ occurs if $k < 0$.

Similarly, plane wave solutions of 1-d CGL, $A = \sqrt{\mu - k^2}\exp[i(kx + \omega t)]$ with $\omega = -\beta\mu + (\beta - \alpha)k^2$, are linearly unstable if

$$1 + \alpha\beta - \frac{2k^2(1+\beta^2)}{\mu - k^2} < 0.$$

This criterion takes into account both the Benjamin-Feir instability – which occurs for solutions with $k = 0$ – and the Eckhaus instability. One can indeed see that setting $k = 0$ in the above formula gives the usual Benjamin-Feir criterion, $1 + \alpha\beta < 0$, whereas one can recover the Eckhaus criterion $\mu - 3k^2 < 0$ by setting $\alpha = \beta = 0$.

2.3. *Phase Equations*

Since the amplitude branch of the continuous spectrum is bounded away from the origin, amplitude perturbations are typically damped out faster than phase perturbations. This separation of time scales suggests that the dynamics in the vicinity of a plane wave may be described in terms of a diffusion equation for only the phase of the solution [17–19]. Such an equation is called a *phase equation* and is formally derived from the corresponding amplitude equation using multiple scales analysis. This process also leads to an expression for the amplitude $|A|$ of the solution in terms of spatial derivatives of the phase of A. One then says that the amplitude of the complex order parameter A is *slaved* to its phase. Recent analytical works [28–31] have confirmed the validity of the phase diffusion equation in the phase stable regime, under appropriate hypotheses, and for special solutions of the Ginzburg-Landau equation with real and complex coefficients. Moreover, Doelman, Sandstede, Scheel and Schneider [32] have very recently proven the validity of the phase-diffusion equation for CGL near a stable plane wave solution.

In the case of the homogeneous solution $A = \sqrt{\mu}\exp(-i\beta\mu t)$ of CGL, one obtains the following phase dynamics to order two in the derivatives of φ,

$$A = \left[\sqrt{\mu} - \frac{1}{2\sqrt{\mu}}\left((\nabla\varphi)^2 + \alpha\nabla^2\varphi\right)\right]\exp[i(-\beta\mu t + \varphi)] \tag{4}$$

$$\frac{\partial\varphi}{\partial t} = (1+\alpha\beta)\nabla^2\varphi + (\beta-\alpha)(\nabla\varphi)^2, \tag{5}$$

where Equation (5) becomes the Kuramoto-Sivashinsky equation [17, 33] (KS),

$$\frac{\partial\varphi}{\partial t} = (1+\alpha\beta)\nabla^2\varphi + (\beta-\alpha)(\nabla\varphi)^2 - \frac{\alpha^2(1+\beta^2)}{2\mu}\nabla^4\varphi, \tag{6}$$

when it is regularized (for $1+\alpha\beta < 0$) by a bilaplacian term. The coefficient of this last term is obtained by looking at fourth-order contributions. Although we will use the Kuramoto-Sivashinsky equation to guide our intuition of the dynamics in phase unstable regimes, is important to note first that the derivation of this equation is formal, and second that it is not typically expected to hold past the phase instability threshold (i.e. when $1+\alpha\beta < 0$). In particular, other fourth order terms missing from this equation actually play an important role [34] when $|1+\alpha\beta|$ gets large (for $1+\alpha\beta < 0$).

One can check that the Benjamin-Feir instability, as described by the Kuramoto-Sivashinsky equation, is supercritical [35]. In the case of the Benjamin-Feir-Eckhaus instability (i.e. for a plane wave solution of CGL with $k \neq 0$), the instability may be supercritical or subcritical, depending on the CGL parameters [36]. Finally, the zig-zag instability of NWS is supercritical, and the Ekhaus instability of 1-dimensional NWS is subcritical [35]. The latter is associated with the following phase-diffusion equation,

$$\frac{\partial\varphi}{\partial t} = \frac{\mu-3k^2}{\mu-k^2}\frac{\partial^2\varphi}{\partial x^2} - \frac{1}{2(\mu-k^2)}\frac{\partial^4\varphi}{\partial x^4} - \frac{4k\mu}{(\mu-k^2)^2}\frac{\partial\varphi}{\partial x}\frac{\partial^2\varphi}{\partial x^2}.$$

2.4. *Phase Turbulence*

Rescaling of the Kuramoto-Sivashinsky equation (6) indicates that its dynamics depends on only one parameter, say the size of the system. When $1 + \alpha\beta < 0$ and for systems of large enough size, KS displays a form of space-time disorder or weak turbulence called *phase turbulence* [37]. The transition to phase turbulence in 1d-KS, as well as the dynamical properties of this regime, have been extensively studied in the literature (see [38–44] and references therein). A discussion of the stochastic and turbulent properties of KS and how they relate to those of the Kardar-Parisi-Zhang (KPZ) equation,

$$\frac{\partial \varphi}{\partial t} = \nu \nabla^2 \varphi + \frac{\lambda}{2} (\nabla \varphi)^2 + \eta(x, y, t),$$

where η is a Gaussian noise, can be found in [45–48].

3. Weak Turbulence in the Complex Ginzburg-Landau Equation

In what follows, we focus on CGL, in one and two spatial dimensions. Consider an initial condition of the form

$$A = \sqrt{\mu} \exp(-i\beta\mu t) + \eta(x, y, t), \tag{7}$$

where η represents some random, uniformly distributed noise of small amplitude. We know that when $1 + \alpha\beta < 0$, the homogeneous solution is phase unstable. Moreover, the Kuramoto-Sivashinsky equation displays phase turbulence in this regime. A natural question to ask is what this implies for the dynamics of CGL, and how the observed behavior depends on the size of the system. Particularly interesting is the question of the persistence of phase turbulence as the size of the system becomes infinitely large.

3.1. *Weak Turbulence in the one-dimensional complex Ginzburg-Landau Equation*

In one spatial dimension, two regimes are commonly seen in systems of finite size [49][a]. Close to threshold, i.e. for $1 + \alpha\beta < 0$ and $|1 + \alpha\beta|$ small enough, phase turbulence is observed. For larger values of $|1 + \alpha\beta|$, phase turbulence does not saturate and leads to the appearance of zeros of the complex

[a] Bichaos and space-time intermittency are also observed, but we do not discuss these here.

order-parameter A [34,50]. Such a regime may be called *hole-mediated turbulence*. The transition between the two regimes may be characterized by measuring the density of zeros of A [51]; the minimum value of $|A|$ - this measure extends to the two-dimensional case [52]; the maximum distance between consecutive maxima of $|\partial\varphi/\partial x|$ [53, 54], where φ is the phase of A; or the maximum value of the winding number - which is the integral of the phase gradient over the whole numerical box [55–58]. Numerical simulations [53, 59] seem to suggest that these two regimes persist as the size of the system becomes large, although such a conclusion has been challenged in the literature [57]. A more recent analysis in terms of modulated amplitude waves [54] also suggests that both phase and hole-mediated turbulence should exist in infinitely large, one-dimensional systems. We now describe these two regimes in more detail.

3.1.1. *Phase Turbulence*

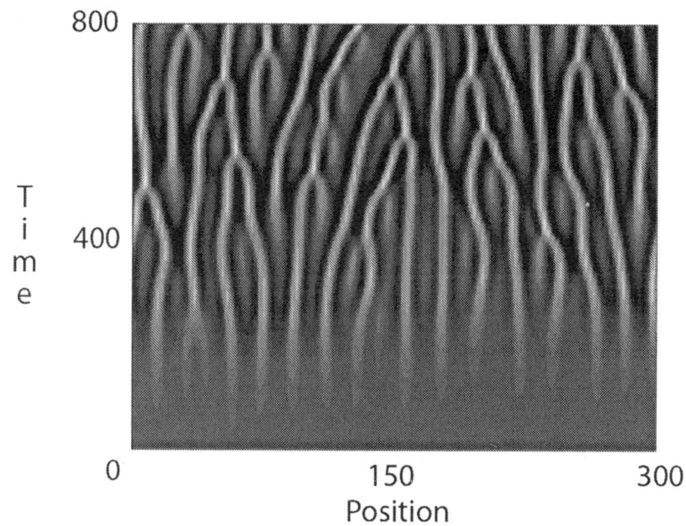

Fig. 2. Space-time diagram illustrating the dynamics of near-modulated amplitude waves in 1d-CGL. Lighter regions correspond to lager values of $|A|$. In this simulation, $0.98 \leq |A| < 1.06$. The parameters are $\mu = 1$, $\alpha = 1.3$, and $\beta = -1$, the boundary conditions are periodic, and the initial condition is given by Eq.(7).

Close to the Benjamin-Feir instability threshold, amplitude modulations, or *near-modulated amplitude waves* (near-MAWs) [53, 54] form, due to the presence of non-uniform phase gradients. Modulated amplitude waves are oscillatory travelling wave solutions of CGL, whose amplitude is periodic in space [53, 54]. In the near-MAW regime, maxima of $|A|$ appear to interact with one-another, and may annihilate in pairs; when two of them are too far apart, a new maximum occurs in between [54]. This dynamics is illustrated in Figure 2, which is a space-time diagram showing the amplitude of the solution A of CGL for $\mu = 1$ and $1 + \alpha\beta = -0.3$, in a system of length $L = 300$. Qualitatively, the appearance of amplitude modulations may be understood as follows: the system is described at lowest order by Eqs. (4) and (5). The latter equation is ill-posed and we know that the phase dynamics is approximately described by the Kuramoto-Sivashinsky equation (6), and is such that non-uniform phase gradients start to develop. The amplitude of A, given by Eq. (4), becomes modulated accordingly. In other words, since $|A|$ is slaved to its phase, the dynamics of the former can be directly linked to the dynamics of the latter.

It turns out that higher-order phase equations are in fact needed to give a good approximation of the dynamics in this regime. In particular, other fourth-order terms should be added to Equation (6) to reproduce some of the MAWs stability features [54]. It is however known from numerical simulations [34] that in such a case the phase equation can exhibit blow-up in finite time. This provides a mechanism for the formation of zeros of A in the hole-mediated turbulent regime, which is described below.

3.1.2. *Hole-mediated Turbulence*

Further away from the Benjamin-Feir threshold, hole-mediated turbulence prevails. Phase turbulence does not saturate and zeros of the complex field A form spontaneously. At this point, it is clear that the amplitude of A is no longer slaved to its phase. Once formed, amplitude holes travel across the system. The minimum of $|A|$ reached at the core of the hole decreases until A vanishes. The phase of A jumps by $\pm 2\pi$ each time such a zero occurs, and the winding number of the solution changes accordingly [58]. This dynamics is illustrated in the space-time diagram of Figure 3, for which the CGL parameters are $\mu = 1$ and $1 + \alpha\beta = -2$, with $L = 300$. A minimal model for the dynamics of holes and defects can be found in [60].

The holes that form as a consequence of the phase instability are reminiscent of a one-parameter family of solutions of CGL called *travelling*

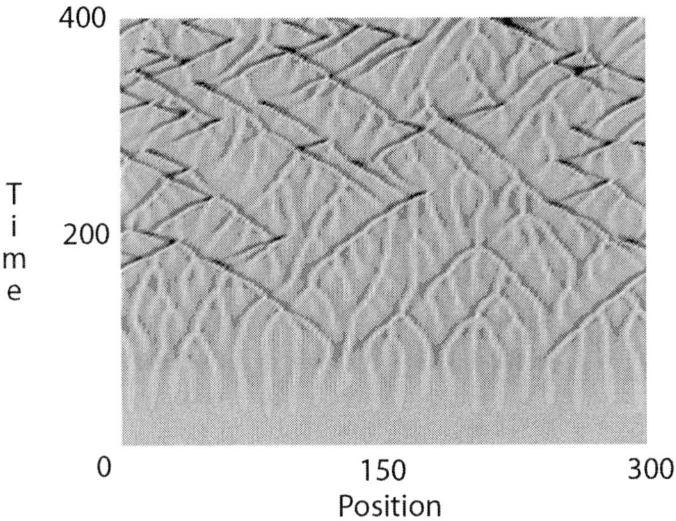

Fig. 3. Space-time diagram illustrating hole-mediated turbulence in 1d-CGL. Dark spots correspond to zeros of A. In this simulation, $0 \leq |A| < 1.25$. The parameters are $\mu = 1$, $\alpha = 2$, and $\beta = -1$, the boundary conditions are periodic, and the initial condition is given by Eq.(7).

holes [61–63]. Other authors have argued that such objects are homoclinic instead of heteroclinic solutions, which have been called *homoclons* [64]. Interestingly, modulated-amplitude waves provide a link between travelling holes and homoclons [54]. A conjecture described in [53, 54] (see also [58]) and well supported by numerical simulations, states that zeros of A arise when the phase-turbulent regime leads to the existence of near-MAWs whose period is larger than that of the saddle-node bifurcation above which MAWs cease to exist.

A different point of view consists in looking at the Fourier transform of the space-time diagram describing the evolution of A. In the (k,ω) space, the dynamics of the system is represented by a collection of points located in the vicinity of the curve $\omega = -\beta\mu + (\beta - \alpha)k^2$, which is the dispersion relation of nonlinear plane waves of CGL. As a consequence, one can also try to describe weak turbulence in CGL as resulting from the competition and interaction of nonlinear waves. A longer discussion of these dual points of view can be found in one of the author's review papers [63].

For $1 + \alpha\beta < 0$, all plane wave solutions of CGL, including the homogeneous solution $A = \sqrt{\mu}\exp(-i\beta\mu t)$ are unstable. The previous discussion, which was for initial conditions close to the homogeneous solution, can be extended to an initial condition corresponding to a plane wave solution. In this case, the transition from phase to hole-mediated turbulence may be described in terms of the winding number [55–58]

$$\nu = \frac{1}{2\pi}\int_0^L \frac{\partial\varphi}{\partial x}\,dx,$$

where φ is the phase of A. When the parameter range for which ν is conserved shrinks to zero, zeros of the complex field A are observed.

The nature of phase and hole-mediated turbulence, as well as that of the transition between these two regimes is fairly well documented in various numerical and theoretical studies reported in the literature and referenced above. In spite of all these efforts, it is however still difficult to give a complete and definite scenario for the formation of zeros of A in the hole-mediated turbulent regime. The topic now calls for mathematical analysis, although such a task is likely to be extremely difficult. But this of course makes this problem particularly interesting.

3.2. *Weak Turbulence in the two-dimensional complex Ginzburg-Landau Equation*

In two-dimensional CGL, *defect-mediated turbulence* [65, 66] is observed when $1 + \alpha\beta < 0$. This is a regime where zeros of A form spontaneously, move across the system and annihilate in pairs. As in the one-dimensional case, the phase instability of a slightly perturbed initial plane wave solution leads to modulations of $|A|$. At this stage, a cellular pattern is observed, as one would expect from the dynamics of the Kuramoto-Sivashinsky equation in two dimensions. However, phase gradients eventually become large enough to drive the system away from the phase approximation, and lead to the spontaneous formation of zeros of A, at least in large systems [65,66]. Such a dynamics is illustrated in Figure 4, which shows a series of gray-scale snapshots of $|A|$, at successive times. Dark regions correspond to low values of $|A|$.

If one starts from such a state and "quenches" the system by suddenly making $1 + \alpha\beta$ positive, the dynamics very quickly relaxes to a situation where the order parameter A vanishes at a few points, which correspond to defect cores. These defects then annihilate in pairs and the solution

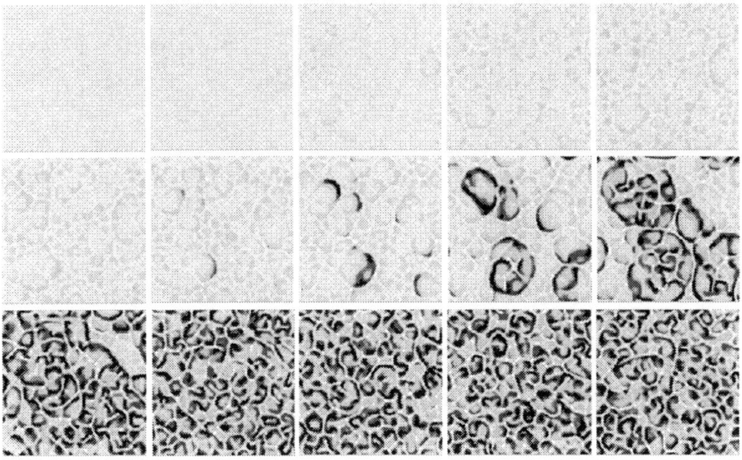

Fig. 4. Sequence of gray-scale snapshots of $|A|$ at times 60, 80, 100, ... 200, from a numerical simulation of 2d-CGL. The parameters are $\mu = 1$, $\alpha = 1$ and $\beta = -1.25$, boundary conditions are periodic, and the initial condition is given by Eq. 7. Lighter regions correspond to larger values of $|A|$. The phase instability leads to the formation first of a cellular pattern in which $|A|$ is modulated, and then to the creation of zeros of A (black spots).

converges toward a state with a few defects arranged in an irregular lattice. This coarse-graining phenomenon is illustrated in Figure 5, which shows a series of gray-scale images of $|A|$, 20 units of time apart. The real part of A corresponding to the last frame of Figure 5 is shown in Figure 6. One can clearly see that the zeros of A are indeed cores of spiral wave defects. It is likely that such an arrangement of defects is metastable. The interaction between defects is exponentially slow due to the screening effect of the shock lines where waves emanating from different spirals meet [67–69]. In the long run, defects should continue to annihilate in pairs, unless a network of spirals is stabilized by the presence of periodic boundary conditions.

Finally, let us note that numerical simulations indicate [52], but again do not prove, that defect-mediated turbulence should occur arbitrarily close to threshold in infinitely large, two-dimensional systems.

4. Conclusions

This chapter gives a brief overview of phase instabilities and weak turbulence in envelope equations associated with common pattern-forming systems. One of the striking features of the complex Ginzburg-Landau equation

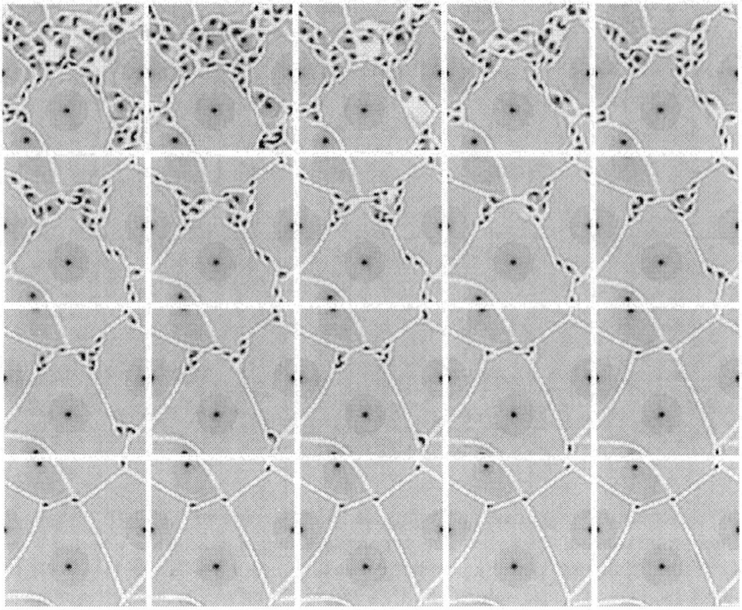

Fig. 5. Sequence of gray-scale snapshots of $|A|$ at times 20, 40, 60, ... 200, from a numerical simulation of 2d-CGL. The parameters are $\mu = 1$, $\alpha = 1$ and $\beta = -0.7$, boundary conditions are periodic, and the initial condition corresponds to a state of defect-mediated turbulence. Lighter regions correspond to larger values of $|A|$. When the value of $1 + \alpha\beta$ is suddenly changed from negative to positive, the system quickly evolves toward an irregular defect lattice. Then, the zeros of A (dark spots) annihilate in pairs, and a coarse-grained structure consisting of a few spiral defects starts to develop.

Fig. 6. Gray-scale picture of the real part of A for the last frame of Fig. 5, illustrating the presence of spiral defects.

is that such a simple partial differential equation can actually describe extremely rich dynamical behaviors. Phase or modulational instabilities lead to phase turbulence, in which the formation of non-zero phase gradients translates into the growth of amplitude modulations. This growth often does not saturate, and a dynamics involving recurring zeros of the amplitude of the complex order parameter A ensues. In one spatial dimension, zeros of A are not structurally stable and occur at points in time and space. In two dimensions, zeros of A are structurally stable and mark the cores of spiral defects.

Since hole- or defect-mediated turbulence results from the non-saturation of phase turbulence, the question of whether phase turbulence may be indefinitely sustained is an important one. It is clear that in a system of finite size, one can always find parameter regimes near the Benjamin-Feir transition for which only phase turbulence exists. As the size of the system - and consequently the number of unstable modes - is increased, the issue is less clear. The current understanding, mostly based on numerical simulations and on some theoretical investigations, is that the result depends on the dimension of the system. In one spatial dimension, there appears to be parameter regimes for which phase turbulence seems to persist indefinitely, even in infinitely large systems. The situation is different in two dimensions, for which it is believed that phase turbulence always leads to defect-mediated turbulence in systems of infinite size. It is however important to keep in mind that these statements are conjectural in nature. An interesting challenge to the mathematical community is now to prove or disprove such results.

References

[1] P. Manneville, *Dissipative Structure and Weak Turbulence* (Academic Press, 1990).

[2] A. C. Newell, T. Passot and J. Lega, *Ann. Rev. Fluid Mech.* **25**, 399-453 (1993).

[3] M.C. Cross, and P.C. Hohenberg, *Rev. Mod. Phys.* **65**, 851-1112 (1993).

[4] D. Walgraef, *Spatio-Temporal Pattern Formation: With Examples from Physics, Chemistry, and Materials Science (Partially Ordered Systems)* (Springer Verlag, 1996).

[5] P. Ball, *The Self-Made Taperstry: Pattern Formation in Nature* (Oxford, 1999).

[6] Y. Kuramoto, *Chemical oscillations, waves and turbulence*, Springer Series in Synergetics, vol. 19 (Springer-Verlag, Berlin, 1984).

[7] A.C. Newell and J.A. Whitehead, *J. Fluid Mech.* **38**, 279-303 (1969).

[8] L.A. Segel, *J. Fluid Mech.* **38**, 203-224 (1969).

[9] P. Collet and J.-P. Eckmann, *Comm. Math. Phys.* **132**, 139-153 (1990).
[10] P. Collet, *European J. Mech., B/Fluids* **10**, No. 2 - Suppl., 119-24 (1991).
[11] A. van Harten, *J. Nonlinear Science* **1**, 397-422 (1991).
[12] P. Collet, J.-P. Eckmann, and H. Epstein, *Helvetica Physica Acta* **65**, 56 (1992).
[13] G. Schneider, *Z. Angev. Math. Phys.* **45**, 433-457 (1994).
[14] G. Schneider, *Comm. Math. Phys.* **164**, 157-179 (1994).
[15] P. Bollerman, A. van Harten, and G. Schneider, in *Nonlinear Dynamics and Pattern Formation in the Natural Environment*, A. Doelman and A. van Harten, Editors, Pitman Research Notes in Mathematics, Vol. 335 (Longman, 1995), pp. 20-36, and references quoted therein.
[16] A. Mielke and G. Schneider, *Nonlinearity* **8**, 743-768 (1995).
[17] Y. Kuramoto and T. Tsuzuki, *Prog. Theoret. Phys.* **55**, 356-369 (1976).
[18] Y. Kuramoto, *Suppl. Progr. Theoretical Phys.* **64**, 346-368 (1978).
[19] Y. Pomeau and P. Manneville, *J. de Physique Lettres* **40**, L609-L612 (1979).
[20] W. Eckaus, *Studies in Nonlinear Stability* (Springer Verlag, Berlin, 1965).
[21] A.C. Newell, in *Lectures in Applied Mathematics* Vol. 15 (Am. Math. Society, Providence, 1974), pp. 157-163.
[22] J.T. Stuart and R.C. DiPrima, *Proc. R. Soc. London* **A 362**, 27-41 (1978).
[23] M.C. Cross and A.C. Newell, *Physica* **10 D**, 299-328 (1984).
[24] N. Ercolani, R. Indik, A.C. Newell, and T. Passot, *J. Nonlinear Sci.* **10**, 223-274 (2000).
[25] L. Kramer and W. Zimmerman, *Physica* **D 16**, 221-232 (1985).
[26] F.H. Busse, *Rep. Prog. Phys.* **41**, 1929-1967 (1978).
[27] T.B. Benjamin and J.E. Feir, *J. Fluid Mech.* **27**, 417-430 (1967).
[28] J. Bricmont and A. Kupiainen, *Comm. Math. Phys.* **150**, 193-208 (1992).
[29] T. Gallay and A. Mielke, *Comm. Math. Phys.* **199**, 71-97 (1998).
[30] I. Melbourne and G. Schneider, *Math. Nachr.* **263-264**, 171-180 (2004).
[31] I. Melbourne and G. Schneider, *J. Differential Equations* **199**, 22-46 (2004).
[32] A. Doelman, B. Sandstede, A. Scheel and G. Schneider, *The dynamics of modulated wave trains*, in preparation.
[33] G.I. Sivashinsky, *Acta Astronautica* **4**, 1177-1206 (1977).
[34] H. Sakaguchi, *Prog. Theor. Phys.* **84**, 792-800 (1990).
[35] S. Fauve, *Large Scale Instabilities of Cellular Flows*, in *Instabilities and Nonequilibrium Structures*, Ed. E. Tirapegui and D. Villarroel (D. Reidel Publishing Company, 1987), p. 63.
[36] B. Janiaud, A. Pumir, D. Bensimon, V. Croquette, H. Richter and L. Kramer, *Physica* **D 55**, 269-286 (1992).
[37] H. Chaté and P. Manneville, *Phase Turbulence*, in *Turbulence: A Tentative Dictionary*, P. Tabeling and O. Cardoso Eds. (Plenum Press, 1994), pp. 67-74.
[38] B. Nicolaenko, B. Scheurer and R. Temam, *Physica* **D 16**, 155-183 (1985).
[39] J.M. Hyman and B. Nicolaenko, *Physica* **D 18**, 113-126 (1986).
[40] Hyman, B. Nicolaenko and S. Zaleski, *Physica* **D 23**, 265-292 (1986).
[41] B. Nicolaenko, B. Scheurer and R. Temam, *Attractors for the Kuramoto-Sivashinsky equations*, Lect. in Appl. Math. Vol. 23 (Am. Math. Society,

Providence, 1986), pp. 149-170.
[42] D. Armbruster, J. Guckenheimer and P. Holmes, *SIAM J. Appl. Math.* **49**, 676-691 (1988).
[43] I.G. Kevrekidis, B. Nicolaenko and J.C. Scovel, *SIAM J. Appl. Math.* **50**, 760-790 (1990).
[44] P. Collet, J.P. Eckmann, H. Epstein and J. Stubbe, *Physica* **D 67**, 321-326 (1993).
[45] V. Yakhot, *Phys. Rev.* **A 24**, 642-644 (1981).
[46] S. Zaleski, *Physica* **D 34**, 427-438 (1989).
[47] C. Jayaprakash, F. Hayot and R. Pandit, *Phys. Rev. Lett.* **71**, 12-15 (1993).
[48] V. L'vov and I. Procaccia, *Phys. Rev. Lett.* **72**, 307 (1994).
[49] B.I. Shraiman, A. Pumir, W. van Saarloos, P.C. Hohenberg, H. Chaté and M. Holen, *Physica* **D 57**, 241-248 (1992).
[50] L. Gil, *Nonlinearity* **4**, 1213-1222 (1991).
[51] D.A. Egolf and H.S. Greenside, *Phys. Rev. Lett.* **74**, 1751-1754 (1995).
[52] P. Manneville and H. Chaté, *Physica* **D 96**, 30-46 (1996).
[53] L. Brusch, M.G. Zimmermann, M. van Hecke, M. Bär and A. Torcini, *Phys. Rev. Lett.* **85**, 86-89 (2000).
[54] L. Brusch, A. Torcini, M. van Hecke, M.G. Zimmermann and M. Bär, *Physica* **D 160**, 127-148 (2001).
[55] R. Montagne, E. Hernández-García, and M. San Miguel, *Phys. Rev. Lett.* **77**, 267-270 (1996).
[56] A. Torcini, *Phys. Rev. Lett.* **77**, 1047-1050 (1996).
[57] R. Montagne, E. Hernández-García, A. Amengual and M. San Miguel, *Phys. Rev.* **E 56**, 151-167 (1997).
[58] A. Torcini, H. Frauenkron and P. Grassberger, *Phys. Rev.* **E 55**, 5073-5081 (1997).
[59] H. Chaté, *Nonlinearity* **7**, 185-204 (1994).
[60] M. van Hecke and M. Howard, *Phys. Rev. Lett.* **86**, 2018-2021 (2001).
[61] N. Bekki and K. Nozaki, 133-135 (1985).
[62] J. Lega, S. Jucquois, B. Janiaud and V. Croquette, *Phys. Rev.* **A 45**, 5596-5604 (1992).
[63] J. Lega, *Physica* **D 152-153**, 269-287 (2001).
[64] M. van Hecke, *Phys. Rev. Lett.* **80**, 1896-1899 (1998).
[65] P. Coullet, L. Gil and J. Lega, *Phys. Rev. Lett.* **62**, 1619-1622 (1989).
[66] P. Coullet, L. Gil and J. Lega, *Physica* **37 D**, 91-103 (1989).
[67] I.S. Aranson, L. Kramer and A. Weber, *Physica* **D 53**, 376-384 (1991).
[68] L. Pismen and A.A. Nepomnyashchy, *Physica* **D 54**, 183-193 (1992).
[69] I.S. Aranson, L. Aranson, L. Kramer, and A. Weber, *Phys. Rev.* **A 46**, 2992 (1992).

CHAPTER 9

PATTERN FORMATION AND PARAMETRIC RESONANCE

Dieter Armbruster* and Tae-Chang Jo†

* *Department of Mathematics, Arizona State University, Tempe, AZ 85287-1804, USA*

† *Mathematics Department, New Mexico Institute of Mining and Technology, Socorro, NM 87801, USA*

The Mathieu partial differential equation is treated as a prototype equation for resonant parametric forcing in an extended system. Two of its pattern forming features are reviewed: i) The onset of pattern formation: Wavenumber selection due to detuning and attractor crowding as a result of high codimension saddle node bifurcations are found. ii) Localized patterns: After averaging and scaling the Mathieu PDE is shown to be equivalent to a perturbed nonlinear Schrödinger (NLS) equation. Localized oscillating patterns (oscillons) are identified as soliton solutions of the NLS. Open problems are identified throughout the paper.

1. Introduction

While the merger of bifurcation and singularity theory has led to a precise notion of a normal form for a bifurcation in a system of ordinary differential equations [10, 15] no equivalently complete theory exists for bifurcations in partial differential equations, which is the realm of pattern formation. The standard approach is to study the onset of bifurcation from a trivial solution through a steady state or Hopf bifurcation, and to use a perturbation analysis to derive a version of the (Complex) Ginzburg Landau equation (CGLE) describing the interaction between the dominant unstable mode and a small waveband of Fourier modes around it. This can be done rigorously in some cases [24, 27] and is done in an adhoc way in many other cases. In almost all cases the resulting CGLE is studied in its own right without any regard to its self-consistency as a perturbation equation. Nevertheless, together with symmetry consideration, this approach has been quite successful describing the evolution and selection of patterns beyond the immediate bifurcation range.

One pattern formation problem that has received comparably little attention is the issue of pattern formation created through a parametric resonant instability in a large number of coupled nonlinear oscillators, or equivalently, in a continuous oscillating medium. There are however many physical examples - first and foremost, all variations of the Faraday problem: [13] A fluid in a container is subjected to vertical oscillations. Surface waves are generated that exhibit all the patterns familiar from other pattern forming systems e.g. the Benard problem, spatio-temporal chaos, and localized oscillations and defects [22, 39]. A variation of the Faraday problem are the experiments by Umbanhowar et al. [37] on vertically shaken sandpiles. Resonantly excited spin waves are another example, the Frenkel-Kontorova [14] chain of coupled pendula is a discrete version of it.

This paper will review recent attempts to analyze a partial differential equation which can be considered the limit of infinitely many coupled Mathieu equations (the Mathieu PDE). Our hope is that this equation will become a "normal form" for parametrically driven pattern formation and that its features capture universal behavior associated with this instability. We will try to illuminate the many open problems that still need to be solved in order to make the Mathieu PDE as successful as the CGLE. Some of these problems are straightforward and may even have been solved, unbeknownst to the authors, while others pose major challenges.

2. The Mathieu Equation

The standard model for the instability of a harmonic oscillator due to parametric resonance is the Mathieu equation which can be written as

$$A_{tt} + A = \varepsilon f \cos(2t) A \tag{1}$$

where $f > 0$ parameterizes the forcing amplitude and the forcing frequency is exactly twice the natural frequency of the harmonic oscillator. It is easy to show [3] that the trivial solution $A = 0$ becomes unstable to an oscillation with frequency 2. With an added cubic nonlinearity, damping and detuning, the Mathieu has become the prototype nonlinear oscillator for the study of parametric resonance,

$$A_{tt} + A = \varepsilon(\delta A - \gamma A_t - \alpha A^3 + f \cos(2t) A). \tag{2}$$

Here, $\gamma > 0$ is the damping coefficient, and α and δ can have either sign parameterizing the nonlinearity and the detuning, respectively. No analytical solutions to Equation (2) exists. However the method of averaging allows to transform the system into an autonomous system which can then be studied using standard bifurcation techniques. Writing $z = A + iA_t$, using a van der Pol transformation $z = u(t)e^{-it}$ and averaging over one period of the forcing we get

$$u_t = \frac{\varepsilon}{2}\left[i(\delta u + \frac{f}{2}\bar{u} - \frac{3}{4}\alpha|u|^2 u) - \gamma u\right]. \tag{3}$$

Converting $u(t)$ into polar coordinates $u = re^{i\varphi}$ leads to

$$r' = \frac{\varepsilon}{2} r(\frac{f}{2}\sin 2\varphi - \gamma) \tag{4}$$

$$\varphi' = \frac{\varepsilon}{2}(\delta - \frac{3}{4}\alpha r^2 + \frac{1}{2} f \cos 2\varphi). \tag{5}$$

This system is most easily analyzed setting $\gamma = 0$ in which case it becomes Hamiltonian. Then we have, in addition to the fixed point at the origin, fixed points at $\varphi_{1,2} = 0, \pi$ and $\varphi_{3,4} = \frac{\pi}{2}, \frac{3\pi}{2}$. Fixed points 1 and 2 exist for $\alpha > 0$ ($\alpha < 0$) and $f > -2\delta$ ($f < -2\delta$), fixed points 3 and 4 exist for $\alpha > 0$ ($\alpha < 0$) and $f < 2\delta$ ($f > 2\delta$). As a result we get the two phase portraits in Figure 1 and 2.

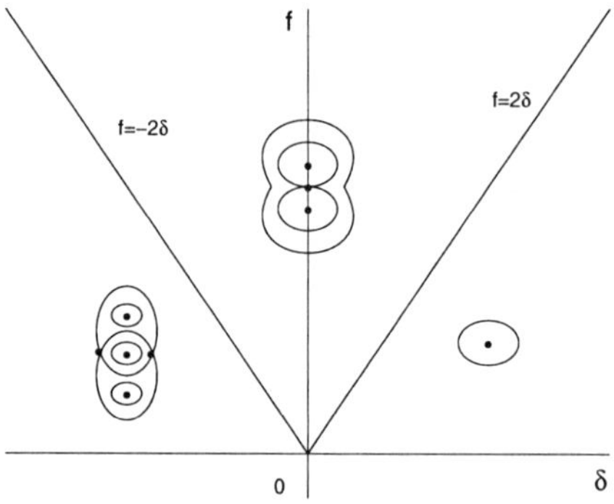

Fig. 1. Schematic phase portraits for the undamped nonlinear Mathieu equation depending on the forcing amplitude f and the detuning δ for $\alpha < 0$.

Notice that, for instance in the case of negative α, the zero solution bifurcates subcritically at the bifurcation parameter $f = 0$ when δ is negative, and bifurcates supercritically at $f = 2\delta$ when δ is positive. In the subcritical regime the system exhibits bistability. Adding damping back again all the closed orbits break and all centers become sinks. In addition the bifurcation diagrams in Figures 1 and 2 are shifted upward by $f = 2\gamma$. I.e., the trivial

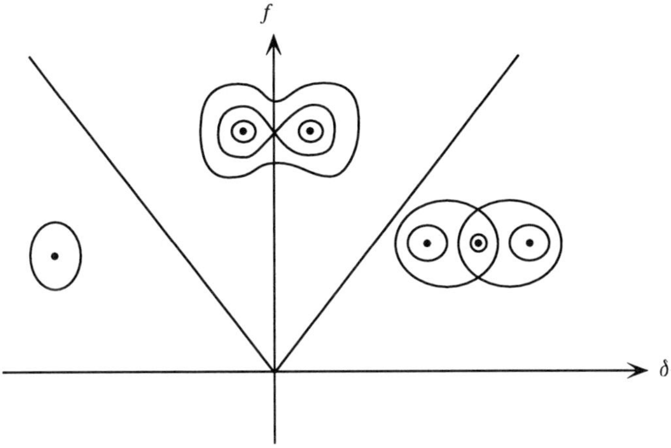

Fig. 2. Same as Figure 1 for $\alpha > 0$.

solution is stable for all detuning as long as $f < 2\gamma$ and we get a saddle node bifurcation at $f = 2\gamma$ for $\alpha\delta > 0$.

Two theorems [34] govern the exact relationship between the averaged equation (3) and the original Mathieu equation (2). By the averaging theorem, solutions to the averaged equation stay ε close to the solutions of the original equation *on a timescale of order* $O(\varepsilon^{-1})$. The timescale improves to *"for all time t"* for trajectories that approach a hyperbolic sink. This has important consequences in the case of the Mathieu equation: The global stability of the averaged system (3) does not depend on the sign of α - for all signs the system is Hamiltonian (if $\gamma = 0$) and globally stable. However, the original Mathieu equation has a restoring force only for $\alpha > 0$ and becomes globally unstable for $\alpha < 0$. Hence the phase portrait in the case $\alpha < 0$ only describes the behavior of the corresponding Mathieu equations for initial conditions close enough to the origin. For larger initial conditions, solutions will, after an time of order ε^{-1} in which they show quasiperiodic behavior, explode and go to infinity. This suggests that there must be an unstable invariant set around the phase portraits of Equation (2) that separates the locally stable from the globally unstable dynamics for the original Mathieu equation.

Open problem 1: Determine the exact relationship between the averaged and the original Mathieu equation for $\alpha < 0$. In particular, verify the last statement.

3. The Mathieu PDE

Coupling Mathieu oscillators via attractive nearest neighbor coupling we get, in the limit of a continuous chain of Mathieu equations, the Mathieu PDE:

$$A_{tt} + A = \varepsilon(\delta A - \gamma A_t - \alpha A^3 + f\cos(2t)A + dA_{xx}). \tag{6}$$

with the coupling coefficient $d > 0$. It is worth pointing out that the limiting behavior of Equation (6) as $\varepsilon \to 0$ is a harmonic oscillator. However, the oscillator is not necessarily spatially homogeneous: Only for initial conditions that are independent of the spatial variable will the oscillation be spatially homogeneous. In the case of many individual oscillators Equation (6) for $\varepsilon = 0$ amounts to a collection of uncoupled oscillators. For the moment, all perturbations are of the same order of magnitude. Other scaling limits will be discussed below. The Mathieu PDE is hyperbolic and shows, in the absence of damping, nonlinearity and forcing, a dispersion relation $\omega^2(k) = 1 - \varepsilon(\delta - dk^2)$. We can formally average Equation (6) to get

$$u_t = \frac{\varepsilon}{2}\left[i(\delta u + du_{xx} + \frac{f}{2}\bar{u} - \frac{3}{4}\alpha|u|^2 u) - \gamma u\right]. \tag{7}$$

Note that for the averaging theorem to hold the term du_{xx} must be $O(1)$. This implies that the following analysis holds for long wavelengths and will break down for short spatial scales. This is a self-consistency condition: As long as only long wavelengths are excited in Equation (7) their dynamics will describe an equivalent dynamics in the original Mathieu PDE. To continue we restrict ourselves to no-flux boundary conditions:

$$u_x(0) = u_x(\pi) = 0. \tag{8}$$

Making an Ansatz of the form $u(x,t) = q(t)\cos kx$ with integer k we get the onset of a steady state bifurcation at the neutral stability curve

$$f = 2\sqrt{\gamma^2 + (dk^2 - \delta)^2}. \tag{9}$$

Figure 3 shows graphs of Equation (9) in the (f, δ) plane for different k. Notice that for $\delta < 0$ the flat solution becomes unstable first, whereas for positive δ the detuning acts as the wavelength selector for the unstable pattern. There exist codimension two bifurcations for δ such that the $k = n$ and the $k = n + 1$ pattern become unstable at the same detuning and the same forcing amplitude. Meron [26] has studied this case and shown the existence of chaotic solutions in the overlap region. Since the nonlinearity in Equation (7) is cubic, only the pattern with the critical wavevector gets

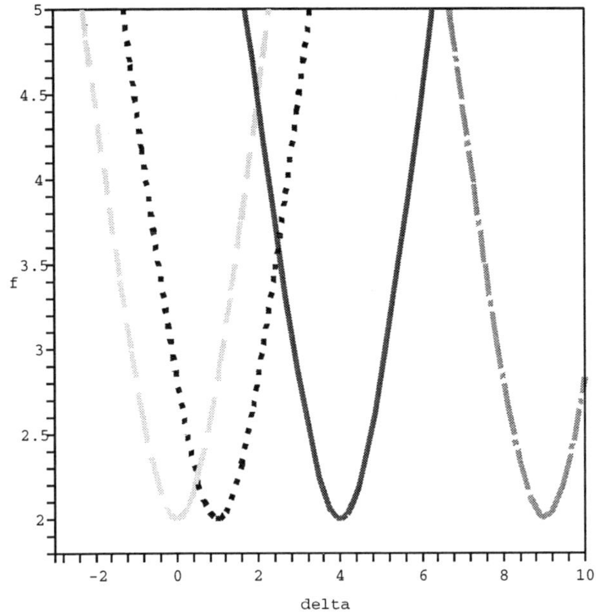

Fig. 3. Neutral stability curves for $k = 0, 1, 2, 3$ in the (f, δ)-plane

excited up to order 3. Hence the resulting center manifold has exactly the structure of the regular Mathieu equation with the detuning δ replaced by $\delta - dk_c^2$. Since there is a saddle-node bifurcation at $f = 2\gamma$ for

$$\delta - dk_c^2 > 0 \tag{10}$$

we have, for a given δ, simultaneous saddle node bifurcations for each mode k that satisfies the condition (10). Asymptotically, for $f \to 2\gamma$ from above, all of these saddle node bifurcations lead to a stable pattern, i.e. for large δ we may have the possibility of many simultaneously stable wave patterns (known as attractor crowding).

Open problem 2: Calculate the dynamics of the center manifold for the codimension n case for a δ that allows simultaneous saddle node bifurcations with wavenumbers $k = 0 \ldots n-1$. Determine a pattern for the secondary bifurcations.

Direct simulations as well as a two parameter path-following analysis for a seven mode Galerkin projection of equation (6) using AUTO lead to the bifurcation diagram shown in Figure 4 [2]. We see that, as f increases, the flat solution is followed by a bistable regime of pure mode solutions generated in a saddle node bifurcation. Further increase in f leads to stable mixed mode and time dependent solutions which terminate in a crises after which the flat solutions is again the only stable solution. A similar "flat desert" has been observed in the Nonlinear Schrödinger Equation [4] and in experiments on granular layers [5].

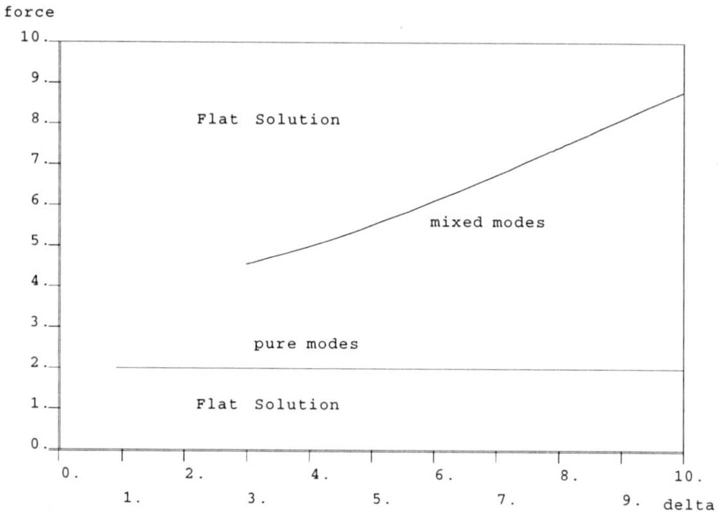

Fig. 4. Schematic bifurcation diagram of equation 6. Both bifurcation curves should intersect at $\delta = 0$ and $f = 2$ (convergence problems in AUTO). We chose $\gamma = 1, \alpha = 1$ and $d = 0.9$.

3.1. *Pattern Formation in 2-d*

Rand and Stubna [36] have extended the discussion of pattern formation in the Mathieu-PDE by replacing $\frac{d^2A}{dx^2}$ in Equation (6) with a 2-d Laplacian. They restrict their analysis to a 3 mode Galerkin projection and determine the resulting 2-d patterns. They found stationary, traveling and rotating patterns.

Open problem 3: Determine the relevant symmetry group for the 2-d Mathieu-PDE and do the bifurcation analysis. How does the symmetry of the bifurcation problem for the parametrically driven pattern formation differ from the symmetry of pattern formation generated by a Hopf bifurcation in an extended system?

4. The Nonlinear Schrödinger Equation

Of special interest in the Faraday type experiments for parametric resonance are spatially localized oscillations. In particular, standing solitary waves were observed in Faraday experiments with fluids in [39]. Similar small, circularly symmetric subharmonic oscillations have been observed in Faraday experiments of granular media [37].

Localized structures typically are explained through a homoclinic or heteroclinic orbit (solitons) in a dynamical system [25, 31], and have been discussed in various models: the nonlinear Schrödinger equation [20, 21], the Ginzburg Landau equation [6, 7, 11, 12, 16, 33], the Swift Hohenberg equation [8] and others.

The averaged Mathieu-PDE (Equation (7)) is closely related to the Nonlinear Schrödinger (NLS) Equation. The transformation

$$t' = \left(\frac{\varepsilon}{2}\right) t\,, \qquad x' = \left(\frac{1}{\sqrt{2d}}\right) x\,, \qquad u' = \left(\sqrt{-\frac{3\alpha}{4}}\right) u \tag{11}$$

leads to the scaled averaged Mathieu PDE,

$$iu_t + \frac{1}{2}u_{xx} + |u|^2 u = -\delta u - i\gamma u - \frac{f}{2}\bar{u}, \tag{12}$$

where we have dropped the prime after the transformation. Equation (12) for $\delta = \gamma = f = 0$ is the NLS Equation and the right hand side of the Equation adds damping, forcing and detuning to it. Note that Equation (12) is equivalent to

$$iu_t + \frac{1}{2}u_{xx} + |u|^2 u = -i\gamma u - \frac{f}{2}\bar{u}\exp(-2i\delta t) \tag{13}$$

through a transformation $u' = \exp(-i\delta t)u$.

The NLS Equation is one of the prototypes of integrable PDEs and as such is especially interesting for its soliton solutions. For the purpose of modelling in pattern formation we consider a soliton as a model for the experimentally found localized solutions.

Open problem 4: (Experimental) Conventional wisdom attributes localized solutions to homoclinic orbits and domain walls to heteroclinic orbits (bright and dark solitons in the NLS Equation). Soliton solutions are characterized by a very distinct exponential decay as $x \to \pm\infty$. To our knowledge there is no experimental study that established convincingly that the experimentally observed localized solutions and domain walls asymptotically decay like solitons.

By setting $u = Re^{i\varphi + i\omega t}$ and assuming a travelling wave solution with a constant speed of c in equation (13) with the perturbations set to zero, one can find a system of ordinary differential equations

$$R\frac{d^2\varphi}{dz^2} + \frac{dR}{dz}\left(2\frac{d\varphi}{dz} - c\right) = 0 \tag{14}$$

$$\frac{d^2R}{dz^2} + cR\frac{d\varphi}{dz} - R\left(\frac{d\varphi}{dz}\right)^2 - \omega R \pm R^3 = 0 \tag{15}$$

where $z = x - ct$. After eliminating φ, the system of equations (14)–(15) can be written as an integrable system of equations

$$\frac{dR}{dz} = Q \tag{16}$$

$$\frac{dQ}{dz} = \left(\omega - \frac{c^2}{4}\right)R + \frac{a}{R^3} \mp R^3 \tag{17}$$

for some positive constant a. Some phase portraits are shown in Figure 5 and corresponding solutions can be found by integrating

$$\left(\frac{dR}{dz}\right)^2 = \left(\omega - \frac{c^2}{4}\right)R^2 - \frac{a}{R^2} \mp \frac{1}{2}R^4 + b \tag{18}$$

for a constant b. There exist both homoclinic and heteroclinic orbits under certain conditions. For $\alpha < 0$ in the original Mathieu equation leading to Equation (13), a soliton solution, the so-called bright soliton, can be found from a homoclinic orbit. It has exponentially decaying zero boundary conditions at $z = \pm\infty$. For $\alpha > 0$, a soliton solution (a "dark soliton") can also be found from a homoclinic or heteroclinic orbit, but it has nonzero boundary conditions. The corresponding spatial structures of the nonlinear Schrödinger equation are shown in Figure 5.

In its most general form the solitons solutions of the unperturbed NLS Equation form a four parameter family of the form [17]

$$u(x,t) = 2\nu\,\mathrm{sech}[2\nu(x - 2\mu t - x_0)]\exp[i2\mu(x - 2\mu t - x_0) + i(2\mu^2 + 2\nu^2)t + i\sigma_0]\,, \tag{19}$$

for constants ν, μ, x_0, and σ_0. For the original Mathieu equation these solutions reflect strongly localized solutions that can travel and show doubly

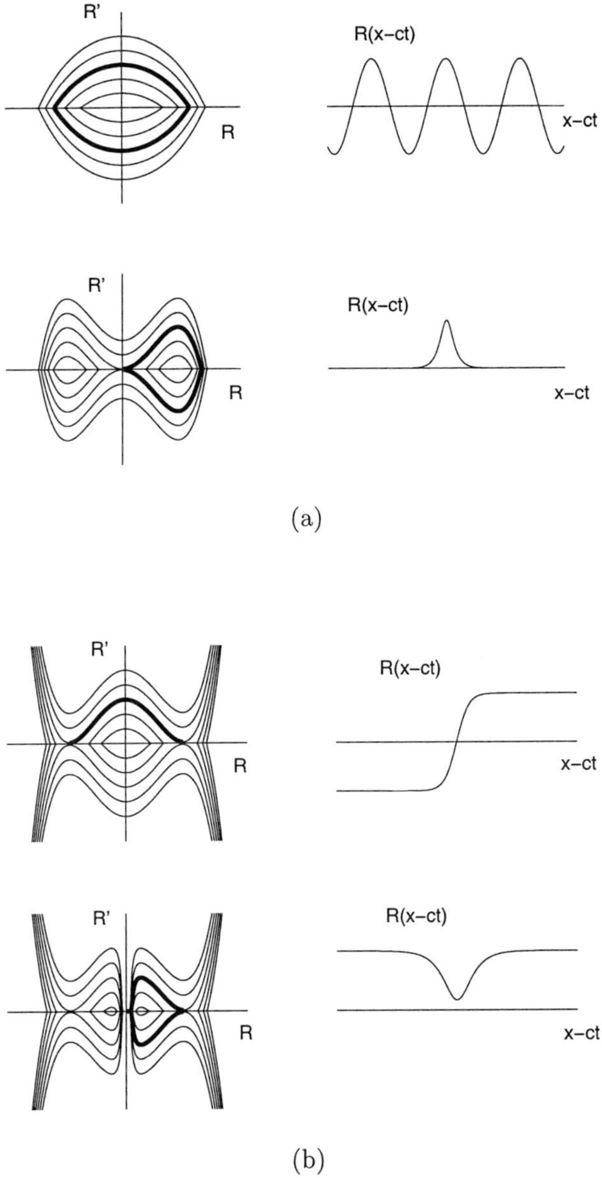

Fig. 5. Phase portraits for the envelopes of solutions to the cubic nonlinear Schrödinger equations and their corresponding spatial structures. The solutions $R(x - ct)$ represent the bold trajectories. (a) Focusing case ($\alpha < 0$) (b) defocusing case ($\alpha > 0$).

periodic oscillations. Equation (13) has been analyzed for $\delta = 1$ numerically by Barashenkov et al in [4] and asymptotically in [1]. Specifically they find i) two exact steady soliton solutions of Equation (13) for $2\gamma < f < 2\sqrt{\gamma^2+1}$, ii) Hopf bifurcations of the steady solitons and iii) period doubling and quasiperiodic transitions to chaos starting from the steady soliton solutions. Laedtke und Spatchek [23] determined, using the theory of periodic Schrödinger operators, that near $f = 2\gamma$ one of the steady solitons is stable while the other one is unstable.

We have recently [18] studied Equation (12) as a perturbation problem of the original NLS Equation using adiabatic perturbation theory [19]. Specifically, we assume that γ, f and δ are of order $O(\varepsilon)$. We look for a soliton pulse solution of Equation (12) of the form

$$u(x,t) = 2\nu \operatorname{sech}\kappa \exp(i\mu\kappa/\nu + i\omega)\ , \quad \kappa = 2\nu(x-\xi)\ . \tag{20}$$

where the parameters ν, μ, ξ, ω now evolve in time. To first order in the adiabatic perturbation theory they evolve according to

$$\frac{d\mu}{dt} = -\frac{f\pi\mu^2}{\nu}\operatorname{csch}(\frac{\pi\mu}{\nu})\sin(2\omega+2\delta t) \tag{21}$$

$$\frac{d\nu}{dt} = -2\gamma\nu + f\pi\mu\operatorname{csch}(\frac{\pi\mu}{\nu})\sin(2\omega+2\delta t) \tag{22}$$

$$\frac{d\xi}{dt} = 2\mu + \frac{f\pi}{4\nu}\operatorname{csch}(\frac{\pi\mu}{\nu})\left[\frac{\pi\mu}{\nu}\coth(\frac{\pi\mu}{\nu})-1\right]\cos(2\omega+2\delta t) \tag{23}$$

$$\frac{d\omega}{dt} = 2\mu^2+2\nu^2 + \frac{f\pi\mu}{2\nu}\operatorname{csch}(\frac{\pi\mu}{\nu})\left[\frac{2\pi\mu}{\nu}\coth(\frac{\pi\mu}{\nu})-1\right]\cos(2\omega+2\delta t) \tag{24}$$

With $\tilde{\omega} = \omega + \delta t$ as a new dependent variable the system becomes autonomous. Fixed points of equations 21 - 24 exist in the region of $f > 2\sqrt{\delta^2+\gamma^2}$ for positive δ and $f > 2\gamma$ for negative δ. There is only one fixed point for $f > 2\sqrt{\delta^2+\gamma^2}$, region A in Figure 6, regardless of the sign of δ. There are two fixed points for $2\gamma < f < 2\sqrt{\delta^2+\gamma^2}$ when $\delta < 0$, region B in Figure 6.

Open problem 5: The bifurcations of these fixed points at $f = 2\gamma$ and $f = 2\sqrt{\delta^2+\gamma^2}$ are saddle node bifurcations and pitchfork bifurcations, respectively. Given the fact that these fixed points represent approximations to soliton solutions of the original Mathieu PDE after averaging and scaling, it should be interesting to determine the exact bifurcation structure in the

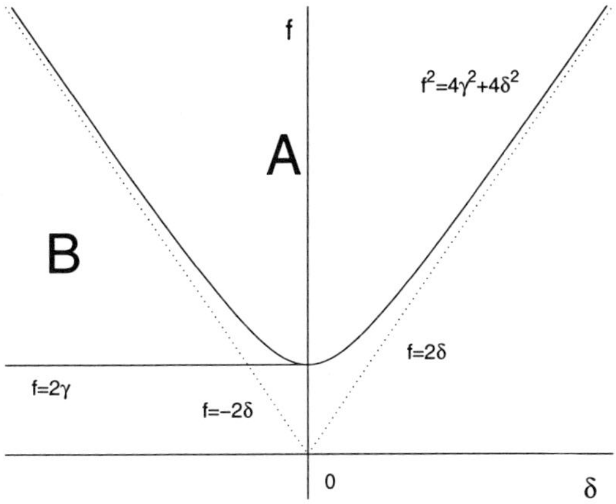

Fig. 6. Parameter regions for the existence of nontrivial fixed points of the ODEs (21)–(24). **A**: One stable nontrivial fixed point. **B**: Two nontrivial fixed points, one stable and one unstable.

original Mathieu PDE near the bifurcation curves in parameter space in dependence on the scaling parameter ε.

As it turns out, the fixed points are the steady, non-oscillating, non-drifting solitons of the NLS Equation determined earlier in [23]. They become an approximation of a localized oscillon solution to the original Mathieu equation after recovering the time scale and the length scale and after a inversion of the van der Pol transformation.

The adiabatic perturbation method is, in contrast to the analysis of [23] and [1], a *global* perturbation analysis as it allows us to study the time evolution of the solution manifold of the solitons of the NLS Equation under the perturbations of forcing, detuning and damping. In addition, by setting those three perturbations to the same order of magnitude we can study a three parameter unfolding of the NLS Equation, allowing us to determine for instance the influence of a variation in detuning from negative to positive values. The price we pay for the unfolding and a more global picture in phase space is the restriction of the validity of our analysis to small perturbations.

Linear stability analysis of the ODEs Eq (21 - 24) around the fixed points shows a single stable fixed point in region A and a stable and an unstable fixed point in region B. However, numerical simulations of the perturbed NLS Equation indicate the fixed point in region A becomes unstable to perturbations that are not modelled by the adiabatic perturbation theory.

Figure 7 presents simulations of the perturbed nonlinear Schrödinger equation for parameters $f = 0.05$, $\delta = 0.01$, and $\gamma = 0.0125$ (region A in Figure 6). We see that, given initial conditions near or a little away from the corresponding nonlinear fixed point, the real parts of the numerical solutions initially move to or stay near the adiabatic approximation of the nontrivial fixed point and then they lose their 'sech' profiles. Tip splitting of the peak of the soliton destroys the shape of a one-soliton solution. As time increases the numerical solution continues to split the tip of the central peak until it develops regularly spaced patterns. Figure 8 shows a simulation of the Mathieu PDE for the same parameters.

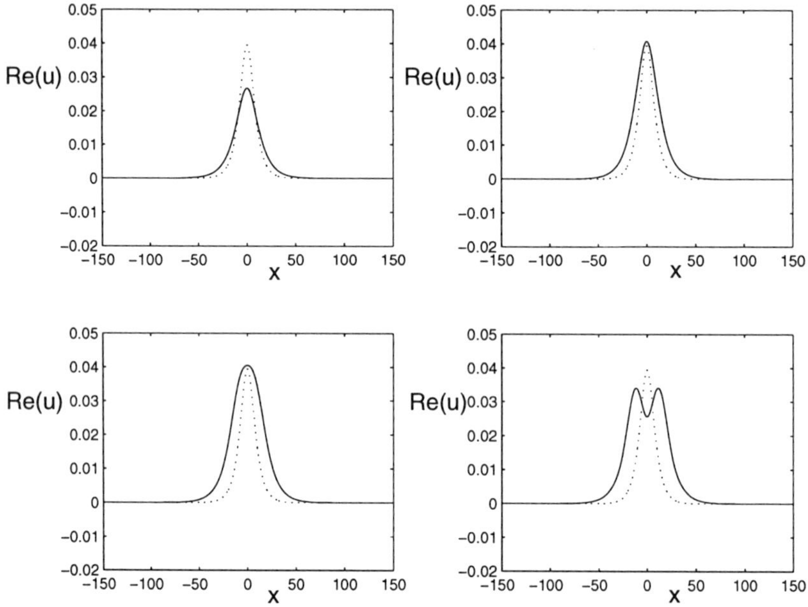

Fig. 7. Soliton solution (real part) corresponding to the nontrivial stable fixed point ($\cdots$) and numerical solution of the perturbed nonlinear Schrödinger equation (—). Shown are real parts of the soliton at $t = 0, 30, 80, 120$ with initial condition of $u(x, 0) = 0.1\,\mathrm{sech}(0.1x)\exp(1.3i)$.

In region B of Figure 6 (parameters $\delta = -0.7$, $f = 0.1$, $\gamma = 0.04$, $\varepsilon = 0.5$, $d = 0.5$, $\alpha = -4/3$), we find a oscillon-like solution that is stable for as long as we care to simulate. Figure 9 shows its simulation for the Mathieu PDE. Typically, amplitude and profile of the oscillon are determined by the system parameters, regardless of initial condition. However, the final

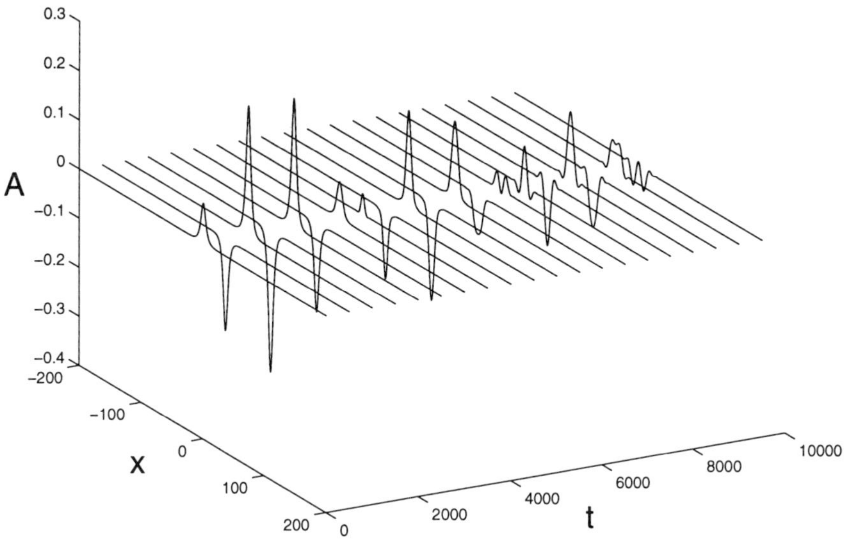

Fig. 8. Time evolution of the unstable oscillon like solution in the Mathieu PDE for an initial condition $A(x,0) = 0.3\ \mathrm{sech}(0.3x)\cos(1.3)$, $A_t(x,0) = 0.3\ \mathrm{sech}(0.3x)\sin(1.3)$.

position of the oscillon is determined by the initial condition which may lead to a transient drift.

We also observe that solutions die out for different initial conditions. This corresponds to the fact that region B is a bistable region for the Mathieu PDE.

Open problem 6: The soliton solutions exist stably for $f > 2\gamma$ in the parameter regime where the Mathieu equation is bistable. The discussion in Section 3 determined that at $f = 2\gamma$ there is a saddle node bifurcation of sinusoidal pattern, with attractor crowding for large δ. How do the stable soliton and the stable sinusoidal pattern interact as the forcing increases? Barashenko et al. [35] have studied a similar problem reducing the interaction to the flat solution.

Open problem 7: Develop a physical understanding for the existence of the "flat desert", i.e. the interval in the forcing amplitude for which the sinusoidal pattern as well as the soliton generated dynamics becomes unstable and the only stable attractor in the NLS Equation and presumably

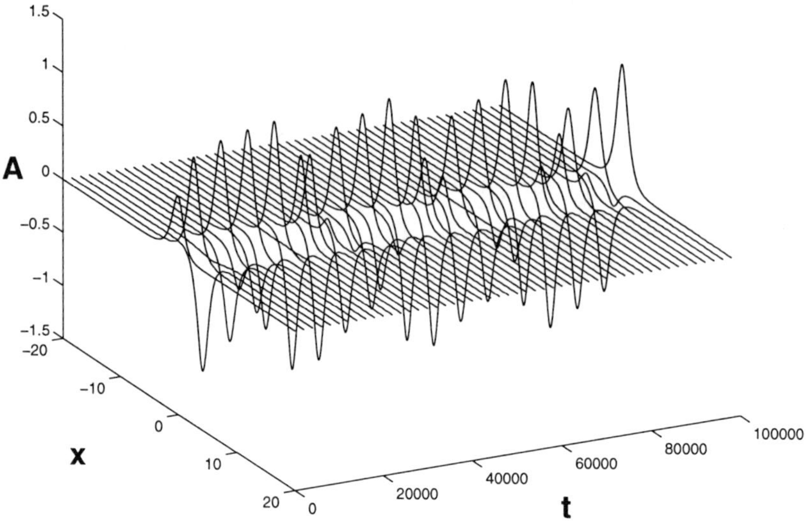

Fig. 9. Time evolution of a stable oscillon like solution in the Mathieu PDE. Initial condition $A(x, 0) = 1.2\ \mathrm{sech}(1.2x)\cos(0.1x + 1.1)$, $A_t(x, 0) = 1.2\ \mathrm{sech}(1.2x)\sin(0.1x + 1.1)$.

the Mathieu PDE becomes the flat solution. A mechanism based on the dynamics of the Mathieu PDE would go a long way to establish it as a prototype equation for pattern formation due to parametric resonance.

Acknowledgment: This work was supported in part through a grant from the National Science Foundation (DMS-0204543).

References

[1] N.V. Alexeeva, I.V. Barashenkov, D.E. Pelinovsky, *Nonlinearity* **12**, 103 (1999)
[2] D. Armbruster, M. George, and I. Oprea, *Chaos* **11**(1), 52-56 (2001)
[3] V.I. Arnold *Mathematical Methods of Classical Mechanics*, Springer Verlag 1989
[4] M. Bondila, I.V. Barashenkov, M.M. Bogdan, *Physica* **D87**, 314 - 320 (1995)
[5] J.R. de Bruyn, C. Bizon, M.D. Shattuck, D. Goldman, J.B. Swift, and H.L. Swinney, *Phys. Rev. Lett.* **81**, 1421, (1998)
[6] P. Coullet, Lega, B.Houchmanzadeh, J.Lajzerowicz, *Phys. Rev. Lett.* **65**, p. 1352 (1990).
[7] P. Coullet K. Emilsson, *Physica* **D 61**, p. 119 (2002).
[8] C.Crawford H.Riecke, *Physica* **D 129**, p. 83 (1999).

[9] J. Eggers, H. Riecke, *Phys. Rev.* **E. 59**(4), pp 4476 - 4483 (1999).
[10] C. Elphick, E. Tirapegui, M.E. Brachet, P. Coullet, G. Iooss, *Physica* **D 29** 95-127 (1987)
[11] C. Elphick A. Hagberg, B. A. Malomed, E. Meron, *Phys. Lett.* **A 230**, p.33 (1997).
[12] C. Elphick, A. Hagberg E. Meron, *Phys. Rev. Lett.* **80**, p.5007 (1998).
[13] M. Faraday, *Phil. Trans. Roy. Soc. London* **121** p. 319 (1831)
[14] A. Gavrielides, T. Kottos, V. Kovanis, and G.P. Tsironis, *Phys Rev* **E 58** 5529 (1998)
[15] M. Golubitsky, D.G. Schaeffer: *Singularities and groups in bifurcation theory*, Springer Verlag 1985
[16] V. Hakim, Y. Pomeau, *European Jr. of Mech.* **B**/*Fluids*, 10, 137 (1991).
[17] A. Hasegawa, Y.Kodama, *Solitons in Optical Communications* Oxford (1995).
[18] T.-C. Jo, D. Armbruster, *Phys. Rev.* **E. 68** 016213 (2003)
[19] V. Karpman E. Maslov, *Sov. Phys. JETP* **46**, p. 281 (1977).
[20] Y. Kivshar B. Malomed, *Review of Modern Physics* **61**, p. 793 (1989).
[21] Y. Kivshar, D. Pelinovsky, *Phys. Reports* **331**, p. 117 (2000).
[22] A. Kudrolli and J.P. Gollub, *Physica* **97D**, 133 (1996)
[23] E. Laedke, K. Spatschek, *J. Fluid Mech.* **223**, p.589 (1991).
[24] I. Melbourne, *J. Nonlin. Sci.* 8 (1998) 1-15
[25] E. Meron, *Phys. Reports* **218**, p. 1 (1992).
[26] E. Meron, *Phys. Rev.* **A 35** (11), 4892 (1987)
[27] A. Mielke, G. Schneider, in *Dynamical Systems and Probabilistic Methods in Partial Differential Equations*, Lectures in Applied Mathematics Vol. 31, Amer. Math. Soc., 1996. Pages 191-216.
[28] J.W. Miles, *J. Fluid Mech.,* **146**, p 285 (1984)
[29] A.H. Nayfeh and D.T. Mook, *Nonlinear Oscillations*, Wiley, 1979
[30] V. Petrov, Q. Ouyang, and H.L.Swinney, *Nature* **388**, 655-657 (1997)
[31] Y.Pomeau, *Physica* **23D**, p. 3 (1986).
[32] R.H. Rand, W.I. Newman, B.C. Denardo, and A.L. Newman, in Proc. for the 1995 Design Engineering Technical Conference, Vol3, Part A, *Vibration of nonlinear, random and time-varying systems*, Boston, 1995, A.S.M.E. **DE-84-1**, 57-68 (1995)
W.I.Newman, R.H. Rand, A.L. Newman, *Chaos* **9**(1), 242- 253, (1999).
[33] W. Saarloos, P.Hohenberg, *Physica* **D 56**, p. 303 (1992).
[34] J.A. Sanders and F. Verhulst: *Averaging Methods in Nonlinear Dynamical Systems*, Springer Verlag 1985
[35] V.S. Shchesnovich, I.V. Barashenkov, *Physica* **D 164** p. 83-109 (2002)
[36] M. Stubna, R.H. Rand, Proceed. of DETC2001,VIB-21401, 2001 ASME Design Engineering Technical Conferences, Pittsburgh, USA
[37] P. Umbanhowar, F. Melo, and H. L. Swinney, *Nature* **382**, 793 (1996)
[38] L.S. Tsimring, I.S. Aranson, *Phys. Rev. Lett.* **79**(2), pp 213-216 (1997)
[39] J. Wu, R.Keolian, and I.Rudnick, *Phys. Rev. Lett.* **52**, 1421 (1984).

CHAPTER 10

MEAN FLOW EFFECTS IN MODEL EQUATIONS FOR FARADAY WAVES

Sten Rüdiger* and José M. Vega†

* *Physikalisches Institut, Universität Bayreuth, 95440 Bayreuth, Germany*
† *E.T.S.I. Aeronáuticos, Universidad Politécnica de Madrid. Plaza Cardenal Cisneros, 3. 28040 Madrid, SPAIN*

We review model equations for parametric surface waves (Faraday waves) in the limit of small viscous dissipation. The equations account for two effects of viscosity, namely damping of the waves and slowly varying streaming and large scale flows (mean flow). Equations for the mean flow can be derived by a multiple scale analysis and are coupled to an order parameter equation describing the evolution of the surface waves. In addition, the equations incorporate a phenomenological damping term due to viscous dissipation. The nonlinear terms, which are undetermined by the derivation of the equation for the surface waves, are chosen so that the primary bifurcation is to a set of standing waves in the form of stripes. Results for the secondary instabilities of the primary waves are presented, including a weak amplification of both Eckhaus and Transverse Amplitude Modulation instabilities due to the mean flow, and a new longitudinal oscillatory instability which is absent without mean flow. Generation of mean flow due to dislocation defects in regular patterns is studied by numerical simulations.

1. Introduction

Parametrically driven surface waves, also known as Faraday waves, are created when a layer of an incompressible fluid is vibrated periodically in the direction normal to the free surface at rest [1]– [4]. Standing surface waves emerge above the primary instability of the planar surface forming a stationary pattern with a symmetry that depends on the fluid parameters and the frequency of the forcing [5]– [8]. While a description of the fluid with conservative equations suffices to explain the excitation of surface waves, consideration of viscosity is necessary to address further characteristics of the waves. A first question concerns the effect of damping on the saturation of amplitudes and on the selection of patterns. Milner [9] has shown how nonlinear terms from an expansion of the energy dissipation due to viscosity in the bulk of the layer mediate nonlinear wave interaction and lead to nonlinear terms in the amplitude equation and therefore to the saturation of amplitudes. Furthermore, the specific form of the coefficients of the nonlinear terms is dominated by triad resonant interactions which govern the selection of patterns [10]– [12].

A further important effect related to viscosity is the existence of slowly varying flows or mean flows. To date, most theories of parametric surface waves near onset have neglected such flows despite the observation [13] that their effect is of the same order as the standard cubic nonlinear and conservative terms which are usually retained. Thus weakly nonlinear corrections to surface waves and mean flows must be considered simultaneously [14]–[18]. A consistent derivation of mean flow equations for Faraday waves requires explicit consideration of special limits, in particular for the physical dimensions of the container. While the relevant coupled amplitude-mean flow equations have been already derived for horizontally one-dimensional waves [13], the extension to two-dimensional waves is not straightforward; a first step in this direction has been made by Vega et al. [19]. Two separate contributions to mean flow were found: An inviscid contribution originating from the slowly varying motion of the free surface, similar to the one appearing in classical Davey-Stewartson models [20], and a viscous one resulting from a slowly varying shear stress produced by time-averaged Reynolds stress in the the oscillatory boundary layer attached to the free surface or the lower plate. The latter describes diffusive or convective vorticity transport from the boundary layer into the bulk [21].

In this article we review the work on mean flow equations derived by a rigorous multiple scale analysis in the limit of weak viscous dissipation coupled to a phenomenological equation. We address the consequences of the mean flows in a laterally unbounded geometry, namely changes in the response of the base pattern of standing waves to perturbations (secondary instabilities) and the generation of mean flows by isolated defects. To simplify the analysis the cubic nonlinear terms of the phenomenological model were chosen such as to lead to a stripe pattern above onset instead of a square pattern which was experimentally observed in the limit of weak viscous dissipation. While it is easy to modify the functional form of the cubic term to produce square patterns, it is natural to first clarify the effect of mean flows on stripe patterns. It is found that the mean flow couples to perturbations of the stripe pattern in different strength depending on the type of perturbation. In general, the incorporation of mean flow is destabilizing with the strongest coupling occurring for longitudinal perturbations. Moreover, the mean flow generates a qualitatively new oscillatory longitudinal perturbation which for small nonlinear damping in the phenomenological equation renders all stripe patterns unstable resulting in time-dependent patterns at onset. By tuning the nonlinear damping to larger values one can generate a range of stable stripe patterns. In this latter situation we consider the effect of mean flows for irregularities in the pattern. Here the simplest case of a pair of dislocations in a pattern of perfect stripes is chosen. Using numerical simulations it is found that the flow generated by a dislocation is a large scale flow consisting of two vortices. In the parameter range considered the mean flow increases the velocity of the dislocation.

The article is organized as follows: In Section 2 we present the nondimensional equations. Section 3 justifies the phenomenological equation for a complex order parameter ψ that describes the evolution of the surface waves. The derivation of this equation is based on the assumption of an irrotational flow in the layer. In Section 4 we sketch the derivation of the mean flow equations and describe the coupling of mean flow and ψ. We then present results of the calculation of secondary instabilities in Section 5. Section 6 contains first results on the numerical simulation of the coupled equations to calculate mean flow effects on patterns with dislocations.

2. Basic Equations

Let a fluid layer of unperturbed depth d^* be supported by a horizontal plate that is vibrating vertically with an amplitude a^* and a frequency $2\omega^*$, where the superscript * denotes dimensional quantities. To obtain nondimensional equations we introduce the characteristic time $1/\omega^*$ and length $1/k^*$, where the wavenumber k^* is related to ω^* by the inviscid dispersion relation

$$\omega^{*2} = g^* k^* + \sigma^* k^{*3}/\rho^*, \tag{1}$$

in terms of the gravitational acceleration g^*, the surface tension σ^* and the density ρ^*, which are all assumed constant. Moreover, we assume that the wavelength k^{*-1} is small compared with the depth of the container. The resulting dimensionless incompressibility and Navier-Stokes equations in a reference frame attached to the vibrating container are

$$\nabla \cdot \boldsymbol{u} + \partial_z w = 0, \tag{2}$$

$$\partial_t \boldsymbol{u} - w(\nabla w - \partial_z \boldsymbol{u}) - \boldsymbol{u}^{\perp} \nabla \cdot \boldsymbol{u}^{\perp} = -\nabla p + \gamma(\nabla^2 \boldsymbol{u} + \partial_{zz}^2 \boldsymbol{u})/2, \tag{3}$$

$$\partial_t w + \boldsymbol{u} \cdot (\nabla w - \partial_z \boldsymbol{u}) = -\partial_z p + \gamma(\nabla^2 w + \partial_{zz}^2 w)/2, \tag{4}$$

in $-d < z < h(x,y,t)$, with the $z = 0$ plane at the unperturbed free surface. The boundary conditions result from no slip at the supporting plate,

$$\boldsymbol{u} = \boldsymbol{0}, \quad w = 0 \quad \text{at } z = -d, \tag{5}$$

and kinematic compatibility and equilibrium of tangential and normal stresses at the free surface,

$$\partial_t h + \boldsymbol{u} \cdot \boldsymbol{\nabla} h = w, \tag{6}$$

$$\partial_z \boldsymbol{u} + \boldsymbol{\nabla} w = (\boldsymbol{\nabla u} + \boldsymbol{\nabla u}^\top) \cdot \boldsymbol{\nabla} h - [2\partial_z w - (\partial_z \boldsymbol{u} + \boldsymbol{\nabla} w) \cdot \boldsymbol{\nabla} h]\boldsymbol{\nabla} h, \tag{7}$$

$$\begin{aligned} &p - (|\boldsymbol{u}|^2 + w^2)/2 - [4a \sin 2t + 1 - \Gamma]h + \Gamma \boldsymbol{\nabla} \cdot [\boldsymbol{\nabla} h/(1 + |\boldsymbol{\nabla} h|^2)^{1/2}] \\ &= \gamma[\partial_z w - (\partial_z \boldsymbol{u} + \boldsymbol{\nabla} w) \cdot \boldsymbol{\nabla} h + \big(\boldsymbol{\nabla} h \cdot (\boldsymbol{\nabla u} + \boldsymbol{\nabla u}^\top)/2\big) \cdot \boldsymbol{\nabla} h]/(1 + |\boldsymbol{\nabla} h|^2) \end{aligned} \tag{8}$$

at $z = h$. Here $\boldsymbol{u} = (u, v, 0)$ and w are the horizontal and vertical velocity components, $\boldsymbol{\nabla} = (\partial_x, \partial_y, 0)$ denotes the horizontal gradient, the superscript $\perp$ over a horizontal vector denotes the result of rotating the vector 90° counterclockwise, namely $\boldsymbol{u}^\perp = (-v, u, 0)$, and the superscript $\top$ over a tensor denotes its transpose; p (=pressure$+(|\boldsymbol{u}|^2+w^2)/2+[1-\Gamma+4a\sin(2t)]z$) is a conveniently modified pressure, and h is the (vertical) free surface deflection. For simplicity we do not consider lateral walls, but impose periodic boundary conditions in the two horizontal directions.

In the nondimensional system the following parameters remain: the dimensionless viscosity $\gamma = 2\nu^* k^{*2}/\omega^*$ (with ν^*= kinematic viscosity), the gravity-capillary contribution $\Gamma = \sigma^* k^{*3}/(\rho^* \omega^{*2})$, the forcing amplitude $a = a^* k^*$, the container depth $d = d^* k^*$ and the aspect ratios L_1 and L_2. According to Eq. (1), $0 \le \Gamma \le 1$ where the extreme cases $\Gamma = 0$ and 1 correspond to the purely gravitational and purely capillary limits respectively.

3. Derivation of Model Equations

Following similar considerations in other systems such as the Rayleigh-Bénard convection [22] a phenomenological model equation was derived that can efficiently be used in analytical and numerical work to describe the evolution of the surface waves. Here we will outline the derivation of the phenomenological model first introduced in Zhang and Viñals [23] (see Zakharov [24] and Crawford et al. [25] for the original use of the method). They assume an inviscid, incompressible, and irrotational fluid that is parametrically driven and add linear viscous damping in a phenomenological way. The governing equation are given by a Laplace equation for the velocity potential φ and boundary conditions on φ and the surface deflection

$h(\mathbf{x},t)$ at the free surface which correspond to the kinematic compatibility (Eq. (2)) and Bernoulli's equation at the free surface.

It is well known that this problem admits a Hamiltonian formulation with Hamiltonian H where the canonically conjugate variables are given by $h(\mathbf{x},t)$ and the velocity potential on the surface $\varphi^s(\mathbf{x},t) = \varphi(\mathbf{x}, z = h(\mathbf{x}))$. Phenomenological damping can be introduced by considering a dissipation function $Q(h(\mathbf{x},t), \varphi^s(\mathbf{x},t))$. The resulting canonical equations of motion are,

$$\partial_t h(\mathbf{x},t) = \frac{\delta H}{\delta \varphi^s(\mathbf{x},t)}, \tag{9}$$

$$\partial_t \varphi^s(\mathbf{x},t) = -\frac{\delta H}{\delta h(\mathbf{x},t)} + Q(h(\mathbf{x},t), \varphi^s(\mathbf{x},t)), \tag{10}$$

where the functional Q determines the rate of viscous dissipation in Eqs. (9)-(10) so that

$$\frac{dH}{dt} - \frac{\partial H}{\partial t} = \int d\mathbf{x}\, Q(h(\mathbf{x},t), \varphi^s(\mathbf{x},t))\partial_t h(\mathbf{x},t). \tag{11}$$

If the fluid is of low viscosity and depth is large compared to wavelength, one can assume that energy dissipation is dominated by the potential flow in the bulk [26]. The functional Q can then be determined by equating the rate of dissipation in Eq. (11) to the rate of energy dissipation due to potential flow,

$$\int d\mathbf{x}\, Q(h(\mathbf{x},t), \varphi^s(\mathbf{x},t))\partial_t h(\mathbf{x},t) = -\frac{\gamma}{2}\int d\mathbf{x} \int_{\infty}^{h(\mathbf{x},t)} dz \nabla^2 (\nabla\varphi)^2. \tag{12}$$

This equation has been used to determine Q order by order in an expansion in the surface wave steepness [9, 27]. To the order relevant here, one finds,

$$\hat{Q}(\mathbf{k},t) = -2\gamma k^2 \hat{\varphi}^s(\mathbf{k},t) + \text{nonlinear terms}, \tag{13}$$

where $\hat{Q}(\mathbf{k},t)$ is the Fourier transform of Q.

The next step in the derivation is the introduction of a complex order parameter field [24],

$$b(\mathbf{k},t) = \sqrt{\frac{\omega(k)}{2k}}\hat{h}(\mathbf{k},t) + i\sqrt{\frac{k}{2\omega(k)}}\hat{\varphi}^s(\mathbf{k},t), \tag{14}$$

where $\hat{h}(\mathbf{k},t)$ and $\hat{\varphi}^s(\mathbf{k},t)$ are the two dimensional Fourier transforms of $h(\mathbf{x},t)$ and $\varphi^s(\mathbf{x},t)$ respectively, and $\omega(k)$ is the inviscid dispersion relation in the dimensionless system. In terms of this new variable, the Hamiltonian system (9,10) can be written as,

$$\partial_t b(\mathbf{k},t) = -i\frac{\delta H}{\delta \bar{b}(-\mathbf{k},t)} + i\sqrt{\frac{k}{2\omega(k)}}\hat{Q}(\mathbf{k},t). \tag{15}$$

Equation (15) can then be expanded in a power series of b where one retains only terms linear in b, as nonlinear terms will be added phenomenologically.

Near onset only amplitudes with wavenumber close to the critical wavenumber are excited, with frequency close to the resonant frequency $\omega = 1$. This facilitates a conventional multiple scale expansion near onset, chosen such that the rotational invariance of the original governing equations is preserved. Expanding

$$b(\mathbf{k},t) = \varepsilon B(\mathbf{k},T_1,T_2)e^{-it} + \mathcal{O}(\varepsilon^2) \tag{16}$$

with $T_1 = \varepsilon t$ and $T_2 = \varepsilon^2 t$ slow time scales corresponding to the time scale of translation of a wave packet and of change in the modulation of the wave packet, respectively, one can use the solvability conditions at orders $\mathcal{O}(\varepsilon^2)$ and $\mathcal{O}(\varepsilon^3)$ to obtain an equation for the evolution of B on the slow time scales. One finally defines a complex order parameter field $\psi(\mathbf{x},t)$ as the inverse Fourier transform of $B(\mathbf{k})$, and finds,

$$\partial_t \psi = -\gamma\psi + \mathrm{i}f\bar{\psi} + 3\mathrm{i}\left(1+\nabla^2\right)\psi/4 + (\mathrm{i}-\gamma\alpha)|\psi|^2\psi, \tag{17}$$

where f is a parameter proportional to the amplitude a of the vibrations. Three simplifications are necessary to obtain this equation. First, the nonlinear functional does not have a closed form representation in real space. As has been done in other systems (cf. Rayleigh-Bénard convection [22]), one introduces phenomenological functional forms for this term thereby choosing the symmetry of the bifurcating pattern at onset artificially. In addition, in the case of Faraday waves the issue of the origin of nonlinear damping and saturation of the waves is sidestepped [12, 28]. In the simplest possible case, the coefficient of the nonlinear term is approximated by an imaginary constant which has already been rescaled to i in Eq. (17). Second, it is also known that linear damping is not sufficient to produce wave saturation in this system [9]. Therefore a phenomenological nonlinear damping coefficient $\alpha\gamma$ is used, where α is a constant assumed to be of order 1.

The positive sign of the imaginary part of the nonlinear coefficient $(i - \gamma\alpha)$ in Eq. (17) is chosen to represent capillary waves [29]. In the opposite limit of gravity waves, the imaginary part of this coefficient has to be negative. We finally note that as a third simplification a further linear term $i\left(1+\nabla^2\right)^2\psi$ has been eliminated in Eq. (17) as this term together with $i\left(1+\nabla^2\right)\psi$ leads to two different wavenumbers becoming critical at threshold, which is an unwanted feature.

4. Mean Flow Equations

We now want to discuss the mean flow equations that have been derived by Vega et al. [19] using a multi-scale analysis in both (horizontal) space and time. The approximation requires that (i) the aspect ratio of the container be large, (ii) the surface waves are weakly damped and (iii) exhibit a small wavelength compared to the container's depth and (iv) a small steepness, which in turn require that $L \gg 1, \quad d \gg 1, \quad \gamma \ll 1, \quad |\nabla h| \ll 1, \quad a \ll 1,$ where $L \leq \min\{L_1, L_2\}$ is a slow horizontal scale in which spatial modulations occur. Furthermore, it is assumed that d is logarithmically large compared to the remaining small parameters (namely γ, a and L^{-1}) and d is treated as a $O(1)$ parameter. The limit is therefore chosen for simplicity such that

$$\gamma^2 \ll \mathrm{e}^{-d} \ll L^{-1} \sim \gamma \sim a \sim \varepsilon^2, \tag{18}$$

where ε is a measure of the surface wave amplitude.

Here we want to sketch the important steps of the method used by Vega et al., and begin with a decomposition of the flow variables and the free surface deflection into oscillatory and time-averaged parts. They are associated with the surface waves and the mean flow (denoted hereinafter with the superscripts o and m), respectively, as

$$(\boldsymbol{u}, w, p, h) = \varepsilon(\boldsymbol{u}^o, w^o, p^o, h^o) + \varepsilon^2(\boldsymbol{u}^m, w^m, p^m, h^m), \tag{19}$$

where ε serves as the small parameter. The oscillatory flow variables associated with the surface waves are required to be such that

$$\langle \boldsymbol{u}^o \rangle^{ts} = \mathbf{0}, \quad \langle w^o \rangle^{ts} = \langle p^o \rangle^{ts} = \langle h^o \rangle^{ts} = 0, \tag{20}$$

with $\langle \cdot \rangle^{ts}$ standing for the time average in the basic oscillating period

$$\langle \psi \rangle^{ts} = (2\pi)^{-1} \int_t^{t+2\pi} \psi \, dt. \tag{21}$$

The variables associated with the mean flow are required to depend weakly on time.

An intermediate step in this analysis is a set of equations for the mean flow and the oscillatory part, with the latter, for the current purposes, ultimately replaced by the phenomenological model equation obtained above.

The intermediate equations for the surface waves are derived by a decomposition using ε, finding at first order an irrotational oscillatory flow in the bulk:

$$\begin{aligned} \boldsymbol{u}^o &= \boldsymbol{\nabla}\varphi + O(\varepsilon^2), \\ w^o &= \partial_z\varphi + O(\varepsilon^2), \quad p = -\partial_t\varphi. \end{aligned} \tag{22}$$

As already mentioned in the introduction the mean flow will be forced in two basic ways. First we must take into account the oscillatory boundary layer attached to the free surface, which provides a slowly varying shear stress at the edge of this layer, that must be imposed as a boundary condition for the mean flow in the bulk. This forcing mechanism was first uncovered by Longuet-Higgins [21], who obtained an explicit expression for the forcing shear stress produced by general boundary layers in 2-D. The counterpart of this expression in 3-D has been only obtained recently by Nicolás and Vega [30]. The general formulae for the shear stress yields

$$\partial_z \boldsymbol{u}^m + \boldsymbol{\nabla} w^m = 2\langle \boldsymbol{\nabla}\left(\boldsymbol{\nabla}\cdot(h^o\boldsymbol{\nabla}\varphi)\right) \tag{23}$$

$$+ (\boldsymbol{\nabla} h^o \cdot \boldsymbol{\nabla})\boldsymbol{\nabla}\varphi + (\boldsymbol{\nabla}^2\varphi)\boldsymbol{\nabla} h^o\rangle^{ts} \quad \text{at } z=0, \tag{24}$$

where only the leading order contribution as $\gamma \to 0$ and $\varepsilon \to 0$ is retained. The boundary layer attached to the free surface has no effect on the other two boundary conditions at the unperturbed free surface, which are obtained from Eqs. (8) and (2,6,22) to be

$$p^m - (1-\Gamma)h^m + \Gamma\boldsymbol{\nabla}^2 h^m = \langle h^o\partial^2_{tz}\varphi + (|\boldsymbol{\nabla}\varphi|^2 + |\partial_z\varphi|^2)/2\rangle^{ts} \tag{25}$$

$$\text{and} \quad \partial_t h^m + \boldsymbol{\nabla}\cdot\left(\int_{-d}^{0} \boldsymbol{u}^m\, dz\right) = -\boldsymbol{\nabla}\cdot\left(\langle h^o\boldsymbol{\nabla}\varphi\rangle^{ts}\right) \quad \text{at } z=0, \tag{26}$$

where we are only taking into account the leading order terms. We are neglecting the effect of a second oscillatory boundary layer attached to the lower plate because its effect is quite small (the oscillatory flow is weak near the lower plate as e^{-d} is small, Eq. (18)). These boundary conditions show that mean flow is forced by surface waves in two ways. The terms on the right hand sides of Eqs. (25) and (26) provide an inviscid forcing mechanism that drives an *inviscid mean flow*, like that appearing in the Davey-Stewartson model [20]. The right hand side of Eq. (24) instead produces a forcing shear stress that generates a *viscous mean flow*, which is absent in the usual inviscid and nearly inviscid theories of Faraday waves. Note that this forcing stress is generically non zero and independent of viscosity at leading order, a fact that is well known but somewhat surprising because this effect is due to the oscillatory boundary layer, and is absent in the strictly inviscid case [16].

The mean flow can be decomposed into its inviscid and viscous parts, as has been done in a number of studies [13, 17, 18]. Alternatively, a decomposition of the mean flow variables into a *short wave* component (oscillatory in the horizontal direction) and a *long wave* component (slowly varying in the horizontal direction) is a convenient choice. Their evolution has been derived by Vega et al. and is given by[a]

$$0 = \nabla \cdot \boldsymbol{U}^{mo}, \tag{27}$$

$$\partial_t \boldsymbol{U}^{mo} = -\varepsilon^2 \nabla Q^{mo} + \gamma(\nabla^2 \boldsymbol{U}^{mo} - \frac{\beta_1^4 \pi^2}{16} \boldsymbol{U}^{mo})/2 + \beta_1 \gamma \langle \mathbf{N}^{vis} \rangle^{ho}, \tag{28}$$

$$\partial_t \boldsymbol{U}^{ms} = -\nabla Q^{ms} - \gamma \frac{\beta_1^4 \pi^2}{32} \boldsymbol{U}^{ms} + \beta_1 \gamma \langle \mathbf{N}^{vis} \rangle^{hs}, \tag{29}$$

$$4Q^{ms} = \beta_1 \pi (1 - \Gamma) h^{ms}, \tag{30}$$

$$\partial_t h^{ms} = -\frac{4}{\beta_1 \pi} \nabla \cdot \boldsymbol{U}^{ms} - \nabla \cdot (\mathrm{i} \langle \psi \nabla \bar{\psi} \rangle^{hs} + c.c.), \tag{31}$$

where $\mathbf{N}^{vis} = \mathrm{i}(\nabla \psi \cdot \nabla) \nabla \bar{\psi} + \mathrm{i}(\nabla^2 \bar{\psi}) \nabla \psi + \text{c.c.}$ denotes the viscous forcing terms and $\langle \cdot \rangle^{hs}$ and $\langle \cdot \rangle^{ho}$ project it onto its horizontal average in the short spatial scales and its short wave component. For a rigorous definition of these projections confer [19], where further details of the derivation can be found. Here we want to mention the following simplifications that are crucial to the method:

- Convective terms in the mean flow equations have been neglected thereby linearizing the mean flow equations. Since for standing waves the associated mean flow is unforced it identically vanishes at large times. Thus this approximation is exact for the linear stability analysis described below. Furthermore, one expects that the neglected convective terms do not play a significant qualitative role in subsequent bifurcated branches, at least near threshold.
- A single mode approximation for the z dependence of the mean flow variables as $g(z)$ is introduced. This function is arbitrary and can be selected to yield the best approximation to the vertical velocity profiles. A reasonable choice is

$$g(z) = \sqrt{2/d} \sin[\pi(z + d)/(2d)]. \tag{32}$$

[a]The coefficient ε^2 in (28) disappears when time is scaled according to the slow time of the mean flow. This has not been done here for convenience.

This function also determines the parameter β_1 in Eqs. (28-31) in terms of the depth d of the layer:

$$\beta_1 = g(0) = \sqrt{2/d}. \tag{33}$$

Finally we replace the complicated equations for the evolution of the oscillatory flow with the model equation (17) given above. To account for the effects of the mean flow on the surface waves we add a convective term $-\boldsymbol{u} \cdot \boldsymbol{\nabla}\psi$ to the right hand side, where $\boldsymbol{u}$ is given by the two parts of the mean flow:

$$\partial_t \psi = -\gamma\psi + \mathrm{i} f \bar{\psi} + 3\mathrm{i}\left(1 + \nabla^2\right)\psi/4 + (\mathrm{i} - \gamma\alpha)|\psi|^2\psi - \beta_1(\boldsymbol{U}^{mo} + \boldsymbol{U}^{ms}) \cdot \boldsymbol{\nabla}\psi. \tag{34}$$

The coefficient β_1 originates from a similar coefficient in the corresponding term in the equation for the surface waves as derived in Vega et al.. Note that there is no dependence on h^{mo} and h^{ms} included, because this is beyond the scope of this phenomenological model.

5. Bifurcations of Periodic Solutions

We review in this section results on the instabilities of regular solutions of the model equations defined by the coupled Eqs. (27-31) and (34). The primary stability from the plane surface to standing waves is a linear stability problem in terms of perturbations in the form of spatially periodic patterns. As can be demonstrated from the form of the forcing terms in Eqs. (28-31) regular stripe patterns do not result in a mean flow, so that the primary instability is completely determined by Eq. (17). As has been described by Zhang and Viñals [23] the trivial solution $\psi = 0$ becomes linearly unstable against a spatially periodic perturbation of ψ of wave number q for $\mu > \mu_c(q) = \sqrt{1 + [3(1 - q^2)/4\gamma]^2} - 1$, the neutral stability curve; μ is the control parameter defined as $\mu = (f - \gamma)/\gamma$. The critical mode with $q = 1$ becomes therefore unstable at $\mu = 0$.

For small $\mu > 0$ stationary and spatially periodic solutions exist that can be approximated by a single Fourier mode $\psi_q(x) = \alpha_q \cos(qx) \exp(\mathrm{i}\Theta_q)$ with

$$\alpha_q^2 = \frac{q^2 - 1 - \frac{4}{3}\alpha\gamma^2 \pm \frac{1}{3}\sqrt{16f^2(1 + \alpha^2\gamma^2) - (4\gamma + 3\alpha\gamma(q^2 - 1))^2}}{1 + (\alpha\gamma)^2}, \tag{35}$$

where the $\pm$ sign stands for $\mathrm{sign}(1 - q^2 + 4\alpha\gamma^2/3)$, and Θ_q satisfies $\sin 2\Theta_q = (1 + 3\alpha\alpha_q^2/4)\gamma/f$, $\cos 2\Theta_q = \frac{3}{4}(q^2 - 1 - \alpha_q^2)/f$. Note that the bifurcation at threshold is subcritical if $q^2 > 1 + 3\alpha\gamma^2/4 \simeq 1$ (recall that γ is small) and supercritical otherwise (a subcritical bifurcation for $q > 1$ as $\gamma \to 0$ has also been found in a direct numerical simulation by Chen and Wu [31].) As mentioned above, this solution for the order parameter leads to vanishing driving terms in the mean flow Eqs. (28)-(31); hence all mean flow variables remain zero for the basic, periodic solution.

The criterion for linear instability yields a range of non-trivial solutions for every $\mu > 0$. These solutions are possibly unstable and their linear stability (secondary stability) can be calculated by introducing small perturbations. Vega et al. [19] have studied the secondary stability using perturbations of general form. In order to address the stability of the primary stationary solution (35) against general longitudinal perturbations, they introduced

$$\psi = A_0[\exp(\mathrm{i}qx) + \exp(-\mathrm{i}qx) + a^{++}\exp(\mathrm{i}(q+k)x) + a^{+-}\exp(\mathrm{i}(q-k)x) \\ + a^{-+}\exp(\mathrm{i}(k-q)x) + a^{--}\exp(-\mathrm{i}(q+k)x)] \tag{36}$$

(where $A_0 = \frac{1}{2}\alpha_q \exp(\mathrm{i}\Theta_q)$), together with the corresponding perturbations of the mean flow variables

$$U_x^{ms} = u^+ \exp(\mathrm{i}kx) + c.c., \quad h^{ms} = c^+ \exp(\mathrm{i}kx) + c.c., \tag{37}$$

and

$$Q^{ms} = d^+ \exp(\mathrm{i}kx) + c.c. \tag{38}$$

where $\boldsymbol{U}^{mo} = 0$ as seen from the incompressibility condition (27) which requires that $U_x^{mo} = 0$. A system of six first order differential equations for the perturbation amplitudes $a^{\pm\pm}$, u^+, and c^+ is derived by inserting the ansatz into Eqs. (29-31,34). The linearization around the basic periodic solution is therefore given by a matrix $A(q, k, \varepsilon, \dots)$ and the basic solution becomes unstable if the real part of an eigenvalue of A becomes positive. Two relevant branches of eigenvalues appear in the vicinity of $k = 0$. One is associated with the broken translational symmetry of the basic state $\psi_q(x)$, generating a mode which becomes marginal at $k = 0$. The other branch is of hydrodynamic nature and corresponds to weakly damped uniform flows away from the quiescent state, with a damping rate of $\gamma\beta_1^4\pi^2/32$ (Eq. (29)). The coupling of the two modes generates a steady long wavelength instability (modified Eckhaus) and a finite wavenumber oscillatory instability described below.

The stability of periodic solutions against transverse amplitude and phase perturbations can be studied similarly. We first note that given that $\partial_y\psi_q = 0$ in the basic state with zero mean flow, terms involving the y

components of the mean flow will be of second order in the amplitudes of the perturbation and hence only the components U_x^{mo} and U_x^{ms} need to be perturbed. Here, in contrast to the case of a longitudinal perturbation, both short and long wavelength components of the mean flow need to be included. However, the equations for U_x^{mo} and U_x^{ms} decouple at linear order and can be analyzed separately.

Considering the short wavelength component of the mean flow velocity U_x^{mo}, and introducing the following perturbation for the order parameter,

$$\begin{aligned}\psi = A_0 \big[&\exp(\mathrm{i}qx) + \exp(-\mathrm{i}qx) + a^{++}\exp(\mathrm{i}(qx+ky)) \\ &+a^{+-}\exp(\mathrm{i}(qx-ky)) + a^{-+}\exp(\mathrm{i}(-qx+ky)) \qquad (39)\\ &+a^{--}\exp(-\mathrm{i}(qx+ky))\big], \qquad (40)\end{aligned}$$

$$U_x^{mo} = v^{++}\exp(\mathrm{i}(2qx+ky)) + v^{+-}\exp(\mathrm{i}(2qx-ky)) + \text{c.c.}, \qquad (41)$$

and

$$Q^{mo} = p^{++}\exp(\mathrm{i}(2qx+ky)) + p^{+-}\exp(\mathrm{i}(2qx-ky)) + \text{c.c.}, \qquad (42)$$

one can derive a linear system of equations for the perturbation amplitudes $a^{+\pm}$, $a^{-\pm}$, and $v^{+\pm}$. Suitable combinations of these modulations lead to the generalizations of the well-known transverse perturbations for Faraday patterns. A transverse amplitude modulation (TAM) is defined by the linear combinations $b_1 = a^{++}+a^{+-}+a^{-+}+a^{--}$ and $v_1 = \mathrm{Im}(v^{++}+v^{+-})$, whereas a transverse phase modulation (zig-zag) is given by $b_2 = a^{++}-a^{+-}-a^{-+}+a^{--}$ and $v_2 = \mathrm{Im}(v^{++}-v^{+-})$.

A similar ansatz has been used for the long wavelength component of the mean flow with perturbations of the form $U_x^{ms} = u^+\exp(\mathrm{i}ky) + \text{c.c.}$, $h^{ms} = c^+\exp(\mathrm{i}ky) + \text{c.c.}$, and $Q^{ms} = d^+\exp(\mathrm{i}ky) + \text{c.c.}$. The order parameter is again given by Eq. (40).

From the eigenvalue equation of each of the resulting systems Vega et al. have demonstrated that modifications to the eigenvalues due to the mean flow are small compared to the case of longitudinal perturbations. Only through the coupling of TAM perturbations with the short wavelength component of the mean flow, the mean flow has a significant modification of the eigenvalues been found. However, its effect on the location of stability curves has been found to be small compared to the longitudinal case.

The numerically obtained results can be summarized in a q-μ diagram containing the different stability curves that together define the region of existence of stable periodic solutions. Figure 1 shows the various stability boundaries for the special cases of $\beta_1 = 0$ (no mean flow) and $\beta_1 = 0.5$.

The other parameter values are $\gamma = 0.1$, $\alpha = 0.5$, and $\Gamma = 0.8$. Except for α, these values correspond approximately to those for the low-viscosity experiments described by Kudrolli and Gollub [5]. For instance, typical experimental values of $\beta_1 = \sqrt{\pi/d}$ (where d is the dimensionless height of the layer) are between 0.5 and 1.2. Figure 1 includes the neutral stability curve of the basic periodic solution and, since the primary bifurcation is subcritical for $q > 1$, we have also included the saddle node curve where the periodic solution bifurcates. The case $\beta_1 = 0$ is shown as a reference, and it agrees with the results of Zhang and Viñals [23].

Basic solutions that are stable against all steady perturbations considered here (Eckhaus, TAM, and zig-zag) exist in a small region close to threshold at $\mu = 0$ between the TAM and zig-zag lines. Periodic solutions are stable against transverse perturbations below the dashed-dotted line in the figure (zig-zag, denoted "Z"), and above the dashed line (TAM). Eckhaus perturbations have negative growth rate below the dotted line. We observe that with increasing β_1 both Eckhaus and TAM curves are shifted so that larger regions in the (μ, q) space become destabilized with respect to Eckhaus or TAM perturbations. As discussed above, the zig-zag line is not affected by the mean flow.

The picture just described changes dramatically if one allows for oscillatory longitudinal perturbations. The oscillatory Eckhaus perturbation leads to an instability provided that $\beta_1 \neq 0$. This finding is consistent with the appearance of an oscillatory instability in the horizontally 1-D, large aspect ratio Faraday system [14], which is also absent when the effect of the mean flow is ignored. The oscillatory nature of the instability is apparent from the non-zero imaginary part of the eigenvalue σ at the critical point in which $\mathrm{Re}(\sigma) = 0$. Furthermore, the instability occurs at small but finite wavenumber k, a fact that has been confirmed by calculating the wave number with largest growth rate both slightly above and below the instability threshold. As mentioned above, the origin of the oscillatory instability can be traced back to the two eigenvalue branches that have small real parts for $k \to 0$. As k increases, the two real eigenvalue branches merge leading to a complex conjugated pair and therefore to the oscillatory nature of the instability at finite k.

Results concerning the location of the oscillatory instability boundary as a function of q for two values of β_1 are shown in Figure 2. Stable regions are located to the right of the plotted curves. As expected, the stability boundary moves toward the steady Eckhaus line with decreasing β_1, presumably merging with it for $\beta_1 \to 0$. Note that for $\beta_1 = 0.5$ the unstable

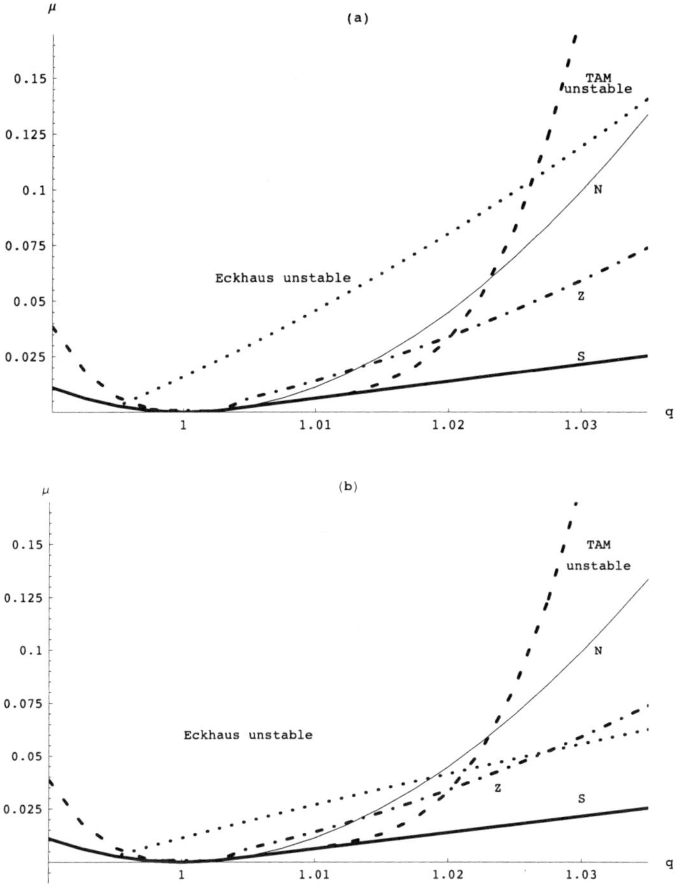

Fig. 1. Bifurcation diagram for (a) the order parameter model without mean flow, and (b) with $\beta_1 = 0.5$. Other values of the parameters are $\gamma = 0.1, \alpha = 0.5$, and $\Gamma = 0.8$. The following stability lines are plotted: Eckhaus (dotted), TAM (dashed), zig-zag (dash-dotted), neutral stability curve of the primary instability (solid line denoted by N), and a saddle node bifurcation (thick solid line marked S). Only the left branch of the Eckhaus line is shown emanating from $(q = 1, \mu = 0)$. The region of stability of the basic solution against an Eckhaus instability is the region below the dotted line. Comparison of (a) and (b) shows that the mean flow decreases the regions of stability against both Eckhaus instability and transverse amplitude modulation.

region covers most of the region of existence of the basic states except for a narrow stripe close to the saddle node bifurcation.

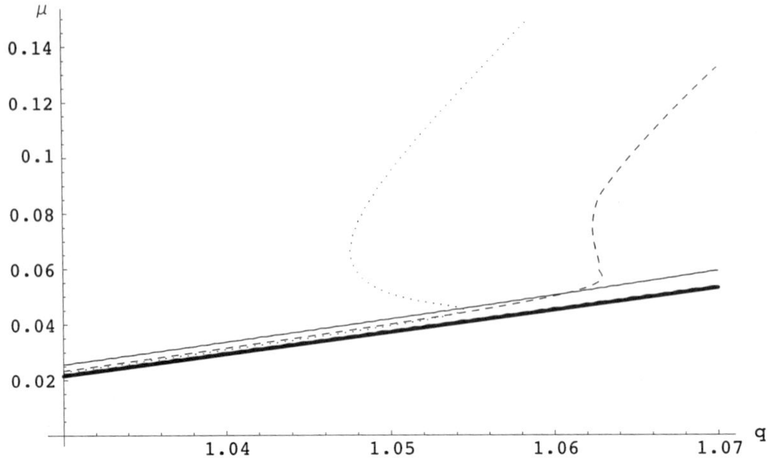

Fig. 2. Stability boundaries of the oscillatory instability for three values of β_1: 0.05 (dotted line), 0.2 (dashed line), and 0.5 (solid line). The thick solid line indicates the saddle node. Periodic solutions are unstable to the left of the curves.

6. Mean Flow Generated by Defects

In the following we will describe numerical simulations that have been performed to estimate the influence of the mean flow on the dynamical evolution of real patterns. Here we will consider the most simple configuration, a sample that contains two isolated dislocations. In general, the presence of a dislocation splits the sample into two regions which possess different wave numbers. If these two wave numbers are taken to be stable against secondary instabilities (according to the stability analysis described above) we can exclude that the defects are dynamically driven by a global instability. However, typically defects are found in motion which consists of two components: a climbing motion along the stripes and a gliding motion parallel to the wave vector of the sample [3, 32]. It is well-known that a climbing motion of the defect can be driven by an evolution of the whole pattern towards an optimal wave number [33]. A second source of climbing motion is the advection term in the order parameter equation due to mean flow and it is the form of the mean flow that we want to study in the following.

Considering the necessity of having a finite range of periodic solutions that are stable with respect to all secondary instabilities, we have chosen a large value for the nonlinear coefficient α. As described by Rüdiger and Viñals [34] this causes a stabilization with respect to the zig-zag and the Eckhaus perturbations. Figure 3 shows the resulting various stability

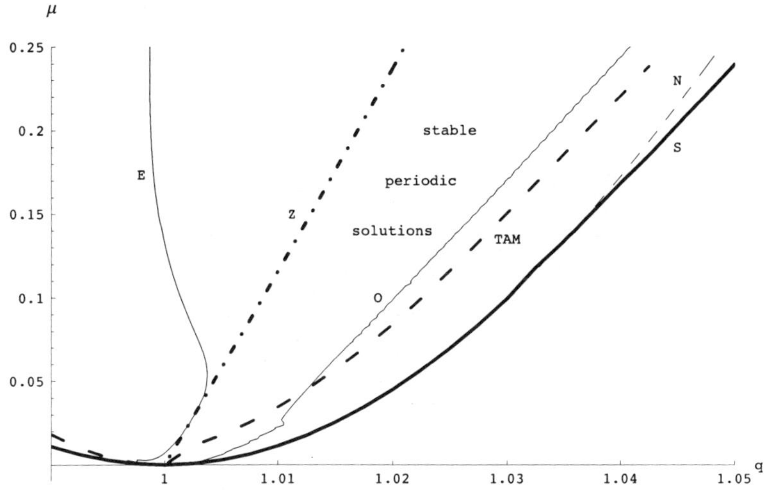

Fig. 3. Stability region for increased nonlinear damping $\alpha = 0.5$ and $\beta_1 = 0.1$. Other values of the parameters used are $\gamma = 0.1$, and $\Gamma = 0.8$. We show the steady Eckhaus line (E), the oscillatory Eckhaus line (O), the zig-zag line (Z), and the TAM line (TAM). Periodic solutions exist above the neutral stability curve (N) or, for $q > 1$ the saddle node bifurcation (S). The region of stability of the basic solution against an Eckhaus instability is the region between the thin solid lines.

boundaries for the case $\beta_1 = 0.1$. The values used for the other parameters are $\gamma = 0.1$ and $\Gamma = 0.8$. The figure also includes the neutral stability curve of the basic periodic solution and, since the primary bifurcation is subcritical for $q > 1$, we have included the saddle node curve where the periodic solution bifurcates. The range of basic solutions that is stable against all the perturbations considered here (Eckhaus, TAM, and zig-zag) is a region close to threshold at $\varepsilon = 0$ between the zig-zag, TAM and oscillatory Eckhaus lines.

The parameters of Figure 3 are thus suitable for the numerical simulation of dislocations. In the simulations we have integrated Eqs. (27-31) and (34) on a two-dimensional domain of length 128π (equivalent to about 60 stripes) using periodic boundary conditions in each direction [34]. The integration code exploits a Fourier function decomposition in a pseudo-spectral method with about 8 points per period of the basic Faraday pattern. A background pattern of wave number 1.0156 was used with two isolated dislocations placed in the sample thus creating a wave number of about 1.031 between the two defects. Although at the value $\varepsilon = 0.15$ used for the simulations the second wave number is located in the unstable region, it turned out to be only slightly unstable and thus no secondary instability that could disturb the defect motion was detected at the time scale of the defect motion.

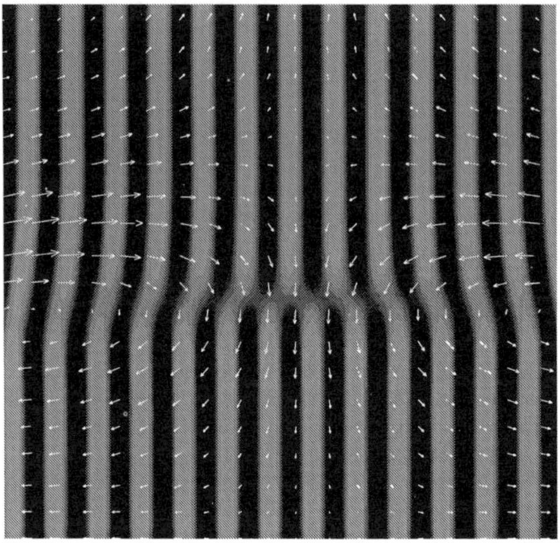

Fig. 4. The "ms" or long wave length component of the mean flow for $\beta_1 = 0.1$ and $\mu = 0.15$ (other parameters as for Figure 3). The amplitude of the velocity is approximately 0.028 (maximal amplitude of the vertical component), and 0.013 (horizontal component). The defect moves downwards.

In the simulations the defects were generally found to be in a climbing motion with a velocity that converges to a steady value after short times. For vanishing mean flow ($\beta_1 = 0$), as well as for all of the tested configurations with mean flow, the two defects move towards each other and annihilate, thus reducing the wave number of the pattern to 1.0156. The velocity fields U^{ms} as generated around a dislocation after a transient time are plotted in Figure 4. The amplitude of the small wave length part U^{mo} is negligibly small. For the long wave length part U^{ms} we found a pattern with two vortices close to each defect. The resulting flow at the defect drives the defect in the same direction as in comparable simulations for $\beta_1 = 0$ and therefore increases the velocity of the defect. The defect velocities for increasing β_1 are plotted in Figure 5.

The dominant advection flow generated by the defect is a long wavelength flow. This flow comprises two contributions: the viscous mean flow determined by Eq. (29) and the inviscid mean flow determined by Eq. (31). By increasing Γ and thus suppressing the effect of the inviscid forcing term on $\boldsymbol{U}^{ms}$ we have found a strongly decreasing component of the mean flow in direction of the climb. Therefore we conclude that at least for the current choice of parameters the main contribution to the advection of the defect originates from the inviscid forcing term in Eq. (31).

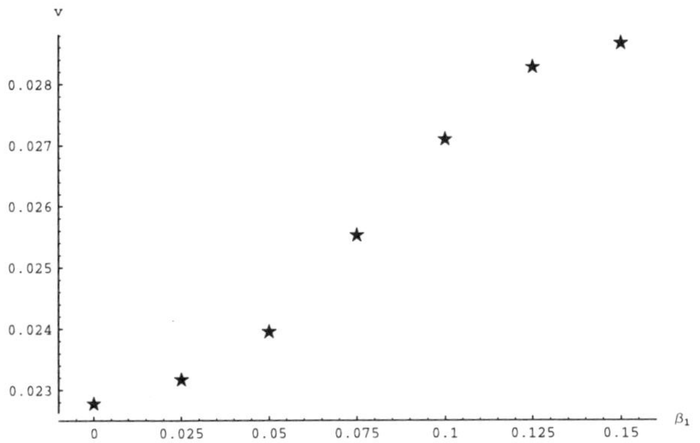

Fig. 5. Velocity of an isolated defect for $\mu = 0.15$ and with increasing coupling to the mean flow (other parameters as for Figure 3).

7. Conclusions

We have reviewed recent work on the generation and action of mean flow for Faraday waves in a horizontally two-dimensional domain. As in similar hydrodynamical pattern forming systems, mean flow effects have been ignored in early studies, but, as has been shown in this article, they have important implications for the stability of patterns. In particular, mean flow couples to perturbations of the stripe pattern and thus changes the domain of stable regular solutions, generally in a destabilizing manner. The strongest coupling, and therefore the strongest destabilization occurs for longitudinal perturbations. More importantly, the mean flow generates a qualitatively new oscillatory longitudinal perturbation, which for small nonlinear damping in the phenomenological equation renders all stripe patterns unstable, resulting in time-dependent patterns at onset. A weaker effect of the mean flow contribution has been found for the TAM instability. It is interesting to note that the TAM perturbations, which are of finite wavelength, couple to the short-wave part of the mean flow, whereas the long wavelength Eckhaus perturbations couple to the large scale mean flow.

Furthermore we have presented preliminary results of numerical simulations of patterns with defects. A background pattern which is stable to secondary perturbations has been found by using large values of the nonlinear damping coefficient. Next, we studied the simplest case of a defect, a pair of dislocations moving in a pattern of perfect stripes. Using numerical simulations we found that the main flow generated by a dislocation is a large scale flow consisting of two vortices. In the parameter range considered here the mean flow increases the velocity of the dislocation. Simulations with different values of the gravity-capillary number Γ showed that the main contribution to the climb of the dislocation originates from the inviscid forcing term.

Owing to the complexity of the derivation of mean flow equations we have limited the analysis to a specific limit in terms of the physical parameters. This concerns in particular the conditions on the dimensions of the layer. Further restrictions of the recent studies are the approximation of the surface waves using a phenomenological equation, and the assumption of a stripe pattern at onset. To consider more general situations and to test the results for the model equations described in this article further work needs to address these limitations.

Acknowledgments

This research was partially supported by the Spanish DGI under Grant BFM2001-2363 and by the U.S. Department of Energy under contract DE-FG05-95ER14566. S.R. would like to thank J. Viñals for helpful discussions and critical comments.

References

[1] M. Faraday, *Philos. Trans. R. Soc. London* **121**, 319 (1831).
[2] J. Miles and D. Henderson, *Annu. Rev. Fluid Mech.* **22**, 143 (1990).
[3] M. Cross and P. Hohenberg, *Rev. Mod. Phys.* **65**, 851 (1993).
[4] J. Gollub and J. Langer, *Rev. Mod. Phys.* **71**, S396 (1999).
[5] A. Kudrolli and J. Gollub, *Physica* **D 97**, 133 (1996).
[6] D. Binks and W. van de Water, *Phys. Rev. Lett.* **78**, 4043 (1997).
[7] D. Binks, M.-T. Westra, and W. van de Water, *Phys. Rev. Lett.* **79**, 5010 (1997).

[8] M.-T. Westra, D. Binks, and W. van de Water, *J. Fluid Mech.* **496**, 1 (2003).
[9] S. Milner, *J. Fluid Mech.* **225** 81 (1991)
[10] W. Edwards and S. Fauve, *J. Fluid Mech.* **278**, 123 (1994).
[11] W. Zhang, Ph.D. thesis, Florida State University (1994).
[12] W. Zhang and J. Viñals, *J. Fluid Mech.* **336**, 301 (1997).
[13] J. M. Vega, E. Knobloch, and C. Martel, *Physica* **D 154**, 313 (2001).
[14] V. Lapuerta, C. Martel, and J. M. Vega, *Physica* **D 173**, 178 (2002).
[15] E. Knobloch, C. Martel, and J.M. Vega, *Ann. N.Y. Acad. Sci.* **974** (2002),201-219
[16] E. Knobloch and J.M. Vega, In *Geometry, Mechanics and Dynamics, Volume in honor of J.E. Marsden*, P. Newton, P. Holmes, and A. Weinstein Eds, Springer-Verlag, (2002), 181.
[17] M. Higuera, J. M. Vega, , and E. Knobloch, in *Coherent Structures in Complex Systems*, edited by L. Bonilla, G. Platero, D. Reguera, and J. Rubi (Springer-Verlag, New York, 2001), p. 328.
[18] M. Higuera, J. M. Vega, and E. Knobloch, *J. Nonlinear Sci.* **12**, 505 (2002).
[19] J. M. Vega, S. Rüdiger, and J. Viñals, to be published.
[20] A. Davey and S. Stewartson, *Proc. R. Soc. London* **A 338**, 101 (1974).
[21] M. Longuet-Higgins, *Phil. Trans. Roy. Soc.* **A 245**, 535 (1953).
[22] J. Swift and P. Hohenberg, *Phys. Rev.* **A 15**, 319 (1977).
[23] W. Zhang and J. Viñals, *Phys. Rev. Lett.* **74**, 690 (1995).
[24] V. Zakharov, *Zh. Prikl. Mekh. Tekh. Fiz.* **9**, 86 (1968), [J. Appl. Mech. Tech. Phys. **9**, 190 (1968)].
[25] D. Crawford, P. Saffman, and H. Yuen, *Wave Motion* **2**, 1 (1980).
[26] L. Landau and E. Lifshitz, *Mechanics*, Pergamon, New York, 1976.
[27] P. Lyngshansen and P. Alstrom, *J. Fluid Mech.* **351**, 301 (1997).
[28] P. Chen and J. Viñals, *Phys. Rev. Lett.* **79**, 2670 (1997).
[29] W. Zhang and J. Viñals, unpublished.
[30] J. Nicolás and J. M. Vega, *Fluid Dyn. Research* **32**, 119 (2003).
[31] P. Chen and K. Wu, *Phys. Rev. Lett.* **85**, 3813 (2000).
[32] A. C. Newell, T. Passot, and J. Lega, *Annu. Rev. Fluid Mech.* **25**, 399 (1993).
[33] T. Walter, W. Pesch, and E. Bodenschatz, to be published.
[34] S. Rüdiger and J. Viñals, to be published.

CHAPTER 11

ROGUE WAVES AND THE BENJAMIN-FEIR INSTABILITY

Constance M. Schober

Department of Mathematics, University of Central Florida,
Orlando, FL 32816, USA

Rogue waves in 2D and 3D are numerically investigated in the framework of dispersive perturbations of the nonlinear Schrödinger (NLS) equation. We find that a chaotic regime greatly increases the likelihood of rogue waves. Enhanced focusing is shown to occur due to chaotically generated optimal phase modulations. A Melnikov analysis indicates persistence of a homoclinic solution $O(\varepsilon)$–close to the optimally phase modulated solution of the NLS. The correlation of the results of the Mel'nikov analysis and of the numerical experiments indicates that the rogue waves are well approximated by homoclinic solutions of the NLS equation. In 3D we find that the mechanisms for rogue waves (BF instability, mode coalescence due to phase modulation) are not destroyed by the tranverse instability but rather interact in a complicated manner.

1. Introduction

Rogue waves, also called freak waves or extreme waves in oceanic applications, are anomalous large amplitude waves whose heights exceed two times the significant wave height of the background sea. On average, approximately two large ships sink at sea every week and rogue waves have been the anecdotal cause behind many of these sinkings [11]. As hard data on these events was lacking or incomplete, statistical models of sea states have been used which predict the occurrence of extreme waves only about once every thousand years [12]. This statistical result is difficult to correlate with predictions from water wave models such as the nonlinear Schrödinger equation [5, 13]. However, very recent satellite data collected by the European Space Agency (ESA) has conclusively established the existence of rogue waves around the globe [11]. In the three weeks of satellite data available, researchers found more than 10 individual rogue waves above 25 meters in height. Significantly, the ESA's data indicates that rogue waves are widespread and are not as rare as previously thought as they occur "in higher numbers than anyone expected".

The ESA's new findings validate the analytical studies on rogue wave phenomena that has been the subject of intense research over the last five years. Various water wave models have been developed which predict rogue waves in different physical settings [12]. From these models we are able to determine some of the mechanisms by which rogue waves are generated. In both linear and nonlinear models, the following mechanisms have been proposed to explain the formation of rogue waves (with of course modifications in the nonlinear case): dispersion enhancement of transient wave groups, geometrical focusing in basins of variable depth, and wave-current interaction [12].

The mechanism applicable to rogue waves in deep water, different in principle from those mentioned above, is focusing due to the Benjamin-Feir (BF) instability. The BF instability is a modulational instability in which a uniform train of surface gravity waves is unstable to a weak amplitude modulation. It has been observed in physical experiments that the BF instability may lead to the generation of high amplitude waves. Since the focusing nonlinear Schrödinger (NLS) equation has been successfully used to model the BF instability in deep water [9, 17], the large amplitude waves can be interpreted as the excitation of homoclinic solutions of the NLS equation [9]. Homoclinic solutions are considered to be simple analytical models of rogue waves since they satisfy the amplitude criterion; the height of the homoclinic solution from trough to crest exceeds that of the modulated wavetrain by many times. Further, as with the anomalous waves being modelled, they are accompanied by deep holes and are detectable only for short periods of time.

In this article we adopt the approach based on the NLS equation, as well as higher order generalizations, to study rogue wave dynamics in deep water. Linearizing to study stability as in the BF instability analysis provides information about the instability only for short times, i.e. for as long as the linearization is approximately correct. For the NLS equation, since exact solutions are known we can follow the instability into the nonlinear regime and determine the long-time behavior. In the next section we consider homoclinic orbits of the Stokes wave (or plane wave) for the NLS equation and discuss their interpretation as rogue waves. When two or more unstable modes are present (see eqn. (1)) we examine the effect of phase modulating the initial wave train on the formation of rogue waves. We obtain explicit formulas for optimally phase modulated homoclinic orbits; i.e.

the modulation is optimal in the sense that the unstable modes are excited simultaneously (also referred to as "nonlinear chirping"), or nearly so, leading to a *significant* amplification beyond the usual BF modulation instability [5].

Water wave dynamics are described only to leading order by the NLS equation. In our study on the long time dynamics of the BF instability we found (both in laboratory and in numerical experiments) that the evolution of periodic modulated deep water gravity waves is generally chaotic [1, 2]. This behavior is well modelled by the broad band modified nonlinear Schrodinger (BMNLS) equation.

In Section 3 we study the generation of rogue waves in the framework of the BMNLS equation. In particular, we investigate how homoclinic chaos affects the formation of rogue waves. As an aid in identifying rogue wave events in the numerical experiments we monitor the evolution of the skewness and kurtosis, the third and fourth statistical moments of the wavefield respectively. We show that i) rogue waves do in fact occur in the BMNLS time series and ii) in a chaotic regime of homoclinic type, the solution can evolve close to a "nonlinearly chirped" wave for a rather general class of initial conditions, i.e. even when we have not prepared the initial data for the phase modulation to occur. In other words, we find that the chaotic dynamics which ensues because of the BF instability and higher order nonlinear effects, rather than diminishing the quantity and/or amplitude of rogue wave events, actually provides a mechanism for further focusing to occur due to chaotically generated optimal phase modulations. In Section 4 we consider the effects of directional spreading and study rogue waves in the 3D NLS equation. We find that the mechanisms for rogue waves (BF instability, mode coalescence due to phase modulation) are not destroyed by the transverse instability but rather they interact in a complicated manner.

In Section 5 we support the numerical study with a Melnikov analysis of equation (5), which indicates persistence of a homoclinic solution in the BMNLS system which is $O(\varepsilon)$–close to the NLS homoclinic solution formed by nonlinear chirping [5]. The Melnikov analysis determines the distinguishing spatial features of the perturbed dynamics which agree with the numerical observations of high amplitude rogue waves in the BMNLS chaotic regime. The correlation of the results of the Mel'nikov analysis and of the BMNLS numerical experiments indicates that in this more general setting rogue waves are still well approximated by homoclinic solutions of the NLS equation. Concluding remarks are provided in Section 6.

2. Rogue Wave Solutions of the 2D (1+1) NLS Equation

One of the simplest nonlinear models for describing the propagation of unidirectional surface waves on deep water is the nonlinear Schrodinger (NLS) equation,

$$iu_t + u_{xx} + 2|u|^2 u = 0.$$

An important physical prediction of the NLS equation is the modulational or Benjamin-Feir instability for periodic boundary conditions $u(x+L) = u(x)$. Modulationally unstable periodic solutions have homoclinic orbits which can be used to model rogue waves.

An example of this is provided by the plane wave solution, $u_0(x,t) = ae^{2i|a|^2 t}$, which has M linearly unstable modes (UMs) with growth rates $\sigma_n^2 = \mu_n^2(\mu_n^2 - 4a^2)$, $\mu_n = 2\pi n/L$, provided

$$0 < (\pi n/L)^2 < |a|^2 \tag{1}$$

is satisfied (the number M of UMs is the largest integer satisfying $0 < M < |a|L/\pi$). That is, the plane wave is unstable with respect to long wavelength perturbations. For each UM there is a corresponding homoclinic orbit; moreover, "combination" homoclinic orbits associated to two or more UMs can be constructed.

A global representation of the homoclinic orbits can be obtained by exponentiating the linear instabilities via Bäcklund transformations [3, 8]. A single homoclinic orbit of the plane wave is given by

$$u(x,t) = ae^{-2ia^2 t}\left(\frac{1 + 2\cos(px)e^{\sigma_1 t + 2i\varphi + \gamma} + A_{12}e^{2\sigma_1 t + 4i\varphi + 2\gamma}}{1 + 2\cos(px)e^{\sigma_1 t + \gamma} + A_{12}e^{2\sigma_1 t + 2\gamma}}\right) \tag{2}$$

where $A_{12} = 1/\cos^2\varphi$, $\sigma_1 = \pm p\sqrt{4a^2 - p^2}$, $\varphi = \sin^{-1}(p/2a)$, and $p = \mu_n = 2\pi n/L$, satisfies (1) for some integer n. Each UM has an associated homoclinic orbit characterized by the mode $p = \mu_n$.

Figure 1 shows the space-time plot of the amplitude $|u(x,t)|$ of a homoclinic orbit ($a = 0.5$, $L = 2\sqrt{2}$ and $p = 2\pi/L$) with one UM. Accompanying all of the surface plots in this paper are two plots of i) the corresponding skewness (dashed line) and kurtosis (solid line) as a function of time and ii) the maximum of the amplitude (UMAX) as well as the mean of the wave-field as a function of time. The interpretation of the skewness and kurtosis is given in the Appendix. The one UM homoclinic solution represents the simplest prototype of rogue wave. As $t \to \pm\infty$, solution (2) limits asymptotically to the plane wave; i.e. the plane wave behavior dominates the motion for most of its lifetime. For $t \approx 15$, UMAX reaches its maximum which is approximately twice that of the background carrier wave (the plane wave). At about $t = 10$ the kurtosis starts to increase indicating the onset of the BF instability and at $t \approx 15$ it reaches it's maximum. However the kurtosis does not exceed three and as a consequence is not as interesting this is not as interesting a case as the following examples from the perspective of rogue waves.

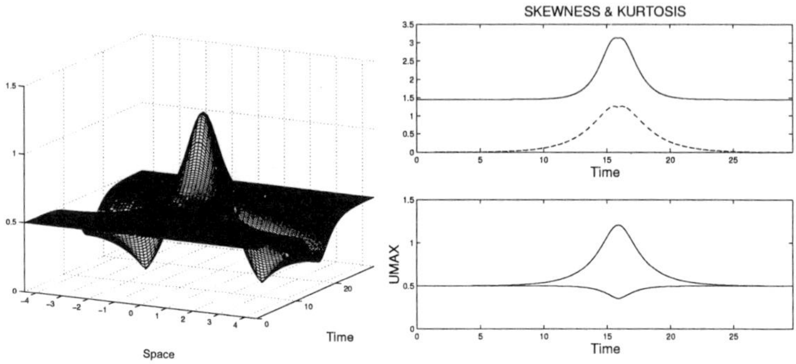

Fig. 1a-b. Rogue wave solution of the NLS corresponding to one unstable mode.

2.1. *Phase Modulated Rogue Waves*

As the number of UMs increases, the space-time structure of the homoclinic solutions becomes more complex. We find that when two or more UMs are present the initial wave train can be phase modulated to produce additional focusing. The family of homoclinic orbits (i.e. rogue waves) associated with two UMs is given by [5]

$$u(x,t) = a e^{2ia^2 t} \frac{g(x,t)}{f(x,t)}, \tag{3}$$

where

$$\begin{aligned} f(x,t) = {} & 1 + 2\cos(p_1 x) e^{\sigma_1 t} + A_{12} e^{2\sigma_1 t} + 2\cos(p_3 x) e^{\sigma_3 t + \gamma} + A_{34} e^{2(\sigma_3 t + \gamma)} \\ & + 2[A_{13}\cos(p_1 + p_3)x + A_{23}\cos(p_3 - p_1)x] e^{(\sigma_1 + \sigma_3)t + \gamma} \\ & + 2A_{12}A_{13}A_{23}\cos(p_3 x) e^{(2\sigma_1 + \sigma_3)t + \gamma} + A_{12}A_{13}^2 A_{23}^2 A_{34} e^{2(\sigma_1 + \sigma_3)t + 2\gamma} \\ & + 2A_{13}A_{23}A_{34}\cos(p_1 x) e^{(\sigma_1 + 2\sigma_3)t + 2\gamma} \end{aligned}$$

$$\begin{aligned} g(x,t) = {} & 1 + 2e^{2i\varphi_1}\cos(p_1 x)e^{\sigma_1 t} + A_{12}e^{4i\varphi_1}e^{2\sigma_1 t} + 2e^{2i\varphi_3}\cos(p_3 x)e^{\sigma_3 t+\gamma} \\ & + A_{34}e^{4i\varphi_3}e^{2(\sigma_3 t+\gamma)} + A_{12}A_{13}^2A_{23}^2A_{34}e^{4i(\varphi_1+\varphi_3)}e^{2(\sigma_1+\sigma_3)t+2\gamma} \\ & + 2A_{13}A_{23}A_{34}e^{2i(\varphi_1+2\varphi_3)}\cos(p_1 x)e^{(\sigma_1+2\sigma_3)t+2\gamma)} \\ & + 2A_{12}A_{13}A_{23}e^{2i(2\varphi_1+\varphi_3)}\cos(p_3 x)e^{(2\sigma_1+\sigma_3)t+\gamma} \\ & + 2e^{2i(\varphi_1+\varphi_3)}[A_{13}\cos(p_1+p_3)x + A_{23}\cos(p_3-p_1)x]e^{(\sigma_1+\sigma_3)t+\gamma)} \end{aligned}$$

$$A_{jk} = \left(\frac{\sin\frac{1}{2}(\varphi_j - \varphi_k)}{\sin\frac{1}{2}(\varphi_j + \varphi_k)}\right)^2, \qquad \sigma_j = \pm\mu_j\sqrt{4a^2 - \mu_j^2}.$$

As in the last example, provided $p_1 = \mu_n$ and $p_3 = \mu_m$ satisfy (1), the solution decays to the plane wave solution, as $t \to \pm\infty$, and the rogue wave lies hidden beneath the background plane wave for most of its lifetime. This rogue wave is characterized by the two modes $\cos(p_1 x)$ and $\cos(p_3 x)$, whose temporal separation is determined by the parameter γ. Figures 2a and 2b illustrate the combination homoclinic orbit (3) obtained when $a = 0.5$, $L = 4\sqrt{2}$, $p_3 = 2p_1 = 2\pi/L$ and $\varphi_j = \sin^{-1}(\mu_j/2a)$, for $\gamma = 0.1$ and $\gamma = 0.2$, respectively. In Figure 2a the modes are well separated and the maximal value of the amplitude is roughly three times the plane wave amplitude (its maximum is 1.4). In this example the kurtosis is well above three.

As γ is varied, the excitation of the two modes occurs at closer values of time and the maximum of the amplitude increases. A homoclinic orbit of maximal amplitude is produced when the two UMs are excited simultaneously. This is an example of "optimal" phase modulation or "nonlinear chirping", where the phases in the initial data have been selected so that the solution chirps up to a maximal height. In Figure 2b the rogue wave rises to a height of approximately 4.1 times the height of the background carrier wave (the maximum height is 2.1). The kurtosis is also significantly larger, $m_4 \approx 14$. We note that Figure 2a shows focusing due to only a weak amplitude modulation of the initial wave train (the growth in UMAX starting at about $t = -5$ and again at about $t = 10$ is due to the BF instability). However, Figure 2b shows focusing due to both an amplitude and a phase modulation (the growth in UMAX starting at about $t = -5$ is due to the BF instability but the additional very rapid focusing at about $t = 3.4$ is due to the phase modulation).

In general it is possible to choose the phases in the homoclinic orbits (general formulas can be found in reference [5]) so that any number of UMs phase lock. Figures 3a and 3b illustrate two members of the family of homoclinic orbits corresponding to three UMs; in Figure 3a the modes are separated and in Figure 3b the modes have coalesced. The rogue wave

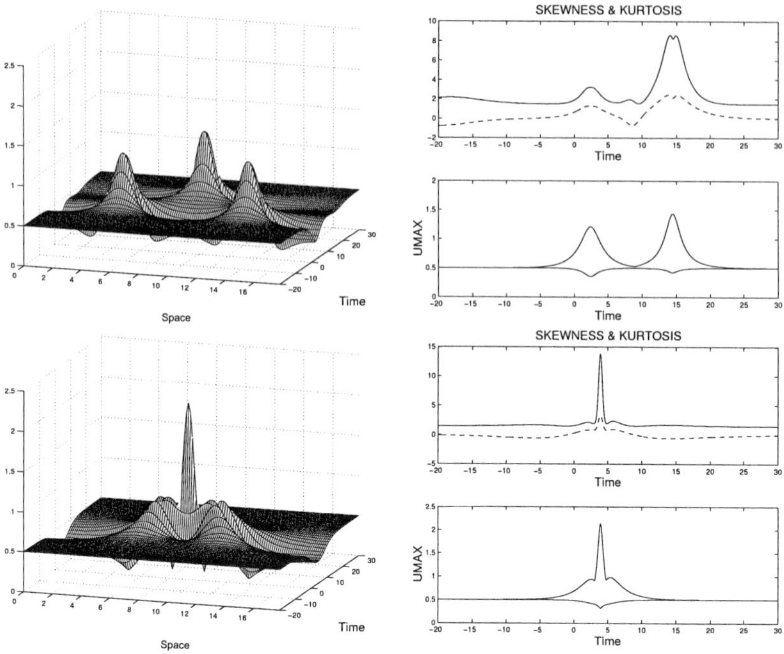

Fig. 2a-b. Rogue wave solutions of the NLS corresponding to two unstable modes without phase modulation (top) and with phase modulation (bottom).

attains a height of approximately three and five times the height of the background carrier wave, respectively. As before the striking difference in wave amplification is due to the additional phase modulation.

3. Rogue Waves in 2D (1+1) higher order NLS Equations

In this section we consider the broader bandwidth modified nonlinear Schrödinger (BMNLS) equation introduced by Trulsen and Dysthe. This equation arises as a higher order asymptotic approximation of slowly modulated periodic wave trains (such as the Stokes wave) in deep water. Assuming the wave slope ka to be $\mathcal{O}(\varepsilon)$ (a is the size of the initial surface displacement), while the bandwidth $|\Delta k|/k$ and $(kh)^{-1}$ (h is the water depth) to be $\mathcal{O}(\varepsilon^{1/2})$, one obtains [7]

$$i\frac{\partial u}{\partial t} + \frac{\partial^2 u}{\partial x^2} + 2|u|^2 u = \sqrt{\varepsilon}\left(-\frac{i}{2}\frac{\partial^3 u}{\partial x^3} + 6i|u|^2\frac{\partial u}{\partial x} - iu^2\frac{\partial u^*}{\partial x} - 2u\left[H(|u|^2)\right]_x\right) + \frac{5\varepsilon}{16}\frac{\partial^4 u}{\partial x^4} + \varepsilon^{3/2}\frac{7i}{32}\frac{\partial^5 u}{\partial x^5}, \qquad (4)$$

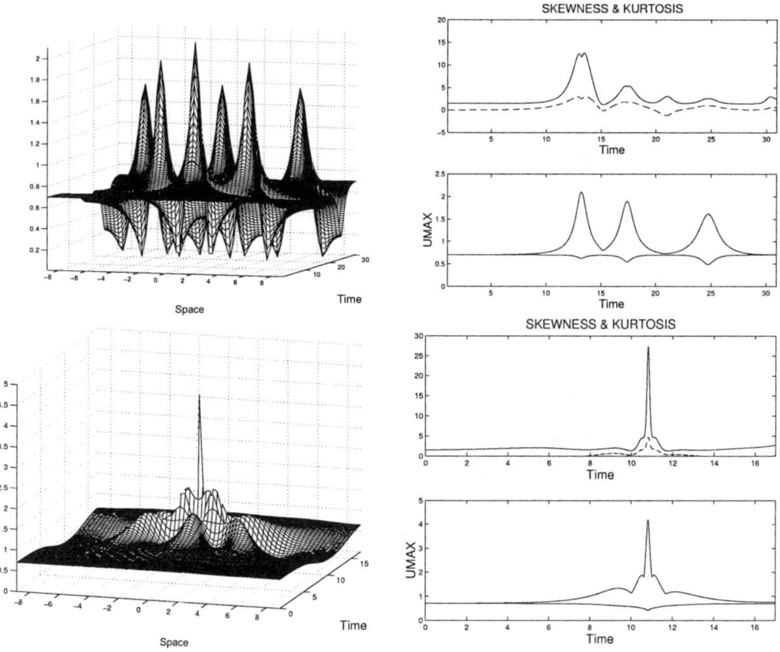

Fig. 3a-b. Rogue waves solutions of the NLS corresponding to three unstable modes without phase modulation (top) and with phase modulation (bottom).

where $H(f)$ represents the Hilbert transform of the function f.

We begin by examining the "even BMNLS" equation which results from considering spatially symmetric wave trains $u(-x,t) = u(x,t)$,

$$i\frac{\partial u}{\partial t} + \frac{\partial^2 u}{\partial x^2} + 2|u|^2 u = \varepsilon \frac{\partial^4 u}{\partial x^4}. \tag{5}$$

Since the even BMNLS dyanmics is chaotic in the presence of two or more UMs [2], we examine cases in the two and three UM regime. Initial data for solutions with two UMs are obtained by linearizing (3) about $t = 0$,

$$u(x,0) = a\left[1 + 4i(\varepsilon_1 \sin\varphi_1 e^{i\varphi_1} \cos p_1 x + \varepsilon_3 \sin\varphi_3 e^{i\varphi_3} \cos p_3 x)\right]$$

where the parameter values are given as in (3). Figure 4a illustrates an interesting rogue wave solution of equation (5) for and $\varepsilon_1 = 10^{-4}$, and $\varepsilon_3 = 10^{-5}$. The solution quickly develops irregularities and becomes chaotic. At about $t = 471.2$ a rogue wave rises up from the plane wave state and develops a crest with an amplitude of approximately four times the background wave height. This BMNLS rogue wave structure is close to the maximal amplitude rogue wave of the NLS obtained when $\gamma = 0.2$ (compare Figure

4a with Figure 2b). Similarly, for numerical experiments in the three UM regime, after the solution becomes chaotic rogue waves develop (see Figure 4b). At $t \approx 208$ a rogue wave develops which is almost five times the height of the background wave. Again, this solution is close to the optimal phase modulated homoclinic solution in the three UM regime.

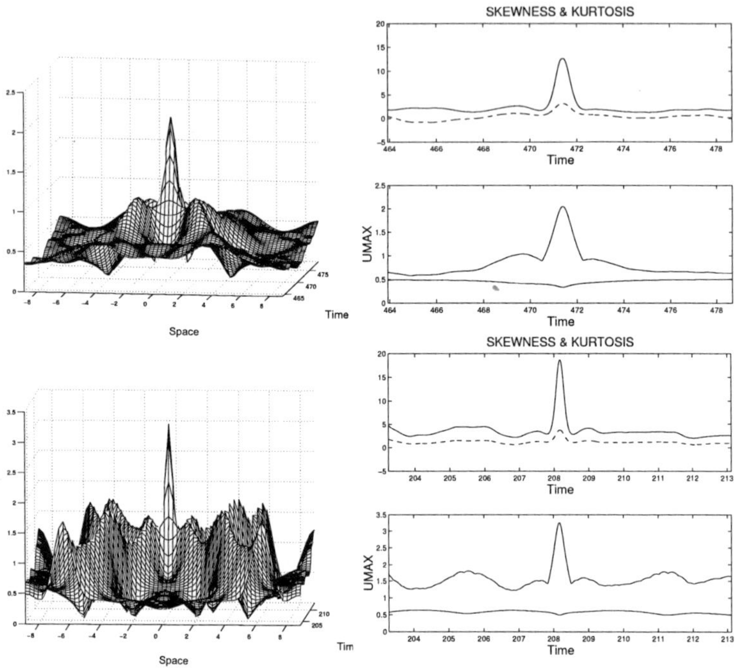

Fig. 4a-b. Rogue waves solutions for the even BMNLS equation when two (top) and three (bottom) unstable modes are present.

The observed wave amplification is only slightly smaller for the full BMNLS equation (4). In this case the symmetry breaking effects of the higher order terms prevent a complete spatial coalescence of the nonlinear modes. In Figures 5a-b loss of spatial symmetry in the rogue wave solution for the full BMNLS (in the two and three UM regime) is clearly visible. The presence of homoclinic orbits in the unperturbed model appears to play a major role both in the perturbed chaotic evolution and in the generation of rogue waves in the full BMNLS equation.

We performed extensive numerical experiments using the BMNLS model in the two and three UM regimes, varying both the perturbation strength

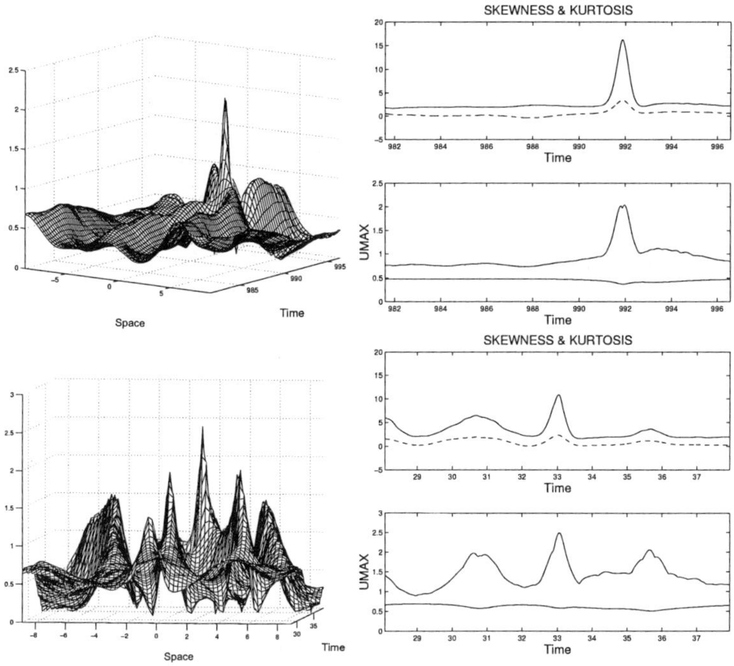

Fig. 5a-b. Rogue wave solutions for the full BMNLS equation when two (top) and three (bottom) unstable modes are present.

ε and the values of the parameters in the initial data. In the three UM regime, we find that rogue wave structures also survive, i.e. are observable within the chaotic regime. We find that a large set of initial conditions can evolve to become a rogue wave. This suggests that the underlying chaotic dynamics favors those high amplitude solutions close to the homoclinic solutions obtained by nonlinear chirping. In other words, homoclinic chaos increases the likelihood of rogue waves.

4. Rogue Waves in the 3D (2 + 1) NLS Equation

In this section we investigate the effect of directional spreading on the development of rogue waves in the framework of the (2+1) NLS equation,

$$i\frac{\partial u}{\partial t} + \frac{\partial^2 u}{\partial x^2} - \frac{1}{2}\frac{\partial^2 u}{\partial y^2} + 2|u|^2 u = 0. \tag{6}$$

The modulations of wave packets in the longitudinal and tranversal directions behave differently. The BF instability condition for the plane wave can be found easily. Linearizing the NLS equation about $u_0(x,t) = ae^{2i|a|^2 t}$ by considering perturbations of the form $u(x,y,t) = u_0(1+\eta(x,y,t))$ (η small), and assuming $\eta(x,y,t) = e^{i(\mu x+\lambda y+\Omega t)}$, one computes the growth rate to be $\Omega^2 = \mu^4 + \frac{1}{4}\lambda^4 - \mu^2\lambda^2 - 4a^2\mu^2 + a^2\lambda^2$. In particular, modulations of the plane wave in only the transverse direction (i.e. $\mu = 0$) are stable. For modulations in only the longitudinal direction (i.e. $\lambda = 0$) the growth rate reduces to 2D case with the instability condition given by (1). In general the instability region is unbounded in the perturbation wave vector plane. As a consequence, it is expected that the modulational instability will be one of the available mechanisms for generating rogue waves in 3D.

In the numerical experiments we use initial conditions which are characterized by a carrier wave plus small perturbations in both the x and y directions,

$$u(x,y,0) = a\left[1 + (\varepsilon_1 \cos p_1 x + \varepsilon_2 \cos p_2 y)\right]. \tag{7}$$

Figure 6 displays typical results for the evolution of a weakly modulated plane wave where the parameters in (7) are chosen so that there are two linearly UMs. Several snapshots in time are provided. The development of the BF instability is clearly visible as well as the later development of transverse instabilities and the formation of a large rogue wave. The initial condition used in Fig. 6 has a stronger modulation in y than in x (see the snapshot at $t = 0.01$). This delays the saturation of the BF instability in x. For fixed values of y, y_i, as t evolves we see the BF instability at work and wave crests developing in x. Since a modulation in y was seeded initially, the closer $u(x,y_i,0)$ is to the plane wave as a function of x, the more unstable and the faster the growth. This explains why at $t = 10.89$ for y_i near the endpoints there is larger growth in x than for y_i in the middle of the period. As the solution evolves away from the plane wave, a transverse instability develops and peaks form which are decoupled in x and y ($t \approx 13$). Finally these peaks merge forming a rogue wave ($t = 14.16$). The amplitude of the rogue wave is approximately 4 times greater than that of the initial wave or 6 times greater than the mean.

The plot of UMAX indicates that the time scales of the two processes are different. The BF instability produces a smooth almost Gaussian growth while the growth due to the interaction of the transverse and BF instability is irregular and in the final stage before the formation of the rogue wave is very rapid. The subsequent decay is equally rapid. This process is also observed in the kurtosis plots. At ($t \approx 10$) the kurtosis is hovering about three, it then rapidly increases to about 15, evolves irregularly, and then

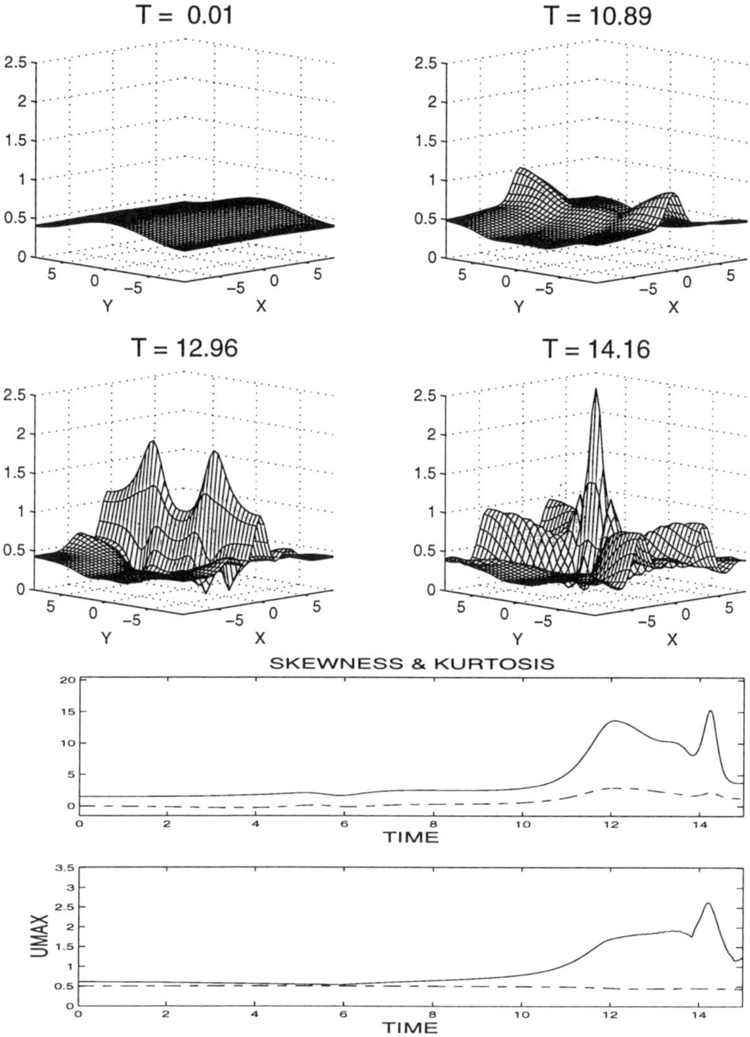

Fig. 6. Evolution of the BF instability in 3D for a weakly modulated plane wave showing the development of a large rogue wave

undergoes a second very rapid increase before it decays to three.

This example illustrates that the mechanisms for rogue waves (BF instability, mode coalescence due to phase modulation) are not destroyed by the transverse instability but rather they interact in a complicated way. In the numerical experiments many rogue waves are observed when the spectrum is narrow-banded near the (1+1) NLS case. This suggests that as a first step the basic mechanisms for the formation of rogue waves can be studied using an analytically tractable one-dimensional model. The Melnikov analysis described in the following section indicates that there is a persistent homoclinic structure which is $\mathcal{O}(\varepsilon)$ close to the NLS phase modulated rogue wave. This result is consistent with the numerical observations of large amplitude rogue waves in Figures 4 and 5.

5. Melnikov Analysis

5.1. *Phase Space Geometry*

The even BMNLS equation (5) can be interpreted as a Hamiltonian system on the Sobolev space $\mathcal{H}_{e,p}$ of even, periodic functions, with Hamiltonian functional

$$H_\varepsilon(u, u^*) = \int_0^L (|u_x| - |u|^4 - \varepsilon |u_{xx}|^2)\, dx.$$

On the invariant plane $\Pi = \{u(x,t) \,|\, \partial_x u(x,t) = 0\}$, the dynamics is simply described in terms of a two-parameter family of plane wave solutions $u_a(t; a, \varphi) = ae^{i(2a^2 t + \varphi)}$. The stability type of these solutions is determined by solving equation (5) linearized about u_a, using a complex Fourier series expansion of the form $\sum_{j=1}^{\infty} a_j e^{\sigma_j t} \cos jx$.

For $\varepsilon = 0$, and $1 < a < 3/2$, one obtains growth rates $\sigma_j = \pm j\sqrt{a^2 - (\pi j/L)^2}$, $j = 1, 2$, corresponding to the two-dimensional stable and unstable eigenspaces of the invariant set $\mathcal{S}_a = \left\{u_a \,|\, 1 < a < \frac{3}{2} \right\}$, as well as an infinite number of center modes characterized by complex conjugate pairs of eigenvalues $\lambda_j = \pm 2i\sqrt{(\pi j/L)^2 - a^2}$, $j \geq 3$.

As seen in Section 2, the invariant set $\mathcal{S}_a$ possesses global two dimensional stable and unstable manifolds, explicitly parameterized in terms of formula (3) for the two-dimensional family of homoclinic solutions (in expression (3), parameter a and the initial phase of the plane wave solutions are coordinates on $\mathcal{S}_a$, while γ_1 and γ_3 parameterize the individual homoclinic orbits.) When $\varepsilon \neq 0$, the perturbed dynamics takes place in the codimension two invariant subspace $\mathcal{F} = \{u \in \mathcal{H}_{e,p} \mid H_\varepsilon(u, u*) = C_H, I(u, u*) = C_I\}$, where $I = \int_0^L |u|^2\, dx$ is the L^2–norm of u, which is preserved by the perturbation.

For equation (5), two main questions need to be addressed. Persistence of invariant manifolds (the singular perturbation makes the proof delicate, though techniques developed by C. Zeng [18] should guarantee at least that much); and persistence of homoclinic structures, such as homoclinic tubes, which explain the features of the irregular dynamics observed numerically.

Persistence of local invariant manifolds is both expected and supported by the numerical evidence; heuristically, we assume this to determine the number of measurements needed to establish transversal intersections of the invariant manifolds. In the two UM regime, the codimension two locally invariant center-stable and center-unstable manifolds $\mathcal{W}_\varepsilon^{cs,cu}(\mathcal{S}_a)$ will intersect transversally along a codimension four locally invariant submanifold $\mathcal{W}_\varepsilon^{cs} \cap \mathcal{W}_\varepsilon^{cu}$. Since splitting does not occur in the directions tangential to the invariant subspace $\mathcal{F}$, it is sufficient to measure the splitting distance between the persistent invariant manifolds along a two-dimensional vector transversal to $\mathcal{W}_\varepsilon^{cs} \cap \mathcal{W}_\varepsilon^{cu}$. In order to define suitable measurements, some results about the integrable structure of the unperturbed NLS equation are needed.

5.2. *Fundamental Invariants*

The integrability of the NLS equation (equation (5) with $\varepsilon = 0$) is established using the Lax pair:

$$\varphi_x = \mathcal{L}^{(x)}\varphi, \qquad \varphi_t = \mathcal{L}^{(t)}\varphi, \tag{8}$$

where

$$\mathcal{L}^{(x)} = \begin{pmatrix} i\lambda & iu \\ iu^* & -i\lambda \end{pmatrix}, \qquad \mathcal{L}^{(t)} = \begin{pmatrix} i[2\lambda^2 - uu^*] & 2i\lambda u + u_x \\ 2i\lambda u^* - u_x^* & -i[2\lambda^2 - uu^*] \end{pmatrix},$$

and λ is the spectral parameter. This system has a common solution $\varphi(x,t;\lambda)$, provided the coefficient $u(x,t)$ satisfies equation (5). Every NLS solution u is characterized by the spectrum $\sigma(u) := \big\{\lambda \in \mathbb{C} \mid \mathcal{L}^{(x)}\boldsymbol{v} = 0, |\boldsymbol{v}| \text{ bounded } \forall x\big\}$ of the associated linear operator $\mathcal{L}^{(x)}$. For periodic boundary conditions $u(x+L,t) = u(x,t)$, the spectrum of u can be written in terms of the transfer matrix $M(x+L;u,\lambda)$ across a period, where $M(x;u,\lambda)$ denotes a fundamental solution matrix of the Lax pair (8). Introducing the Floquet discriminant $\Delta(u,\lambda) := \text{Trace}\,[M(x+L;u,\lambda)]$, one obtains

$$\sigma(u) := \{\lambda \in \mathbb{C} \mid \Delta(u,\lambda) \in \mathbb{R},\ -2 \le \Delta(u,\lambda) \le 2\}\,.$$

The Floquet spectrum of a typical solution consists of bands of continuous spectrum with simple periodic eigenvalues as endpoints (for example, finite-genus solutions are those with a finite number of bands of continuous spectrum). *Simple critical points*, $\{\lambda_j^s \mid \Delta(\lambda,u) = \pm 2,\ d\Delta/d\lambda \ne 0\}$

and *double points* $\{\lambda_j^d \mid \Delta(\lambda,u)=\pm 2,\ d\Delta/d\lambda = 0,\ d^2\Delta/d\lambda^2 \neq 0\}$ are distinguished points of the periodic spectrum. Complex double points in the Floquet spectrum of a given potential u typically reflect the instability of u and label its homoclinic orbits.

The Floquet discriminant of a plane wave solution $u(x,t) = a\, e^{2ia^2 t}$ is readily computed to be $\Delta(a,\lambda) = 2\cos(\sqrt{a^2+\lambda^2}L)$ Then, the associated Floquet spectrum consists of the continuous bands $\mathbb{R}\bigcup\, [-ia, ia]$, and a discrete part containing the simple periodic/antiperiodic eigenvalues $\pm ia$, and the infinite sequence of double points

$$\lambda_j^2 = (j\pi/L)^2 - a^2, \qquad\qquad j \in \mathbb{Z}.$$

Of these, if $[aL/\pi] = M$ (where $[p]$ = largest integer$\leq p$, $p > 0$), $2M$ are complex (pure imaginary) double points, while the remaining λ_n's for $|n| > M$ are real.

The Floquet discriminant functional is invariant under the NLS evolution, and thus encodes the infinite family of constants of motion (in fact, parameterized by $\lambda \in \mathbb{C}$). We briefly describe the results of Y.Li and D.W. McLaughlin [10], which show how the critical structure of the Floquet discriminant (as a functional on $\mathcal{H}_{e,p}$) is in one-to-one correspondence with critical level sets of solutions of the unperturbed PDE.

Given a solution u_c with a purely imaginary critical point λ^c, regarding λ^c as a functional on a neighborhood $\mathcal{U}$ of u_c, one has

$$\left.\frac{\partial}{\partial\lambda}\Delta(\lambda;u)\right|_{\lambda^c(u)} = 0; \qquad \lambda^c(u_c) = \lambda^c.$$

The following invariant functional $\boldsymbol{F} : \mathcal{U} \to \mathbb{R}$, $\boldsymbol{F} := \Delta\left(\lambda^c(u); u\right)$ is locally smooth, provided $\frac{d^2}{d\lambda^2}\Delta(\lambda,u) \neq 0$,$\forall\, u \in \mathcal{U}$. Then, the sequence $F_j(u) = \Delta(\lambda_j^c(u), u)$, generated as λ_j^c varies among the critical points of the potential u, defines a natural sequence of constants of motion in the following sense. First, it provides a Morse-function type approach to the description of the topology of level sets of solutions of the integrable PDE (i.e., the stratification is described by identifying the critical level sets, labelled by the double points of the associated Floquet spectrum). Second, the sequence $\{F_j\}_{j=1}^{\infty}$ provides a local description of the strata which are close to critical level sets of saddle type (this is important when considering the effects of perturbations).

The local analysis is encoded in the first and second variation of the $F_j's$. The gradient of F_j has the following explicit representation [10]; if $\vec{\psi}^{\pm}(x,\lambda)$ are the Bloch eigenfunctions (common eigenfunctions of the Lax operator

$\mathcal{L}^{(x)}$ at (u, λ) and the shift operator $(\mathcal{S}\psi)(x) = \psi(x+L))$, then

$$\frac{\delta F_j}{\delta \vec{u}}(u) = i\frac{\sqrt{\Delta^2 - 4}}{W[\psi^+, \psi^-]} \left[\begin{array}{c} \psi_2^+ \psi_2^- \\ -\psi_1^+ \psi_1^- \end{array} \right] \Bigg|_{\lambda=\lambda^c}. \tag{9}$$

We observe that $\delta F_j / \delta \vec{u}\,(u_c) = 0$, that is the functional is critical at the critical level set. On the other hand, if u^H is the homoclinic orbit in the isospectral level set of u_c, then

$$\frac{\delta F_j}{\delta \vec{u}}(u^H) = i\frac{\sqrt{\Delta^2 - 4}}{W[\Psi^+, \Psi^-]} \left[\begin{array}{c} \Psi_2^+ \Psi_2^- \\ -\Psi_1^+ \Psi_1^- \end{array} \right] \Bigg|_{\lambda=\lambda^d} \tag{10}$$

does not vanish. Therefore $\delta F_j / \delta \vec{u}\,(u^H)$, $j = 1, \ldots M$ (M being the number of complex double points) define directions transversal to the homoclinic manifold and can be used to construct Melnikov-type measurements.

5.3. *The Melnikov Integrals*

In this section we generalize the formulas for the Melnikov integrals obtained in [10] to the case of 2 complex double points (ν_1, ν_2) and compute them for the conservative perturbation (5).

Denoting with $f(\vec{u}) = (u_{xxxx}, u^*_{xxxx})^T$ the vector of the perturbation, the components of the distance vector between the persistent invariant manifolds $W^{cs}_\varepsilon(\mathcal{S}_a)$ and $W^{cu}_\varepsilon(\mathcal{S}_a)$ along the directions ∇F_j, $j = 1, 2$, are given by

$$d_j = \varepsilon M_j + \mathcal{O}(\varepsilon^2), \qquad M_j = \int_{-\infty}^{+\infty} \langle \nabla F_j, f \rangle |_{u=U_H}\, dt,$$

where $< , >$ is the standard complex inner product. The integrand is evaluated along the homoclinic orbit of the plane wave solution $U_H(x,t)$, and ∇F_j is given by

$$\nabla F_j = -i\frac{\sqrt{\Delta^2 - 4}}{W[\vec{\Psi}^{(+)}, \vec{\Psi}^{(-)}]} \left(\begin{array}{c} \Psi_2^{(+)} \Psi_2^{(-)} \\ \Psi_1^{(+)} \Psi_1^{(-)} \end{array} \right) \Bigg|_{\lambda=\nu_j}, \tag{11}$$

with $\vec{\Psi}^{(+)}, \vec{\Psi}^{(-)}$ Bloch eigenfunctions at the new potential U_H evaluated at $\lambda = \nu_j$. The new Bloch basis can be obtained by means of a Bäcklund transformation; one also obtains the compact representation for the Wronskian $W[\vec{\Psi}^{(+)}, \vec{\Psi}^{(-)}] = (\lambda - \nu)(\lambda - \nu^*) W[\vec{\psi}^{(+)}, \vec{\psi}^{(-)}]$, which allows one to effortlessly evaluate the limit $\lambda \to \nu$ in equation (11).

In order to construct the Melnikov integrals, we first consider the limit $\lambda \to \nu_2$ and obtain

$$\nabla F_2 = C_2 \frac{W[\psi^{(+)}, \psi^{(-)}](\nu_2)}{(|\xi_1|^2 + |\xi_2|^2)^2} \begin{pmatrix} (\xi_1^*)^2 \\ (\xi_2^*)^2 \end{pmatrix}, \tag{12}$$

with $C_2 = i d_+ d_- (\nu_2 - \nu_2^*)(\nu_2 - \nu_1)(\nu_2 - \nu_1^*)\sqrt{\Delta(\nu_2)\Delta''(\nu_2)}$, and with $\xi(\nu_2) = G(\nu_2, \nu_1; \varphi(\nu_1))\chi(\nu_2)$. Here $G(\nu_2, \nu_1; \varphi(\nu_1))$ is the gauge matrix which carries eigenfunctions $\varphi(\lambda)$ of the plane wave solution to eigenfunctions $\xi(\lambda)$ of u_1, the potential obtained after one iteration of the Bäcklund formula, and $\chi(\nu_2) = d_+\psi^{(+)}(\nu_2) + d_-\psi^{(-)}(\nu_2)$. This leads to the following Melnikov integral

$$M_2(\gamma) = C_2 \int_{-\infty}^{+\infty} \int_0^1 W[\psi^{(+)}, \psi^{(-)}] \frac{(\xi_1^*)^2 (U_H)_{xxxx} - (\xi_2^*)^2 (U_H^*)_{xxxx}}{(|\xi_1|^2 + |\xi_2|^2)^2} \, dx dt. \tag{13}$$

Similarly, computing the limit $\lambda \to \nu_1$ leads to the following expression for ∇F_1:

$$\nabla F_1 = C_1 \frac{W[\psi^{(+)}, \psi^{(-)}](\nu_1)}{(|\zeta_1|^2 + |\zeta_2|^2)^2} \begin{pmatrix} (\zeta_1^*)^2 \\ (\zeta_2^*)^2 \end{pmatrix}, \tag{14}$$

with $C_1 = i c_+ c_- (\nu_1 - \nu_1^*)(\nu_1 - \nu_2)(\nu_1 - \nu_2^*)\sqrt{\Delta(\nu_1)\Delta''(\nu_1)}$, and with $\zeta(\nu_1) = G(\nu_1, \nu_2; \chi(\nu_2))$. The corresponding Melnikov integral is

$$M_1(\gamma) = C_1 \int_{-\infty}^{+\infty} \int_0^1 W[\psi^{(+)}, \psi^{(-)}] \frac{(\zeta_1^*)^2 (U_H)_{xxxx} - (\zeta_1^*)^2 (U_H^*)_{xxxx}}{(|\zeta_1|^2 + |\zeta_2|^2)^2} \, dx dt. \tag{15}$$

The numerical evaluation of the two Melnikov integrals, using expression (3) for the homoclinic orbit of the unstable plane wave solution, indicate that M_1 and M_2 are mutually proportional functions of the parameter γ. The presence of a unique nondegenerate zero of $M_1 \sim M_2$ suggests (together with numerical evidence of irregular behavior) a single measurement is sufficient for establishing persistence of invariant hyperbolic sets.

The value $\gamma = \gamma_0$, at which a nondegenerate zero of M_1 occurs is $O(\varepsilon)$–close to the value of γ for which the homoclinic solution (3) achieves maximal amplitude. Figure 2b shows an amplitude plot of the homoclinic orbit with $\gamma = \gamma_0$. Remarkably, this solution can be identified with the rogue wave observed intermittently (see e.g. Figure 4a) throughout the numerical simulations. While typically a Melnikov analysis only indicates persistence of hyperbolic structures (which are the source of irregular dynamics), for the even BMNLS equation it also supports the thesis that chaotic dynamics leads to optimal phase modulation and increases the likelihood of rogue wave events.

We finally remark that, due to the imposed spatial symmetry, the imaginary parts of the gradients ∇F_j vanish identically, while for the full BMNLS equation the presence of symmetry breaking terms in the perturbation makes the Melnikov integrals complex quantities. Recently, we studied a simple symmetry breaking perturbation of the NLS equation, for which the existence of nondegenerate zeros of the real part of the Melnikov integral provides the condition for the onset of chaotic dynamics, while the existence of nondegenerate zeros of the imaginary part provides the condition for spatially asymmetric chaotic wave forms [4]. Using this fact, the Mel'nikov analysis of the even BMNLS also indicates chaotic dynamics and persistence of a homoclinic structure for the full BMNLS equation, the only difference is that the waveform is asymmetric [6].

6. Conclusions

In this paper we have investigated the modulational instability as a mechanism for the development of rogue waves using simplified model equations such as the NLS and BMNLS equations. Using the integrable theory of the NLS, we obtained formulas for optimally phase modulated homoclinic orbits where the UMs are simultaneously excited leading to a significant amplification beyond the usual BF modulation instability.

The numerical results indicate that rogue waves are robust to perturbations of the evolution equations. They are shown to exist in the BMNLS and the 3D NLS equation. In fact, enhanced wave amplification is shown to occur in the chaotic regime since the solution evolves close to a "nonlinearly chirped" wave for a rather general class of initial conditions. Significantly, we find that the wavefield is chaotic before the chirped rogue waves appear, i.e. the chaos provides a mechanism for further focusing to occur due to chaotically generated optimal phase modulations.

The Mel'nikov analysis provides necessary conditions for the persistence of a homoclinic solution in the BMNLS system which is $O(\varepsilon)$–close to the NLS homoclinic solution formed by nonlinear chirping. The Melnikov analysis determines the distinguishing spatial features of the perturbed dynamics which agree with the numerical observations of high amplitude rogue waves in the BMNLS chaotic regime. The correlation of the results of the Mel'nikov analysis and of the BMNLS numerical experiments indicates that in this more general setting rogue waves are still well approximated by homoclinic solutions of the NLS equation.

Acknowledgments

This work was partially supported by grant NSF-DMS0204714 of the National Science Foundation, U.S.A.

Appendix: Statistical Diagnostics

In the numerical experiments we monitor the evolution of the third and fourth statistical moments of the (probability density function of the wave amplitude) wavefield

$$M_3(a) = \sum_{j=1}^{N} \frac{(a_j - \bar{a})^3}{N}, \qquad M_4(a) = \sum_{j=1}^{N} \frac{(a_j - \bar{a})^4}{N},$$

where $\bar{a}$ is the mean height of the wavefield, and N is the number of data points sampled. The skewness, m_3, and the kurtosis, m_4, of the wavefield are then given by

$$m_3(a) = \frac{M_3}{\sigma^3}, \qquad m_4(a) = \frac{M_4}{\sigma^4},$$

where σ is the standard deviation.

Skewness is a measure of the vertical asymmetry of the wavefield. Positive values indicate the wavefield is skewed above average height, i.e. the crests are bigger than the troughs. Negative values indicate that the wavefield is skewed below average height.

The kurtosis is a measure of whether the distribution for the wavefield is peaked or flat, relative to a Gaussian distribution and defines the contribution of large waves to the wavefield. Wavefields with high kurtosis tend to have a distinct peak near the mean, decline rapidly, and have heavy tails. That is the larger the value of the kurtosis, the large/wider the tails of the pdf. The kurtosis for a Gaussian distribution is three. For this reason, excess kurtosis much above three indicates that the contribution of large waves is significant and corresponds to a higher probability of a rogue wave event.

References

[1] M.J. Ablowitz, J. Hammack, D. Henderson, and C.M. Schober, *Phys. Rev. Lett.* **84** (2000) 887–890.

[2] M.J. Ablowitz, J. Hammack, D. Henderson, and C.M. Schober, *Physica **D*** **152-153**, 416 (2001).

[3] A. Calini, N.M. Ercolani, D.W. McLaughlin, and C.M. Schober, *Physica **D*** **89**,227–260 (1996).

[4] A. Calini and C.M. Schober, *J. Math. and Comp. Sim.* **55**, 351–364 (2001).

[5] A. Calini and C.M. Schober, *Phys. Lett.* ***A*** **298**, 335 (2002).

[6] A. Calini and C.M. Schober, *preprint* submitted *Physica* **D**, 2004.

[7] K. Trulsen and K.B. Dysthe, *Wave Motion* **24**, 281 (1996).

[8] N.M. Ercolani and D.W. McLaughlin, Toward a topological classification of integrable PDE's. *MSRI Proc. Workshop on Symplectic Geometry,* Eds. R. Devaney, H. Flaschka, W. Meyer and T. Ratiu (1990).

[9] K.L. Henderson, D.H. Peregrine and J.W. Dold, *Wave Motion* **29**, 341 (1999).

[10] Y. Li and D.W. McLaughlin, *Comm. Math. Phys.* **612**, 175–214 (1994).

[11] News realease of the European Space Agency, http://www.esa.int, July 21 2004.

[12] Rogue waves 2000, Eds. M. Olagnon and G. Athanassoulis, *Ifremer* **32**, (2001).

[13] M. Onorato, A. Osborne, M. Serio, S. Bertone, *Phys. Rev. Lett.* **86**, 5831 (2001).

[14] A. Osborne, M. Onorato, M Serio, *Phys Lett* ***A*** **275**, 386 (2000).

[15] E. Pelinovsky, T. Talipova and C. Kharif, *Physica **D*** **147**, 83 (2000).

[16] P.G. Saffman and H.C. Yuen, *Phys. Fluids* **21**, 8 (1978).

[17] Trulsen, K., Dysthe, K, Freak Waves –A three-dimensional wave simulation. *Proc. 21st Symp. on Naval Hydrodynamics*, National Academy Press, 550-560 (1997).

[18] C. Zeng, *Comm. Pure Appl. Math.* **53**, 1222–1283 (2000).

CHAPTER 12

HETEROGENEOUS PACEMAKERS IN OSCILLATORY MEDIA

Michael Stich[†*] and Alexander S. Mikhailov[*]

[†] *Instituto Pluridisciplinar, Universidad Complutense de Madrid, Paseo Juan XXIII 1, 28040 Madrid, Spain*

[*] *Abt. Physikalische Chemie, Fritz-Haber-Institut der Max-Planck-Gesellschaft, Faradayweg 4-6, 14195 Berlin, Germany*

We discuss analytical and numerical results related to target patterns produced by heterogeneous pacemakers in oscillatory reaction-diffusion systems. The model consists of a complex Ginzburg-Landau equation where the oscillation frequency is space-dependent. Analytical calculations are performed in the framework of the phase dynamics approximation. We present results on stable inward and outward travelling target waves in one and two space dimensions. Furthermore, we consider large heterogeneities, positive and negative wave dispersion, localized wave sink patterns, and the case where phase slips develop.

1. Introduction

Reaction-diffusion systems can display a rich variety of complex regimes, including oscillations, wave propagation, and spatiotemporal chaos [1]. One of the most prominent wave patterns is the target pattern, which consists of concentric waves that are periodically emitted from a small central region, called pacemaker [2]. Target patterns have been observed in many chemical, physical, and biological systems [3–6].

Stable pacemakers and target patterns may already be found in uniform oscillatory media, as a result of the intrinsic nonlinear dynamics [7,8]. Nonetheless, the great majority of target patterns observed in chemical reaction-diffusion systems are associated with the presence of a heterogeneity, e.g. a dust particle, that locally modifies the properties of the medium and gives rise to a *heterogeneous pacemaker*. Although such patterns have been investigated in a number of studies [9–15], target patterns have received much less attention than, e.g, rotating spiral waves.

As a theoretical model to study heterogeneous pacemakers in oscillatory reaction-diffusion systems, we choose the complex Ginzburg-Landau equation. The medium is nonuniform, possessing a localized region where the oscillation frequency is modified. This area may form a pacemaker which generates a spatially extended wave pattern. In a previous publication, we focused on such pacemakers and their wave patterns in one-dimensional media [16]. In addition, localized wave sink patterns were considered. In this article, we review the results presented there and extend the discussion to large heterogeneities and two-dimensional media.

2. The Complex Ginzburg-Landau Equation

Reaction-diffusion systems can display different types of oscillatory dynamics. However, in the vicinity of a supercritical Hopf bifurcation, all such systems are described by the complex Ginzburg-Landau equation (CGLE) [15,17–19]. The equation reads

$$\partial_t A = (1 - \mathrm{i}\omega)A - (1 + \mathrm{i}\alpha)|A|^2 A + (1 + \mathrm{i}\beta)\nabla^2 A, \tag{1}$$

where A is the complex oscillation amplitude, ω is the linear frequency parameter, α is the nonlinear frequency parameter, and β is the linear dispersion coefficient. The frequency of stable uniform oscillations $A(t) = \rho_0 \exp(-\mathrm{i}\omega_0 t)$ in this system is $\omega_0 = \omega + \alpha$ and their amplitude is $\rho_0 = 1$. The parameter ω is inversely proportional to the distance from the Hopf bifurcation. Therefore, close to the bifurcation point $\omega \gg 1$ is satisfied. Although ω can be scaled out of the CGLE, we keep it to preserve the qualitative comparability with reaction-diffusion systems. The parameters α and β are generally of order unity.

We assume that the frequency ω is changed by an amount $\Delta\omega$ within a small region of radius R (located in the center of the medium), so that

$$\omega(x) = \begin{cases} \omega & \text{for} \quad |x| > R \\ \omega + \Delta\omega & \text{for} \quad |x| \le R. \end{cases} \tag{2}$$

This region may become either a source or a sink of travelling waves.

Target patterns are two-dimensional patterns. Although no real reaction-diffusion system is strictly two-dimensional, the extension of the system in the third dimension is usually small and not essential to the pattern-forming processes. Therefore, it can be neglected in most cases. In terms of spatial polar coordinates r and θ, target waves are described by

$$A(r,t) = \rho_{\mathrm{tp}}(r) \exp(\mathrm{i}\psi_{\mathrm{tp}}(r) - \mathrm{i}\omega_k t), \tag{3}$$

where ρ_{tp} is the amplitude, ω_k the frequency, and $\psi_{\mathrm{tp}}(r)$ the phase of the target waves. Target waves do not depend on θ. Far from the center of the pattern, the direct influence of the wave source and the curvature effects become small and the target waves can approximately be described as one-dimensional plane waves, given by

$$A(x,t) = \sqrt{1-k^2}\exp(\mathrm{i}kx - \mathrm{i}\omega_k t), \tag{4a}$$

$$\omega_k = \omega_0 + (\beta - \alpha)k^2. \tag{4b}$$

If $\beta - \alpha > 0$, the frequency increases with the wavenumber and the waves have *positive* dispersion. In the opposite case, wave dispersion is *negative*.

As will be demonstrated below, pattern formation in this system is determined by the heterogeneity. Since moreover the size L of the medium is assumed to be significantly larger than the wavelength of the pattern and the size of the heterogeneity, the impact of the boundaries can be neglected in first approximation. The boundary conditions become important if no heterogeneity is present [20,21] or if wave instabilities occur [22,23].

3. Pacemakers in the Phase Dynamics Approximation

Introducing phase φ and real amplitude ρ as $A = \rho\exp(-\mathrm{i}\varphi)$ and substituting this into the CGLE, we obtain

$$\partial_t\rho = (1-\rho^2)\rho + \nabla^2\rho - \rho(\nabla\varphi)^2 + \beta\rho\nabla^2\varphi + 2\beta\nabla\varphi\nabla\rho, \tag{5a}$$

$$\partial_t\varphi = \omega(x) + \alpha\rho^2 + (2/\rho)\nabla\rho\nabla\varphi + \nabla^2\varphi - (\beta/\rho)\nabla^2\rho + \beta(\nabla\varphi)^2. \tag{5b}$$

Wave patterns can be described using the phase dynamics approximation if phase perturbations are smooth, i.e. if they have large characteristic lengths [15]. Then, the amplitude ρ follows adiabatically the dynamics of the phase, so that $\rho^2 \approx 1 - (\nabla\varphi)^2 + \beta\nabla^2\varphi$ in first approximation. Consequently, Eqs. (5) are reduced to a single equation for the phase,

$$\partial_t\varphi = \omega(x) + \alpha + (\beta-\alpha)(\nabla\varphi)^2 + (1+\alpha\beta)\nabla^2\varphi. \tag{6}$$

We assume that uniform oscillations are modulationally stable, i.e. that the Benjamin-Feir-Newell condition $1 + \alpha\beta > 0$ is satisfied.

After applying the Cole-Hopf transformation $\varphi = (1+\alpha\beta)\ln Q/(\beta-\alpha)$ to Eq. (6) (requiring $\alpha \neq \beta$), the new variable Q obeys the linear equation

$$\partial_t Q = \frac{\beta-\alpha}{1+\alpha\beta}(\omega(x)+\alpha)Q + (1+\alpha\beta)\nabla^2 Q, \tag{7}$$

which formally is equivalent to a Schrödinger equation for a quantum particle in a potential well with width $2R$ and depth $d = (\beta-\alpha)\Delta\omega/(1+\alpha\beta)^2$.

The next step is to derive expressions for the frequency ω_k and the wavenumber k of the waves. These quantities correspond to the largest eigenvalue and its eigenfunction associated with the operator appearing in Eq. (7). For a detailed discussion, we refer the reader to [15,16].

The solution of Eq. (7) depends on the dimensionality of the system. In one- and two-dimensional media, a heterogeneity acts as a wave source or *pacemaker*, creating an extended target pattern if and only if the condition

$$(\beta - \alpha)\Delta\omega > 0 \tag{8}$$

is fulfilled. If Eq. (8) does not hold, i.e. if

$$(\beta - \alpha)\Delta\omega < 0, \tag{9}$$

a heterogeneity cannot lead to an extended target wave pattern and is called a *wave sink*. A wave sink does not entrain the system and creates "only" a localized wave pattern. In terms of the interpretation as a Schrödinger equation, a pacemaker corresponds to a potential well, while a wave sink corresponds to a potential barrier.

The direction of propagation of the (extended or localized) waves is determined by the sign of the local frequency shift. For $\Delta\omega > 0$, the wave is moving outward, for $\Delta\omega < 0$ inward.

4. Stable, Extended and Localized Wave Patterns

4.1. *Positive Wave Dispersion*

If $\beta - \alpha > 0$ and when the oscillation frequency is locally *increased* ($\Delta\omega > 0$), the pacemaker condition (8) is fulfilled and a target pattern of outward propagating waves is formed. In one-dimensional media, the wavenumber k of generated waves is determined by the equation

$$k = \sqrt{k_{\max}^2 - k^2}\,\tan\left(R\frac{\beta - \alpha}{1 + \alpha\beta}\sqrt{k_{\max}^2 - k^2}\right), \tag{10}$$

where

$$k_{\max} = \sqrt{\Delta\omega/(\beta - \alpha)} \tag{11}$$

is the maximum value of the wavenumber (as $R \to \infty$). The pacemaker frequency is given by Eq. (4b). In Figure 1, the wavenumber and frequency of a pacemaker are shown as a function of the radius of the heterogeneity. As a result of the wave propagation, the medium becomes entrained by the wave source and the effective oscillation frequency approaches the same value ω_k everywhere in the system. An example is shown in Figure 2(a).

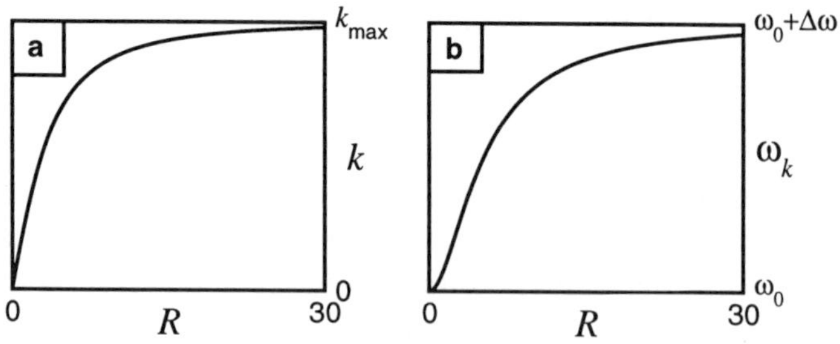

Fig. 1. Wavenumber (a) and frequency (b) of a 1d-pacemaker. The parameters are $\alpha = 0.5$, $\beta = 1.0$, $\Delta\omega = 0.2$. The curves do not depend on ω.

When the oscillation frequency is locally *decreased* ($\Delta\omega < 0$), the heterogeneity represents a wave sink and creates a localized pattern [Figure 2(b)]. Waves initiate near the core boundary and propagate inward. The frequency of the oscillations inside the wave sink adjusts to the value of uniform oscillations ω_0, giving rise to a pattern where a constant phase shift between the center of the wave sink and the rest of the medium is established. Hence,

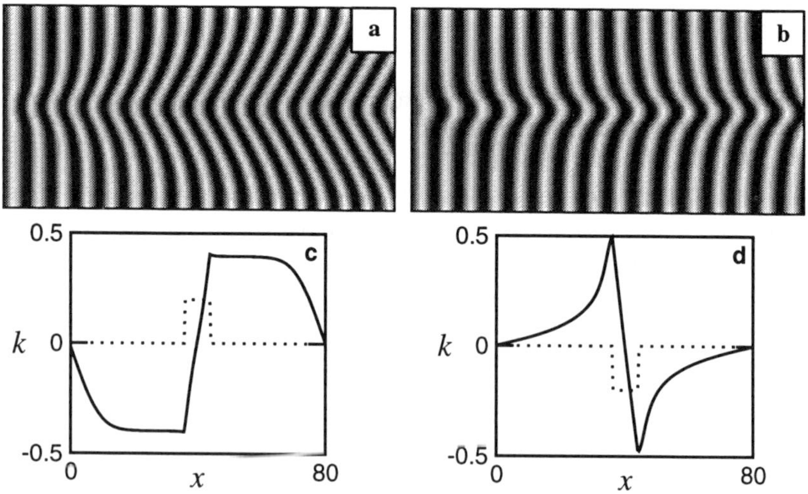

Fig. 2. Wave patterns for positive dispersion. The heterogeneity ($R = 4$) is placed in the center of the medium ($L = 80$). (a,b) Space-time diagrams of ReA (in gray scale, black corresponding to -1.1 and white to $+1.1$) are shown for time intervals $\Delta t = 200$. In all space-time diagrams, space is displayed along the vertical axis and time along the horizontal axis. The parameters are $\omega = 0, \alpha = 0.5, \beta = 1.0$. (a) Pacemaker; $\Delta\omega = 0.2$. (b) Wave sink; $\Delta\omega = -0.2$. (c,d) Spatial distributions of the local wavenumber for (a,b). In (c,d), the local frequency shifts $\Delta\omega$ are displayed as dotted lines.

the heterogeneity is entrained by the uniform oscillations.

Figure 2(c,d) display the spatial distributions of the local wavenumber, defined as $k = -\nabla\varphi$ and evaluated numerically for the patterns shown in Figures 2(a,b). For a pacemaker, the wavenumber is constant outside the core (except at the no-flux boundary, where it approaches zero). For a wave sink, the local wavenumber rapidly falls down to zero, indicating that no travelling waves are present. Since the value of k is determined numerically, it has different signs on both sides of the heterogeneity, and different signs for outward and inward travelling waves on the same side of the core.

4.2. *Negative Wave Dispersion*

If $\beta - \alpha < 0$ and the oscillation frequency is locally *increased* ($\Delta\omega > 0$), the heterogeneity creates a localized wave pattern with frequency ω_0 [Figure 3(a)]. Waves are initiated inside the core and propagate outward,

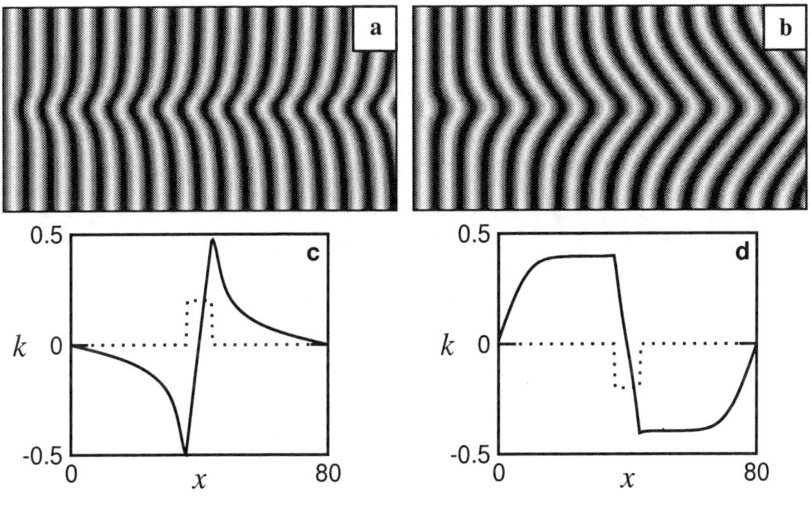

Fig. 3. Wave patterns for negative dispersion. The system is as in Figure 2. The parameters are $\omega = 1, \alpha = -0.5, \beta = -1.0$. (a,c) Wave sink; $\Delta\omega = 0.2$. (b,d) Pacemaker; $\Delta\omega = -0.2$.

but decay close to the core boundary. The heterogeneity represents a wave sink.

If the local oscillation frequency is locally *decreased* ($\Delta\omega < 0$), an extended wave pattern is formed [Figure 3(b)]. Outside the core, the medium is filled with propagating waves. The wavenumber k is again given by

Eq. (10) where $(\beta - \alpha)$ is replaced by $(\alpha - \beta)$ (to assure positive k). The frequency ω_k of generated waves is given by Eq. (4b). Note that ω_k is smaller than the frequency ω_0 of uniform oscillations. Despite the fact that waves are moving inward, the heterogeneity entrains the medium and thus represents a pacemaker. Figure 3(c,d) show the corresponding spatial distributions of the local wavenumber.

4.3. *Large Cores*

For a wave sink, a constant phase shift is present between the pattern in the center of the sink and the uniform oscillations outside. As R increases, the phase shift becomes larger and the propagation of waves *inside* the core becomes possible.

In Figure 4, simulations for the four different wave patterns distinguished in context with Figures 2 and 3 are displayed. While uniform oscillations take place in the cores of the pacemakers [Figures 4(a,d)], wave propagation is observed inside the cores of the wave sinks [Figures 4(b,c)]. This means that inward travelling target patterns may also be observed in systems with positive dispersion if large heterogeneities are present [Fig-

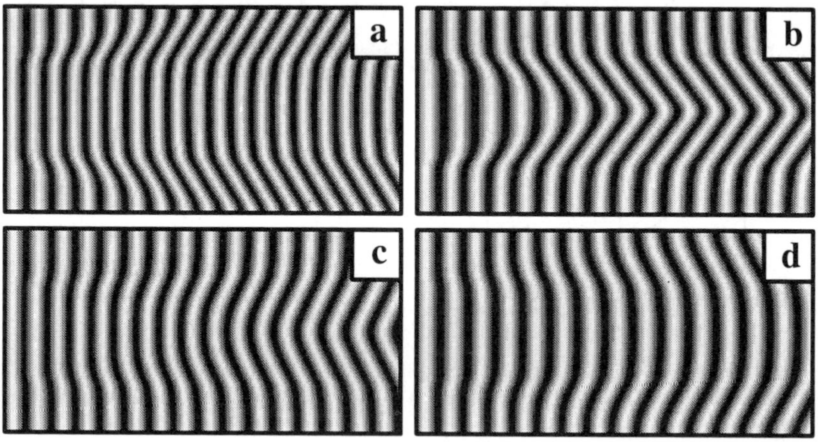

Fig. 4. Wave patterns for large heterogeneities. Space-time diagrams of ReA are shown for time intervals $\Delta t = 100$, system size $L = 80$, and $R = 20$. (a) Uniform oscillations inside the heterogeneity (positive dispersion); $\Delta\omega = 0.2$, $\omega = 2$, $\alpha = -1$, $\beta = 0$. (b) Inward travelling waves inside the heterogeneity (positive dispersion); $\Delta\omega = -0.2$, $\omega = 2$, $\alpha = -1$, $\beta = 0$. (c) Outward travelling waves inside the heterogeneity (negative dispersion); $\Delta\omega = 0.2$, $\omega = 0$, $\alpha = 1$, $\beta = 0$. (d) Uniform oscillations inside the heterogeneity (negative dispersion); $\Delta\omega = -0.2$, $\omega = 0$, $\alpha = 1$, $\beta = 0$.

ure 4(b)]. Accordingly, outward travelling waves may also be seen in systems with negative dispersion [Figure 4(c)].

4.4. *Two-Dimensional Waves*

In two spatial dimensions, the eigenfunctions of the potential well are different from the case of one dimension and involve Bessel functions of first kind, zeroth order, J_0 (inside the well) and modified Bessel functions of second kind, zeroth order, K_0 (outside the well). The equation, from which $k(R)$ can be determined numerically, is given by

$$u\frac{J_1(u)}{J_0(u)} = v\frac{K_1(v)}{K_0(v)}, \tag{12}$$

with $u = R\frac{\beta-\alpha}{1+\alpha\beta}\sqrt{k_{\max}^2 - k^2}$ and $v = R\frac{\beta-\alpha}{1+\alpha\beta}k$. The qualitative behavior of $k(R)$ is similar to the one-dimensional case, i.e. a heterogeneity serves as a pacemaker even for small $\Delta\omega$ [24]. In Figure 5 we show a snapshot of a two-dimensional simulation in a medium with negative dispersion displaying a pacemaker emitting stable target waves and a wave sink modulating locally the wave field. The pacemaker is located in the center of the target pattern, while the sink is close to the right lower corner. The wave sink slightly compresses the waves locally and the amplitude decreases which can be seen in Figure 5(b) as a dark spot. Note that the wave field remains continuous, i.e. no phase slips occur here.

5. Unstable Wave Patterns and Phase Slips

The phase dynamics approximation is valid only when phase gradients, i.e. (local) wavenumbers, remain sufficiently small. Since the maximal possible wavenumber is given by $k_{\max} = \sqrt{\Delta\omega/(\beta-\alpha)}$ (Eq. (11)), the analytical results presented above hold only for pacemakers and sinks with small frequency shifts, i.e. $|\Delta\omega| \ll 1$. As $\Delta\omega$ is increased, the wavenumber of emitted waves increases and pattern instabilities, which involve the generation of phase slips, develop. These phenomena cannot be described in the framework of the phase dynamics approximation.

5.1. *One-Dimensional Media*

Plane waves in the CGLE become unstable with respect to the Eckhaus instability if their wavenumber exceeds the threshold given by

$$k_{\mathrm{EI}} = \sqrt{\frac{1+\alpha\beta}{2(1+\alpha^2)+1+\alpha\beta}}. \tag{13}$$

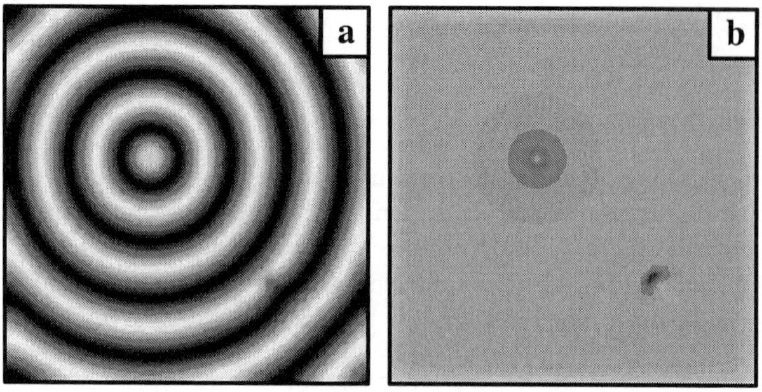

Fig. 5. Two-dimensional pacemaker with a wave sink. (a) shows an image of ReA, (b) an image of the $|A|$ at $t = 500$. The dark spot in (b) denotes the wave sink, where $|A|$ is decreased to its minimum value 0.67 (black does not denote zero here). The parameters are: $\omega = 0$, $\alpha = 1$, and $\beta = 0$. The size of the medium is $L_x = L_y = 100$. The pacemaker is characterized by $\Delta\omega = -0.6$ and $R = 1.6$, and the sink by $\Delta\omega = 0.7$ and $R = 1.6$. In 2d-simulations, heterogeneities have a square shape with half-width R.

If the instability is subcritical, long-wavelength perturbations do not saturate. Then, the wave is compressed until locally the amplitude drops down to zero (defect) at some time moment corresponding to a phase singularity [25, 26]. Such an event is associated with a phase slip of 2π, the disappearance of one wave from the wave train and a subsequent readjustment of the wavenumber on both sides of the defect. In two spatial dimensions, the Eckhaus instability may lead to spiral breakup [23].

An example for the formation of the phase slips as a result of the Eckhaus instability is shown in Figure 6(a,b) for a medium with positive dispersion. The initial condition consists of a pacemaker emitting stable waves. Then, the frequency difference $\Delta\omega$ in the core region is increased above the threshold. The travelling wave is now unstable and collapses finally, associated with a phase slip. After a short transient, a wave pattern with periodically generated phase slips is established. The wavenumber in the far region is small and lies below the Eckhaus threshold while near the core a wave pattern with a large k is established. In general terms, the oscillations inside a certain area around the core are *desynchronized* from the rest of the medium.

Eckhaus-unstable waves are also possible for inward propagating waves ($\beta - \alpha < 0$, $\Delta\omega < 0$), as shown in Figure 6(c,d). Again, phase slips occur at a finite distance from the core boundary and the wavenumber k is decreased far from the pacemaker. However, since the medium has negative dispersion,

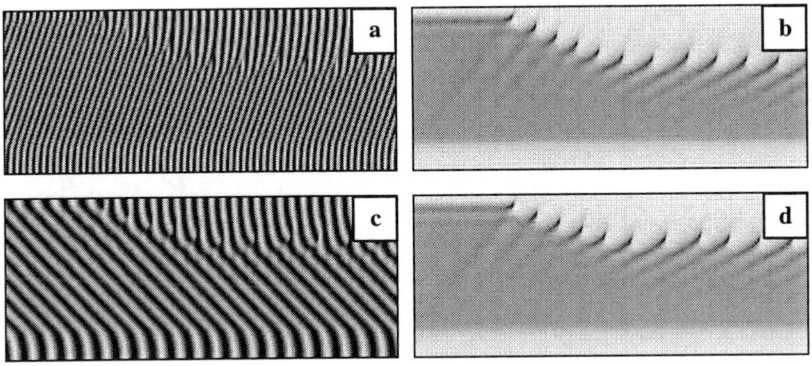

Fig. 6. Formation of phase slips through the Eckhaus instability in media with positive (a,b) and negative dispersion (c,d). Space-time diagrams of ReA (a,c) and $|A|$ (b,d) are shown. A linear gray scale color map is chosen where the minimum refers to black and the maximum to white. For $|A|$, black corresponds to zero, and white to $|A| \approx 1.1$. Only half of the medium ($L = 160$) is displayed, with the core located in the lower part of the figure. (a,b) The parameters are $\Delta\omega = 0.207$, $\omega = 0$, $\alpha = 0.55$, $\beta = 1.0$, $R = 14.8$, $\Delta t = 500$ (for $\Delta\omega = 0.2069$ no phase slips develop). (c,d) The parameters are $\Delta\omega = -0.2072$, $R = 14.8$, $\omega = 1$, $\alpha = -0.55$, $\beta = -1.0$, $\Delta t = 250$.

the frequency ω_k is smaller close to the pacemaker than in the far field. Note that the modulations of the amplitude, which finally lead to the phase slips, are directed outward [Figure 6(d)] like in the case of positive dispersion.

For each pacemaker with radius R and frequency shift $\Delta\omega$, it is possible to compute k and to check whether it is smaller or larger than the Eckhaus wavenumber k_{EI}. For a more general view, however, k_{EI} can be compared with the maximal wavenumber $k_{\max}$ that a core with fixed $\Delta\omega$ may emit. The results are displayed in Figures 7(a,b).

Figure 7(a) shows for the case $\Delta\omega > 0$ in which part of the parameter plane spanned by α and β outgoing target patterns (I,II) and in which part outgoing localized waves (III) are present. The regions (II) and (III) are separated by the line of vanishing dispersion (VD) $\beta = \alpha$. The Benjamin-Feir-Newell (BFN) lines $\beta = -1/\alpha$ indicate where uniform oscillations become unstable. Hence, target waves exist in the region between the VD line and the upper BFN line. However, the Eckhaus instability restricts the parameter region, where target waves are stable, to the area (I) between the dotted lines, where $k_{\max} < k_{\mathrm{EI}}$. In region (III), no extended waves are present, so that the Eckhaus instability criterion, derived for plane waves, is not applicable. However, other desynchronization phenomena are possible, as shown below.

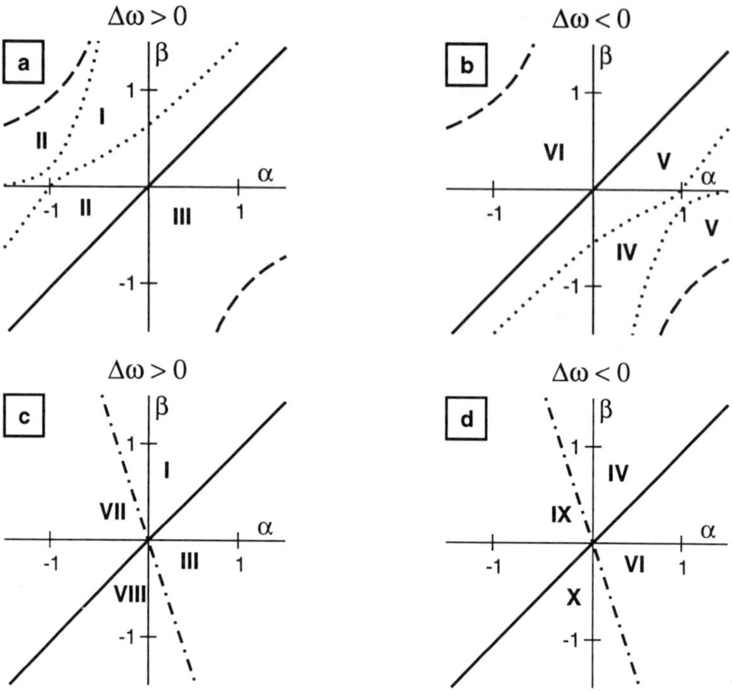

Fig. 7. Parameter space (α, β) for $\Delta\omega > 0$ (a,c) and $\Delta\omega < 0$ (b,d). In (a) and (b), $\omega \gg 1$ is assumed. In (c) and (d), ω is scaled out. The following lines are plotted: VD line ($\beta = \alpha$, solid line), BFN line ($\beta = -1/\alpha$, dashed line), the lines where $k_{\mathrm{EI}} = k_{\mathrm{max}}$ (plotted for $\Delta\omega = 0.2$, dotted lines), and the line of vanishing phase velocity ($\beta = \alpha(1-k^{-2})$, plotted for $k = 0.5$, dot-dashed line). More information in the text.

In Figure 7(b), the respective cases are displayed for $\Delta\omega < 0$, i.e. ingoing target patterns correspond to (IV,V), ingoing localized waves to (VI), and between the two dotted lines (IV), target waves are stable with respect to the Eckhaus instability.

We can conclude from Figures 7(a,b) that the Eckhaus instability restricts significantly the parameter region where stable target patterns can be expected. Furthermore, the figure illustrates a special symmetry of the system, i.e. the pacemaker condition and the BFN and EI criteria are invariant with respect to simultaneous changes of the signs of $\Delta\omega$, α, and β.

Figures 7(c,d) show how the direction of propagation of target waves is affected if the parameter ω is scaled out of the CGLE, $A \to A\exp(\mathrm{i}\omega t)$. In this coordinate frame, inward travelling target waves are possible even for positive dispersion and positive frequency shift. Although this is not observed in reaction-diffusion systems close to the Hopf bifurcation where ω is large, this case is presented here since much work is devoted to the rescaled CGLE in general. In order to keep the figure simple, we do not discuss the Eckhaus instability since that has been done above.

Figure 7(c) shows for the case $\Delta\omega > 0$ in which parts of parameter space target patterns (I,VII) and wave sinks (III,VIII) are present. In (I), the usual outgoing target patterns for positive dispersion are found while in (VII) target waves move inward for positive dispersion. The regions (I) and (VII) are separated by the line of vanishing phase velocity $v_{\rm p} = \omega_k/k$, here drawn for waves with wavenumber $k = 0.5$. For $k \to 1$, this line approaches $\beta = 0$, for $k \to 0$, its asymptotics is $\alpha = 0$. Although this line is valid for plane waves, it also describes qualitatively the change of wave direction of localized waves around a wave sink.

If the frequency shift $\Delta\omega$ in the core is further increased, the location where phase slips occur moves closer to the core. Then, the formation of phase slips cannot be interpreted as a result of an Eckhaus instability for plane waves, but as a *local* desynchronization phenomenon. For examples of such patterns in one-dimensional media and of the formation of phase slips at wave sinks, we refer the reader to [16].

5.2. *Two-Dimensional Media*

In two spatial dimensions, target pattern formation proceeds in a similar way as in one dimension and only selected examples are shown for this case.

In the simulation displayed in Figure 8, the interaction of target waves with a large wave sink is studied. It can be seen that a large wave sink (with sufficiently large $\Delta\omega$) is able to break the waves that are emitted by a pacemaker. These broken waves have open ends which, in the absence of the pacemaker, would actually curl in and form a spiral (not shown here). The open ends constitute amplitude defects that can be easily detected in Figure 8(b) as dark spots. Since the pacemaker periodically emits waves, the broken waves move outward and annihilate at the no-flux boundary.

As the magnitude of $\Delta\omega$ increases, phase slips are expected to develop in the wave pattern. A simulation confirming this conjecture is shown in Figure 9. There, a strongly negative frequency shift in a medium with negative dispersion leads to the emission of unstable, inward travelling target waves. In analogy with the one-dimensional case, phase slips are observed at a certain distance from the wave source. A phase slip is associated with

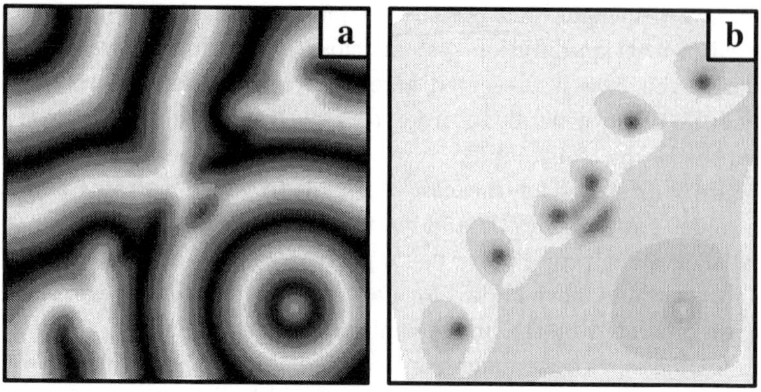

Fig. 8. Two-dimensional pacemaker generating open wave ends at a wave sink. (a) shows an image of ReA, (b) an image of $|A|$ at $t = 500$. The pacemaker ($R = 3.2$, $\Delta\omega = 0.6$) is located near the lower right corner while the sink ($R = 7.2$, $\Delta\omega = -0.6$) is located in the center. The dark spots in (b) denote open wave ends where the amplitude is vanishing (defects). The parameters are $\omega = 0$, $\alpha = -1$, $\beta = 0$, $L_x = L_y = 100$.

an amplitude defect which has a circular geometry, as seen in Figure 9(b). In addition, phase slips also occur at the boundary of the pacemaker, which may be due to the strong curvature of the waves in the center of the pattern. Both types of phase slips are also seen in the space-time diagram for a cross section through the center of the system [Figure 9(c)].

6. Discussion

Our study demonstrates that a large variety of patterns can be produced in one- and two-dimensional oscillatory media by introducing a heterogeneity, i.e. a localized region where the oscillation frequency is modified. Depending on the parameters of the medium, which determine the dispersion of waves, and the properties of the heterogeneity, such as size, sign, and magnitude of the frequency shift, target patterns with outward and inward propagating waves or localized wave sink patterns can be observed.

We emphasize the possibility of inward travelling waves, which have been regarded as a curiosity until experiments demonstrated that they may be observed in real reaction-diffusion systems [27, 28]. Our study shows that they can be found either in a medium with negative dispersion and $\Delta\omega < 0$ or in a medium with positive dispersion and $\Delta\omega < 0$ where the heterogeneity is large. If the medium is not close to the Hopf bifurcation, inward travelling waves may be found even for positive dispersion and $\Delta\omega > 0$. Furthermore, we have shown that a heterogeneous wave sink, a pattern

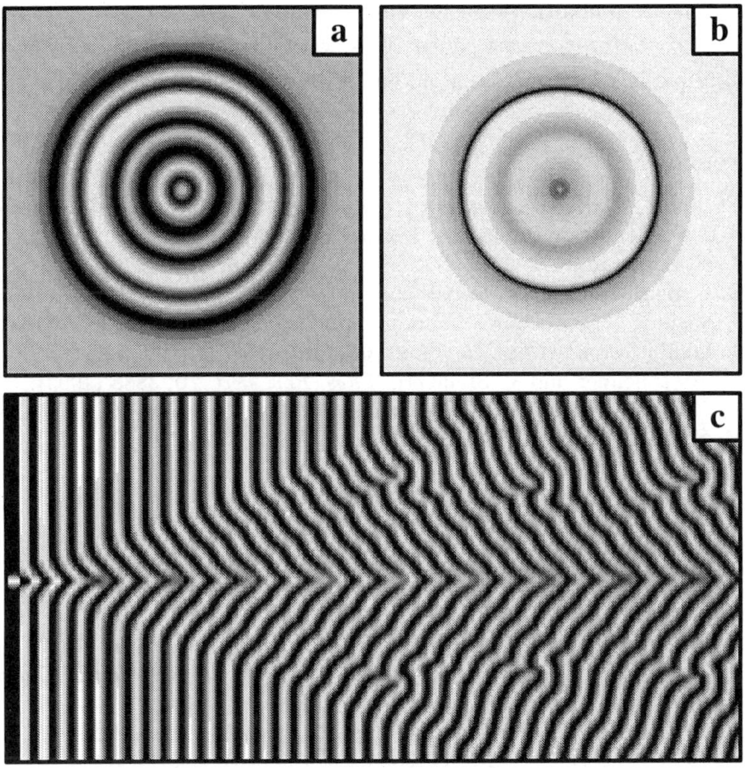

Fig. 9. Two-dimensional pacemaker with unstable target waves in a medium with negative dispersion. (a) shows an image of ReA, (b) an image of $|A|$ at $t = 131$. The black line in (b) denotes a circular defect line. A strong decrease of $|A|$ is also present close to the core. (c) shows a space-time diagram of ReA for a cross section through the center of the system parallel to the x-axis for a time interval $\Delta t = 250$. The parameters are $\omega = 0$, $\alpha = 1$, $\beta = 0$, $\Delta\omega = -0.7$, $R = 1.6$, and $L_x = L_y = 150$.

which has been poorly investigated, can lead to an interesting dynamic behavior, when it is strong (by breaking the waves) or large (by resembling a pacemaker).

The formation of phase slips is observed, when the shift of the oscillation frequency in the core is increased. Phase slips occur if the medium is no longer able to compensate the frequency shift through the propagation of waves: desynchronization takes place and oscillations in the region close to the core become decoupled from the rest of the medium.

Our study is based on a general model for oscillatory media. Therefore, we expect that the results qualitatively hold for many systems, even if the oscillations are relaxational rather than harmonic. Given the rich dynamics produced by heterogeneous pacemakers, we believe that they can be used as building block for controlling and engineering reaction-diffusion systems.

References

[1] *Chemical Waves and Patterns*, edited by R. Kapral and K. Showalter (Kluwer Academic Publishers, Dordrecht, 1995).
[2] A. N. Zaikin and A. M. Zhabotinsky, *Nature* **255**, 535 (1970).
[3] G. Gerisch, *Roux Arch. Entwicklungsmech. Organismen* **116**, 127 (1965).
[4] S. Nasuno, M. Sano, and Y. Sawada, *J. Phys. Soc. Jpn.* **58**, 1875 (1989).
[5] S. Jakubith *et al.*, *Phys. Rev. Lett.* **65**, 3013 (1990).
[6] M. Assenheimer and V. Steinberg, *Phys. Rev. Lett.* **70**, 3888 (1993).
[7] M. Stich, M. Ipsen, and A. S. Mikhailov, *Phys. Rev. Lett.* **86**, 4406 (2001).
[8] M. Stich, M. Ipsen, and A. S. Mikhailov, *Physica* **D 171**, 19 (2002).
[9] N. Kopell, *Adv. Appl. Math.* **2**, 389 (1981).
[10] P. S. Hagan, *Adv. Appl. Math.* **2**, 400 (1981).
[11] J. J. Tyson and P. C. Fife, *J. Chem. Phys.* **73**, 2224 (1980).
[12] A. E. Bugrim, M. Dolnik, A. M. Zhabotinsky, and I. R. Epstein, *J. Phys. Chem.* **100**, 19017 (1996).
[13] H. Mahara *et al.*, *J. Phys. Soc. Jpn.* **69**, 3552 (2000).
[14] M. Hendrey, K. Nam, P. Guzdar, and E. Ott, *Phys. Rev.* **E 62**, 7627 (2000).
[15] Y. Kuramoto, *Chemical Oscillations, Waves, and Turbulence* (Springer, Berlin, 1984).
[16] M. Stich and A. S. Mikhailov, *Z. Phys. Chem.* **216**, 521 (2002).
[17] M. Ipsen, F. Hynne, and P. G. Sorensen, *Chaos* **8**, 834 (1998).
[18] M. Ipsen, L. Kramer, and P. G. Sorensen, *Phys. Rep.* **337**, 193 (2000).
[19] I. S. Aranson and L. Kramer, *Rev. Mod. Phys.* **74**, 99 (2002).
[20] M. Golubitsky, E. Knobloch, and I. Stewart, *J. Nonlin. Sci.* **10**, 333 (2000).
[21] J. A. Sherratt, *SIAM J. Appl. Math.* **63**, 1520 (2003).
[22] B. Sandstede and A. Scheel, *Physica* **D 145**, 233 (2000).
[23] M. Bär and L. Brusch, N. *J. Phys.* **6**, 5 (2004).
[24] B. Simon, *Ann. Phys.* **97**, 279 (1976).
[25] B. Janiaud *et al.*, *Physica* **D 55**, 269 (1992).
[26] J. Eckmann, T. Gallay, and C. E. Wayne, *Nonlinearity* **8**, 943 (1995).
[27] V. K. Vanag and I. R. Epstein, *Science* **294**, 835 (2001).
[28] J. Wolff, M. Stich, C. Beta, and H. H. Rotermund, *J. Phys. Chem.* **B** (2004), in press.

PART III

Modelling and Characterization of Spatio-Temporal Complexity

CHAPTER 13

A FINITE-DIMENSIONAL MECHANISM RESPONSIBLE FOR BURSTS IN FLUID MECHANICS

Edgar Knobloch

Department of Applied Mathematics, University of Leeds,
Leeds LS2 9JT, U.K.

A simple finite-dimensional mechanism is described that appears to be responsible for the presence of bursts near onset of primary instability in a number of fluid systems. The manifestation of the mechanism differs dramatically from system to system, but the mechanism is common to all of them. The geometrical ideas behind the mechanism are discussed, and the results applied to three systems, (a) binary fluid convection, (b) natural doubly diffusive convection, and (c) Faraday oscillations in an almost circular container. Where possible the results are confronted with the results of direct numerical simulations in two or three dimensions.

1. Introduction

Bursts are ubiquitous in fluid mechanics, and are both of theoretical and practical importance. In recent years, with the advent of dynamical systems techniques, it has become clear that many distinct mechanisms are capable of generating bursts. The question therefore arises as to which are commonly observed in nature. Although as posed such a question is ill-defined, the present paper may be seen as an attempt to identify a mechanism that is sufficiently simple and robust that one might expect to see it commonly in applications.

Bursts can be divided into two broad classes: fundamentally low-dimensional bursts, and high-dimensional bursts. The former form the focus of the present article; the latter occur in fully developed, perhaps turbulent flows and involve many degrees of freedom. A recent review by Knobloch and Moehlis [1, 2] divided the former further, into bursts of finite dynamic range (such as those observed in the breakdown of a turbulent boundary layer [3]) and bursts of large dynamic range such as certain bursts observed in large aspect ratio binary fluid convection [4]. In addition bursts can occur near onset or far from onset [5] but remain fundamentally low-dimensional.

Among the ideas pursued in these reviews is the notion that bursts near onset are easily generated as a result of forced symmetry-breaking [6]. In systems with symmetry such forced symmetry breaking may destroy the degeneracy between multiple symmetry-related modes and produce complex transitions among these modes in situations where, in the absence of the forced symmetry-breaking, the system would behave in a simple 'laminar' fashion. The present article focuses on different issues and there is almost no overlap with the reviews cited above. In particular symmetries in the present article play a different role, although the presence of a subcritical instability is a common theme. Our idea is simple: we seek parameter regimes in which a secondary bifurcation destabilizes the solutions that would be stable above the saddle-node bifurcation at which the subcritical branch turns around. This requires the presence of a competing mode but this mode need not be one of a family of nearly degenerate modes. If such a mode is present a *stability gap* is created near onset: at the primary bifurcation the system cannot find a simple stable state, and instead the instability may result in a transition to burst-like behavior, either periodic or chaotic. We show explicitly how such stability gaps may be created, and illustrate their consequence by direct numerical simulation of binary fluid [7] and natural doubly diffusive convection [8]. Related ideas are then applied to nearly inviscid Faraday waves in an almost circular cylinder. In each case we provide a geometrical interpretation for the presence of complex dynamics in the stability gap. Finally, in the convection systems we are able to explore the evolution of the bursts with increasing system size. In Alan Newell's memorable expression, the resulting 'wimpy' turbulence consists of temporally and spatially confined 'events', which may be viewed as spatially localized bursts.

2. Binary Fluid Convection

Cross-diffusion effects are important in binary fluid mixtures. In liquids the Soret effect dominates, and the sign of this effect determines the behavior of the mixture in response to an applied temperature gradient. For mixtures with a negative Soret coefficient the heavier component migrates towards the lower (hotter) boundary, i.e., a concentration gradient is set up that opposes the destabilizing temperature gradient that produced it. Under these conditions the onset of convection may take the form of growing oscillations, and these may lead to a variety of states with complex time-dependence in the nonlinear regime [9]. This behavior was first observed in salt-water mixtures [10] and was subsequently studied in great detail in water-ethanol mixtures.

The system is described by the nondimensional equations [11]

$$\mathbf{u}_t + (\mathbf{u}\cdot\nabla)\mathbf{u} = -\nabla P + \sigma R[\theta(1+S) - S\eta]\hat{\mathbf{z}} + \sigma\nabla^2\mathbf{u}, \tag{1}$$

$$\theta_t + (\mathbf{u}\cdot\nabla)\theta = w + \nabla^2\theta, \tag{2}$$

$$\eta_t + (\mathbf{u}\cdot\nabla)\eta = \tau\nabla^2\eta + \nabla^2\theta, \tag{3}$$

together with the incompressibility condition

$$\nabla\cdot\mathbf{u} = 0, \tag{4}$$

where $\mathbf{u}$ is the velocity field, P is the pressure, and θ denotes the departure of the temperature from its conduction profile, in units of the imposed temperature difference ΔT. The variable η is defined such that its gradient represents the dimensionless mass flux. Thus $\eta \equiv \theta - C$, where C denotes the concentration of the heavier component relative to its conduction profile in units of the concentration difference that develops across the layer as a result of the Soret effect. In the following we describe solutions to these equations in a two-dimensional rectangular container $D \equiv \{x, z | 0 \le x \le \Gamma, -\frac{1}{2} \le z \le \frac{1}{2}\}$ subject to the boundary conditions

$$\mathbf{u} = \mathbf{n}\cdot\nabla\eta = 0 \text{ on } \partial D, \tag{5}$$

and

$$\theta = 0 \text{ at } z = \pm 1/2, \qquad \theta_x = 0 \text{ at } x = 0, \Gamma. \tag{6}$$

These boundary conditions approximate well the experimental setup [11]. The system is therefore specified by five dimensionless parameters: the Rayleigh number R providing a dimensionless measure of the imposed temperature difference ΔT, the separation ratio S that measures the resulting concentration contribution to the buoyancy force due to the Soret effect, and the Prandtl and Lewis numbers σ, τ, in addition to the aspect ratio Γ. In this report all calculations are performed for a $\Gamma = 10$ container, with $\sigma = 0.6$, $\tau = 0.03$, values appropriate for ^{3}He-^{4}He mixtures above the λ point [12]. The results are reported in terms of a *reduced* Rayleigh number $\varepsilon \equiv (R - R_c)/R_c$, where R_c is the critical Rayleigh number for the primary instability.

Equations (1)-(6) are equivariant with respect to the operations

$$R_x: \qquad (x, z) \to (\Gamma - x, z), \qquad (u, w, \theta, \eta) \to (-u, w, \theta, \eta), \tag{7}$$

$$\kappa: \qquad (x, z) \to (x, -z), \qquad (u, w, \theta, \eta) \to (u, -w, -\theta, -\eta), \tag{8}$$

where $\mathbf{u} \equiv (u, w)$. These two operations generate the symmetry group D_2 of a rectangle. It follows that even solutions, i.e., solutions invariant under R_x, satisfy $(u(x,z),\ w(x,z),\ \theta(x,z),\ \eta(x,z)) = (-u(\Gamma - x, z),\ w(\Gamma - x, z),\ \theta(\Gamma - x, z),\ \eta(\Gamma - x, z))$ at each instant in time, while odd solutions are invariant under κR_x and satisfy $(u(x,z), w(x,z), \theta(x,z), \eta(x,z)) = -(u(\Gamma - x, -z),\ w(\Gamma - x, -z),\ \theta(\Gamma - x, -z),\ \eta(\Gamma - x, -z))$, again at each instant of time. Primary bifurcations from the conduction state $u = w = \theta = \eta = 0$ necessarily produce states of one or other type [13, 14]. In the following we refer to these as *chevron* states since they consist of rolls that travel out from the center of the container towards the lateral walls where they disappear. In an even chevron the outgoing waves are in phase; in an odd chevron they are exactly out of phase. A *blinking* state results when these states undergo a symmetry-breaking Hopf bifurcation.

When $S = -0.1$ convection sets in at $R_c = 1972.13$ via a Hopf bifurcation with frequency $\omega_c = 4.918$. This bifurcation creates an even chevron state but this state is unstable since the bifurcation is subcritical, and the first stable nonlinear state instead takes the form of a burst-like state called a *repeated transient* (Figure 1). States of this type were studied in detail by Kolodner [15] in experiments on water-ethanol mixtures, and their origin is discussed by Batiste et al [11]. Figure 1 suggests that these states are *three-frequency* states, in which ω_1 is the fast chevron frequency, ω_2 represents the blinking frequency, while the third frequency ω_3 represents the slow modulation frequency. Figure 1 shows that these states consist of long intervals consisting of a slowly growing (even) chevron state. Instead of saturating this state becomes unstable to the onset of blinking, which leads to a collapse of the state to a small amplitude chevron, followed by a slow regrowth. In the time series shown these collapse events are periodic with period $T_3 = 2\pi/\omega_3$. Figure 2 shows that as R decreases the period $T_3 = 2\pi/\omega_3$ increases rapidly and apparently diverges at $R \approx 1971.5$. Such a divergence suggests that the three-frequency states are created via a global bifurcation. Since no stable nonzero solutions are present for smaller values of R for these parameter values the three-frequency repeated transient state represents the *first* nontrivial state of the system.

In contrast, as R increases the modulation frequency ω_3 increases but drops out from the time series between $R = 1980.0$ and $R = 1981.0$ (Figure 3). We identify this transition with a Hopf bifurcation [11], and note that this bifurcation is *supercritical*, i.e., viewed in the direction of decreasing R this bifurcation creates a *stable* three-frequency state from a *stable* two-frequency state, with no observable hysteresis. Figure 4 shows

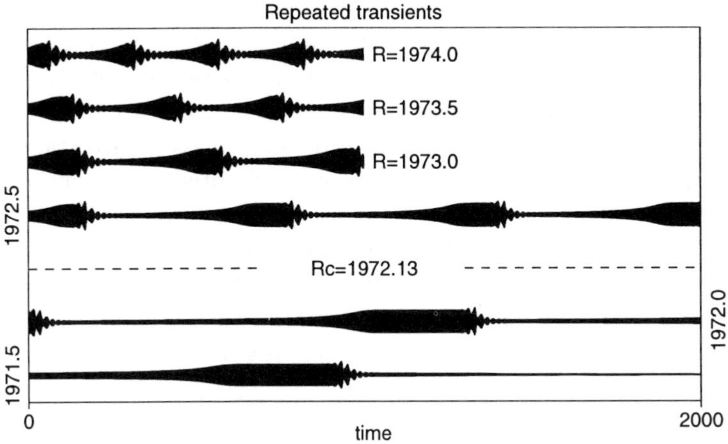

Fig. 1. The repeated transient state in a ^{3}He-^{4}He mixture with $S = -0.1$ at several values of R. The time series $w(x = 0.87\Gamma, z = 0, t)$ suggest the presence of three frequencies, with the lowest frequency ω_3 increasing from zero as R increases from $R \approx 1971.5$ ($\varepsilon \equiv \varepsilon^* \approx -3.2 \times 10^{-4}$). Courtesy O. Batiste.

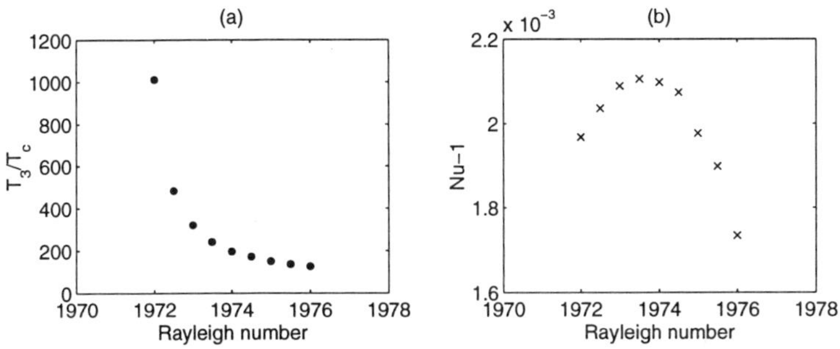

Fig. 2. (a) The modulation period $T_3 \equiv 2\pi/\omega_3$ of the repeated transients when $S = -0.1$ in units of $T_c \equiv 2\pi/\omega_c$ as a function of the Rayleigh number R. (b) The corresponding Nusselt number. Courtesy O. Batiste.

that this two-frequency state is a *symmetric* blinking state, and traces the evolution of this state towards larger values of R, and in particular to a time-dependent state referred to as the *fish* state, followed by a transition to stationary convection. Figure 5 shows an example of the fish state just prior to this transition. An examination of the spatial structure of the waves (Figure 6) shows that there is a brief phase of the oscillation during which small amplitude counterprogagating waves fill the container. This state is highly unstable, however, with the waves at one end growing at the expense of those at the other. Once the amplitude of the growing state is large enough the system shifts into a new and larger amplitude state in which the waves are spatially confined towards one side, with *no* waves at the other, and forms a wall-attached state. The wall-attached state is not stable, however, and continues to decrease in amplitude and contract until it becomes so weak and confined that waves start to regrow at the other sidewall. At this point the wall-attached state disintegrates, and the small amplitude extended chevron-like state is restored. All these transitions are quite easily distinguished in the time trace shown in Figure 5 and in the space-time plot shown in Figure 6, and suggest that the formation of the initial pulse-like state is triggered by a nonlinear focusing effect reminiscent of the nonlinear Schrödinger equation. Many of these details resemble behavior observed in experiments on water-ethanol mixtures [16].

When $S = -0.5$ convection sets in at $R_c = 2643.43$ with $\omega_c = 12.836$. The primary instability is again to an even chevron, but this time we find strongly irregular dynamics already quite close to onset (Figure 7). The time series in Figure 7 is best described as an intermittent repeated transient, in which the final collapse event may be preceded by several spatially symmetric bounces before the onset of the symmetry-breaking instability that disrupts the state and leads to the temporary formation of a confined state towards one side, much as already described for $S = -0.1$. This state then drifts towards the nearest wall and shrinks in lateral extent, until it triggers another collapse event that permits waves to grow at the other sidewall. The decaying symmetric blinking state that results reestablishes a small amplitude chevron state which then regrows on a much longer timescale. The spatially symmetric bounces are associated with relatively sharp peaks in the Nusselt number, while the symmetry-breaking collapse events produce bursts in the Nusselt number that are markedly asymmetric, much as in the fish state discussed above.

The frequency of the burst-like events in the Nusselt number increases with R (Figure 8). Perhaps the most remarkable time series of all is shown in Figure 9 for $R = 2750$. As seen from Figure 10 this state corresponds to a spatially and temporally intermittent state. The state is not 'random', however, but is characterized by well-defined episodes during which a spatially confined state forms and collapses. Indeed the time series shows an irregular switching between two states, a large amplitude state with a relatively low ω_1, and a small amplitude state with a large ω_1. The former is a spatially confined slowly drifting wave, while the latter is an extended

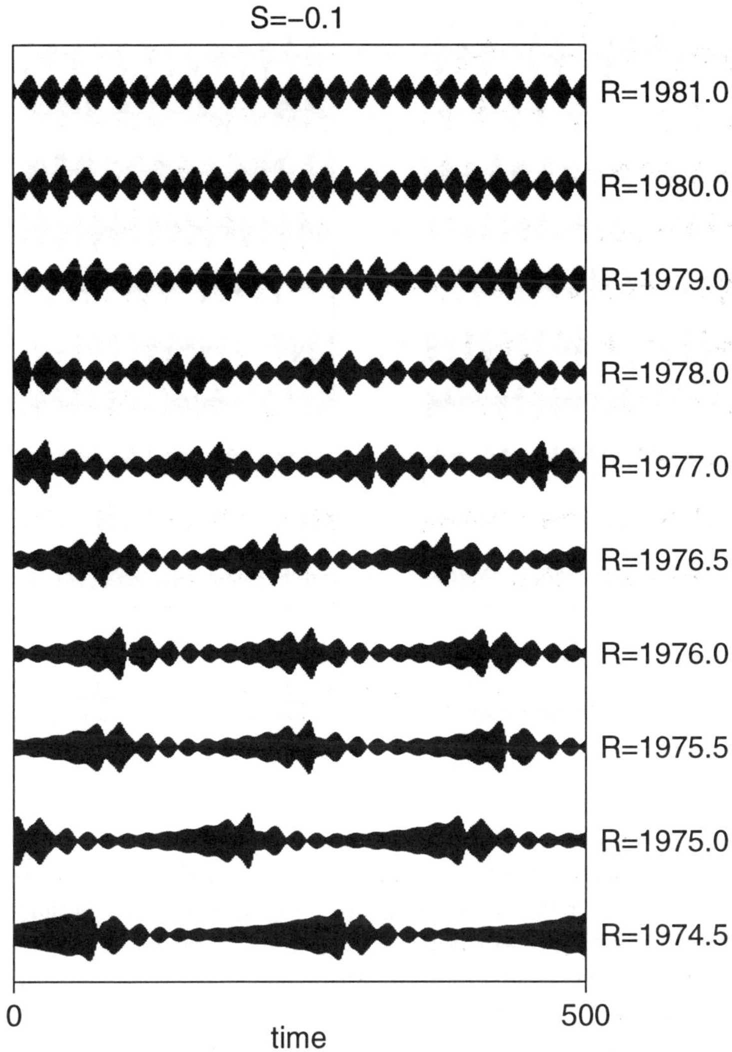

Fig. 3. The time series $w(x = 0.87\Gamma, z = 0, t)$ for a ^{3}He-^{4}He mixture with $S = -0.1$ at several values of R showing the transition from the repeated transient state in Figure 1 to a symmetric periodic blinking state at $R = 1981$ ($\varepsilon = 4.5 \times 10^{-3}$). Courtesy O. Batiste.

more-or-less symmetric pulsating chevron-like state. The latter is unstable to a symmetry-breaking blinking instability which amplifies the waves near one of the sidewalls at the expense of those near the opposite wall.

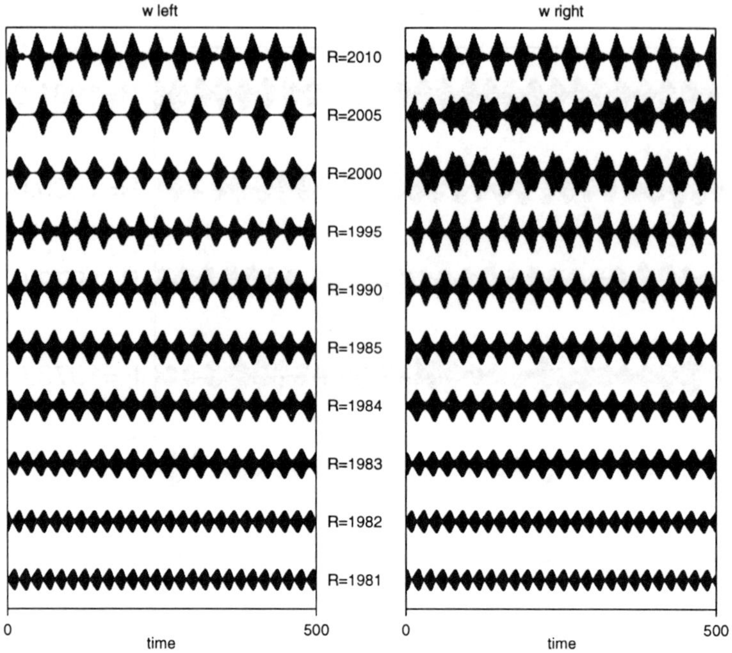

Fig. 4. Time series $w(x = 0.13\Gamma, z = 0, t)$ (left column) and $w(x = 0.87\Gamma, z = 0, t)$ (right column) for different values of R increasing upwards when $S = -0.1$. Courtesy O. Batiste.

The amplified waves are then reflected from the wall but continue to grow, slowing down markedly. This asymmetric state then abruptly collapses (by a mechanism that remains unclear) into a highly nonlinear spatially confined state consisting of slowly travelling waves. These waves propagate in the direction of the original reflected wave, and speed up as they approach the opposite wall. The whole pulse gradually retracts towards that wall, ultimately leaving much of the cell free of convection. Once the peak of the pulse reaches the wall its amplitude drops dramatically and it disintegrates into smaller amplitude counterpropagating waves that invade the convection-free part of the container, permitting the regrowth of the original higher frequency small amplitude chevron-like state filling the domain (Figure 10). Overall this behavior resembles that observed for $S = -0.1$ (see Figure 6) but here it is much more dramatic, and we may think of it as back-and-forth 'sloshing'. The resulting state is reminiscent of a state computed in doubly-diffusive convection [17, 18].

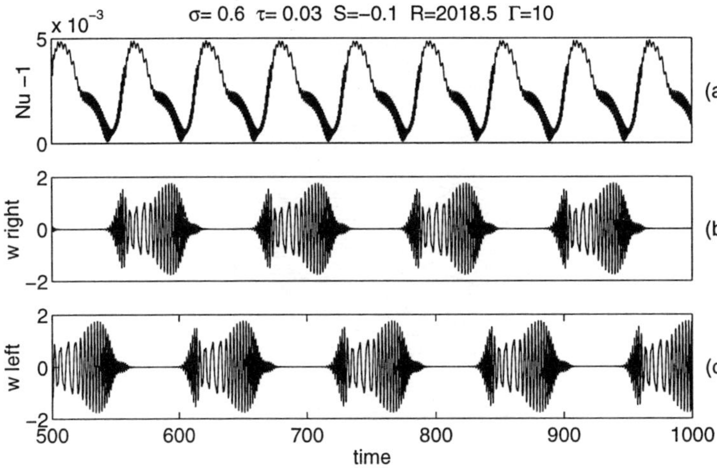

Fig. 5. (a) The Nusselt number Nu(t), (b) $w(x = 0.87\Gamma, z = 0, t)$, (c) $w(x = 0.13\Gamma, z = 0, t)$ for the *fish* state at $R = 2018.5$ ($\varepsilon = 0.024$) when $S = -0.1$. Courtesy O. Batiste.

2.1. *Origin of Blinking States and Repeated Transients*

In this Section we summarize the properties of a simple model system, based on normal form theory for the interaction of two Hopf modes with opposite parity, that accounts for essentially all the properties revealed in our simulations when $\varepsilon \ll \Gamma^{-2} = 10^{-2}$. We do not have a corresponding understanding for the dynamics observed at larger Rayleigh numbers. The model is based on the observation that, at specific aspect ratios Γ_c, the odd and even chevrons bifurcate simultaneously, albeit with different frequencies [19]. For nearby aspect ratios they come in in close succession. Under these conditions we may expect the chevrons to interact in the nonlinear regime, and to do so already at small amplitude. We write each of the fields u, w, θ, C in the form

$$\theta(x,z,t) = \mathrm{Re}\{z_+(t)e^{i\omega_+ t} f_+(x,z) + z_-(t)e^{i\omega_- t} f_-(x,z)\} + \dots, \qquad (9)$$

where $f_\pm(x,z)$ are the (complex) eigenfunctions of the even and odd chevrons, and $z_\pm(t)$ are their amplitudes. Standard normal form theory now yields the following equations for the (real) amplitudes $r_\pm \equiv |z_\pm|$:

$$\dot{r}_+ = (\mu + a_+ r_+^2 + b_+ r_-^2 - r_+^4 + c_+ r_+^2 r_-^2 + d_+ r_-^4) r_+ + \dots, \qquad (10)$$

$$\dot{r}_- = (\mu - \delta + a_- r_-^2 + b_- r_+^2 - r_-^4 + c_- r_+^2 r_-^2 + d_- r_+^4) r_- + \dots, \qquad (11)$$

Fig. 6. Space-time plot of the fish state in Figure 5 with time increasing upwards. The state is periodic in time with period T, and is invariant under the spatial reflection R_x followed by evolution for a time $T/2$. Courtesy O. Batiste.

with a pair of decoupled equations for the associated (nonlinear) frequencies $\omega_{\pm}$. Here the coefficients are all real, and we take $a_{\pm} > 0$ so that both chevrons bifurcate subcritically, the even ones (given by $(z_+, z_-) = (r_+, 0) \exp i\omega_+ t)$ at $\mu = 0$ and the odd ones (given by $(z_+, z_-) = (0, r_-) \exp i\omega_- t)$ at $\mu = \delta > 0$. It is now a simple matter to show that the chevrons undergo saddle-node bifurcations at $\mu = -a_{\pm}^2/4 < 0$, and steady state bifurcations to a mixed parity state (z_+, z_-), $r_+ r_- > 0$, at $\mu = \delta - b_- r_+^2 - d_- r_+^4$ (even chevrons) and $\mu = -b_+ r_-^2 - d_+ r_-^4$ (odd chevrons). The bifurcations to the mixed parity states are to be identi-

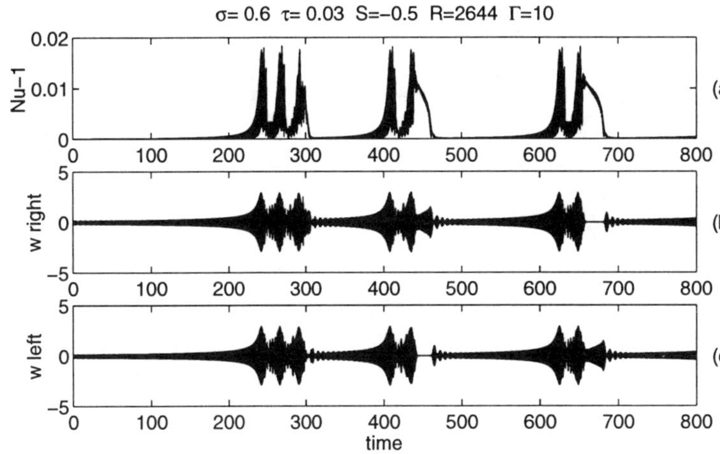

Fig. 7. (a) The Nusselt number Nu(t), (b) $w(x = 0.87\Gamma, z = 0, t)$, (c) $w(x = 0.13\Gamma, z = 0, t)$ at $R = 2644$ ($\varepsilon = 2.1 \times 10^{-4}$), $S = -0.5$, showing irregular bursts. Courtesy O. Batiste.

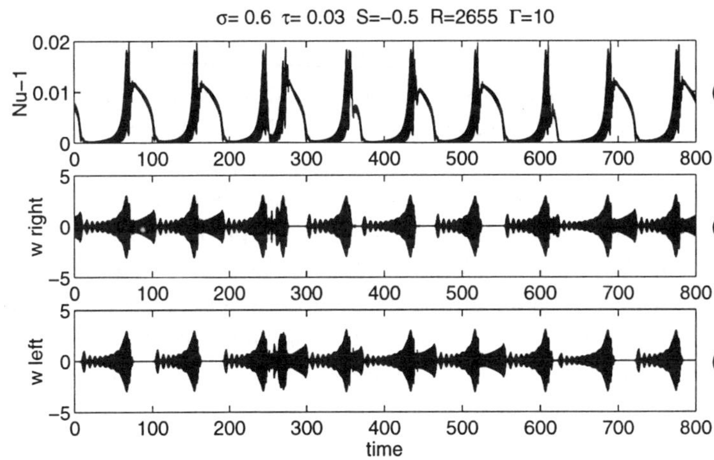

Fig. 8. As for Figure 7 but showing the time series corresponding to $R = 2655$ ($\varepsilon = 4.4 \times 10^{-3}$). Courtesy O. Batiste.

fied with Hopf (more precisely torus) bifurcations from the chevron states to *blinking* states. This is because near this bifurcation on the even chevron

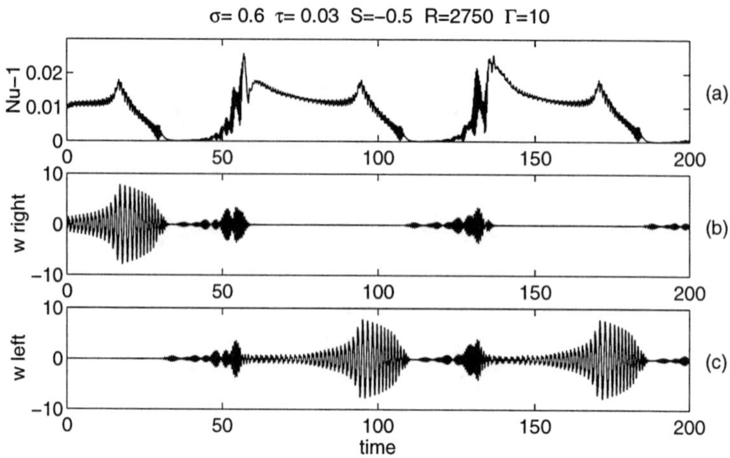

Fig. 9. Irregular switching between the fish state and a blinking state at $R = 2750$ ($\varepsilon = 0.040$). Courtesy O. Batiste.

branch the dynamics take the form

$$\theta(x,z,t) = \mathrm{Re}\, r_+ e^{i\omega_+ t}\{f_+(x,z) + (r_-/r_+)e^{i(\omega_- - \omega_+)t} f_-(x,z)\} + \dots, \quad (12)$$

describing an even chevron with a periodically oscillating odd parity contribution. The second term amplifies (reduces) waves in the left half of the container at the same time as it reduces (amplifies) waves in the right half, and shows that the blinking frequency at leading order is simply the beat frequency $\omega_2 = \omega_+ - \omega_-$. We emphasize that these frequencies are the *nonlinear* frequencies, not the onset frequencies predicted by linear theory.

When the bifurcation to blinking states falls below the saddle-node bifurcation on the chevron branch the resulting blinking states are initially unstable but acquire stability with increasing amplitude at a *tertiary* Hopf bifurcation [20,21]. This bifurcation introduces a *third* frequency ω_3 into the dynamics of the system. We identify the resulting states with the observed repeated transients.

The key transitions involving the one-, two- and three-frequency states are captured by a simplified model system obtained from the above equations by dropping inessential terms. In particular, we drop the term $-r_-^5$ and mimic its effect by taking $a_- < 0$. The resulting model [11]

$$\dot{r}_+ = (\mu + a_+ r_+^2 - r_+^4 - r_-^2) r_+ \quad (13)$$

$$\dot{r}_- = (-\nu + a_- r_-^2 + b_- r_+^2) r_- \quad (14)$$

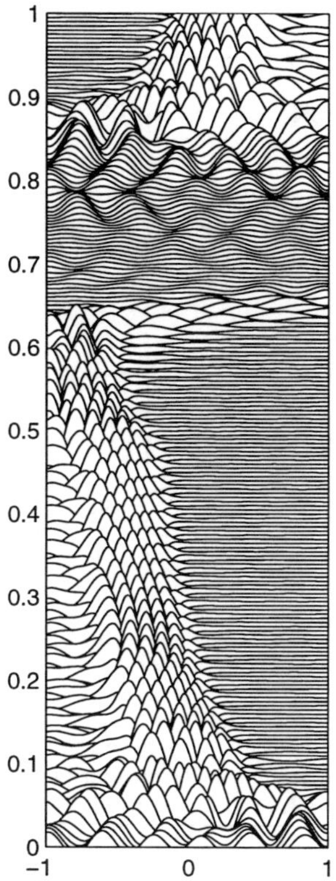

Fig. 10. Space-time plot of the state in Figure 9 with time increasing upwards. Courtesy O. Batiste.

with $a_+ > 0$, $b_- > 0$ is the simplest set of equations capable of describing correctly the stability properties of the even chevrons and the mixed parity states observed in the partial differential equations for moderate values of $|S|$. The model removes the primary bifurcation to odd chevrons but leaves the secondary bifurcations from the even chevrons fundamentally unchanged. In the following we think of μ, ν as proportional to $R - R_c(\Gamma)$, $\Gamma - \Gamma_c(R)$, respectively.

Figure 11 summarizes the properties of the model in the case in which the three-frequency state created from the blinking states is stable. This is always the case when $b_- = 1$, $a_- = 0$ and $a_+ > 0$, and hence for sufficiently small negative values of a_- as well. The figure shows the loci of the primary (H_1), secondary (H_2) and tertiary (H_3) Hopf bifurcations,

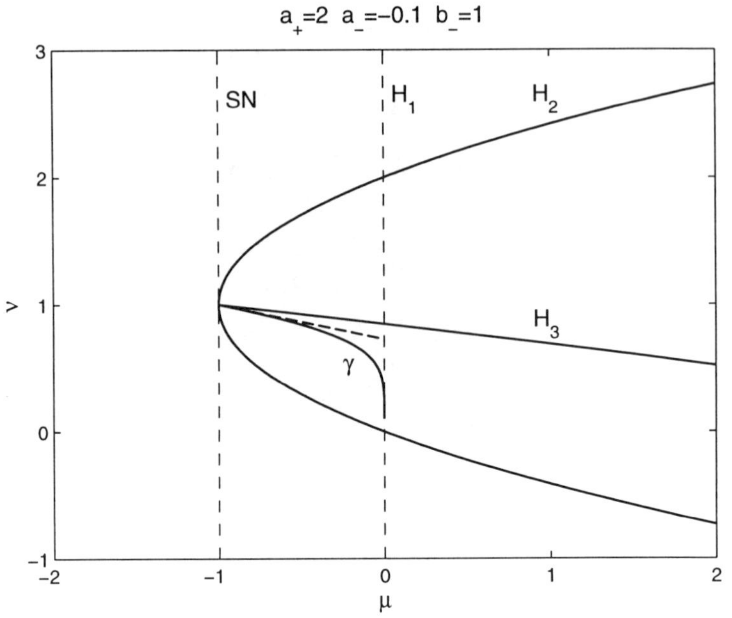

Fig. 11. Codimension-one bifurcation surfaces in the (μ, ν) plane for equations (13,14) with $a_+ = 2.0$, $a_- = -0.1$ and $b_- = 1.0$. H_1: primary (Hopf) bifurcation to the chevron state $(r_+, 0)$, SN: saddle-node bifurcation on the chevron branch, H_2: (secondary) Hopf bifurcation to blinking states (r_+, r_-), H_3: (tertiary) Hopf bifurcation from (r_+, r_-) responsible for the appearance of the three-frequency states, and γ: global bifurcation at which these states disappear. The heavy broken line represents the asymptote to γ.

as well as the locus of the saddle-node bifurcations (SN) on the chevron branch. It should be remembered that in the (r_+, r_-) variables only the bifurcation H_3 remains a Hopf bifurcation, with H_1 and H_2 represented by pitchfork bifurcations. In addition the figure shows the curve $\gamma : \mu = \mu^*(\nu)$ of global bifurcations at which the limit cycle (corresponding to the three-frequency states) created at H_3 disappears by simultaneous collision with small and large amplitude chevron states. The location of this line must be determined numerically. An asymptotic calculation of this curve near the codimension-two point at which H_2 and SN coincide yields the heavy broken line; this line is tangent to γ at the codimension-two point, as it must.

Figure 12a shows the bifurcation diagram obtained by traversing the (μ, ν) plane in Figure 11 along the line $\nu = 1.6$. The figure shows a small interval of subcritical but stable chevrons, followed by a supercritical pitch-

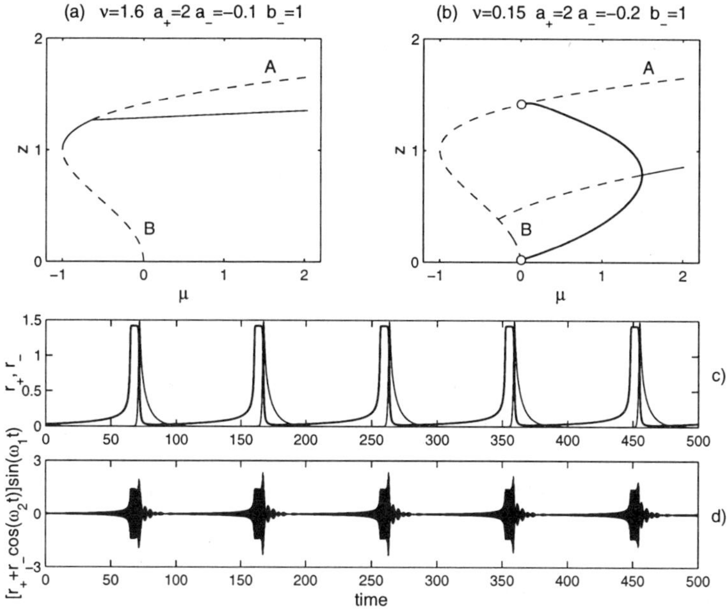

Fig. 12. Bifurcation diagrams along (a) the line $\nu = 1.6$ when $a_+ = 2.0$, $a_- = -0.1$, $b_- = 1.0$, and (b) the line $\nu = 0.15$ when $a_+ = 2.0$, $a_- = -0.2$, $b_- = 1.0$. The open circles indicate the global bifurcation; this bifurcation occurs very close to $\mu = 0$ for a large range of values of ν. (c) The time series $r_+(t)$ (thick line) and $r_-(t)$ (thin line) corresponding to case (b) with $\mu = 0.02$. (d) The quantity $[r_+(t)+r_-(t)\cos(\omega_2 t)]\sin(\omega_1 t)$ corresponding to (c) when $\omega_1 = 20.0$, $\omega_2 = 0.8$ for comparison with Figure 1. Note the exponential growth during the chevron phase, followed by an overshoot when the blinking instability sets in, and the ringing down during the subsequent collapse phase.

fork bifurcation to a state with $r_- \neq 0$ that represents a blinking state in physical variables. In the example shown this bifurcation occurs at $\mu < 0$ so that the first stable state just above onset ($\mu = 0$) is a finite amplitude blinking state. This case is typical of the behavior of the partial differential equations when $S = -0.01$ in appropriate ranges of Γ (not shown). In contrast, when $\nu < a_+b_-/2 = 1$ the first stable state encountered as μ increases is a periodic state $(r_+(t), r_-(t))$ corresponding to the three-frequency repeated transient state with the frequencies ω_1 and ω_2 filtered out (Figure 12b). This state appears in a global bifurcation at $\mu = \mu^* < 0$ [22], at which $\omega_3 = 0$. Figure 12c shows the time series corresponding to this state when $\mu = 0.02$, $\nu = 0.15$, $a_+ = 2.0$, $a_- = -0.2$, $b_- = 1.0$. For these parameter values $\mu^* \approx 0$ (open circles in Figure 12b), and Figure 11 shows that this situation persists for a large range of values of ν. In this case the chevron state regrows from very small amplitude, and the resulting oscillation resembles closely the state shown in Figure 1. In particular there is almost no hysteresis between this state and the conduction state, and the system behaves *as if* the primary instability at $\mu = 0$ were directly responsible for generating repeated transients. Observe that during the growth phase of the variable r_+ the variable r_- vanishes, indicating that the growing state is a pure chevron; r_- becomes nonzero only during the collapse phase, indicating that the collapse is triggered by a symmetry-breaking instability of the growing even chevron. The amplitude and the period $2\pi/\omega_3$ of the limit cycle in Figure 12c decrease with increasing μ, with the oscillations disappearing at H_3. As already mentioned we interpret this transition as the transition from the repeated transient state to the (symmetric) periodic blinking state with increasing Rayleigh number seen in Figure 3. For the model parameters this transition is supercritical, indicating absence of hysteresis, as in the figure.

Within the model the repeated transient state $(r_+(t), r_-(t))$ has all the properties of this state observed in the experiments, except for the (apparent) absence of oscillations during the collapse phase. In fact, if the frequencies ω_1 and ω_2 are restored, and the pointwise quantity $[r_+(t) + r_-(t)\cos\omega_2 t]\sin\omega_1 t$, cf. Eq. (12), plotted instead of $r_+(t)$ or $r_-(t)$, these oscillations *are* present (Figure 12d), and their amplitude depends on the chevron amplitude r_+ in the manner observed in the simulations. In fact, the time series shown in Figure 12d bears a number of qualitative features, including the pointed overshoot at maximum as the mode r_- begins to grow and the "ringing down" due to the fact that the variable r_+ decays more rapidly than r_-, that are documented in experiments as well [15].

Despite its remarkable simplicity the model (13,14) captures completely the two scenarios for generating blinking states identified in the simulations of both ^{3}He-^{4}He and water-ethanol mixtures, and the origin and properties of the repeated transients. Extensions of the model [11] indicate the possibility that repeated transients may, under appropriate circumstances, be chaotic. In Figure 13 we show an example of such a chaotic repeated transient state. We believe that this state is associated with the global bifurcation in which the repeated transients first appear, cf. Figure 12b. As already noted the frequency ω_3 decreases to zero as $\mu \downarrow \mu^* < 0$, i.e., as $\varepsilon \downarrow \varepsilon^* < 0$ in the partial differential equations. As this occurs the three-frequency states approach simultaneously the unstable large and small amplitude chevron states, hereafter A and B, respectively (see Figure 12). The character of the repeated transient when $\varepsilon \approx \varepsilon^*$ is determined by the leading eigenvalues of A and B in the $(r_+, 0)$ direction, hereafter $-\lambda_A < 0$ and $\lambda_B > 0$, and the leading eigenvalues in the $(0, r_-)$ direction. If the latter are real, $\alpha_A > 0$ and $-\alpha_B < 0$, say, and $\rho \equiv \alpha_B \lambda_A / \alpha_A \lambda_B > 1$, the repeated transients will remain periodic and stable all the way to ε^*, where the period T_3 diverges and the global bifurcation takes place [11]. In contrast, when $0 < \rho < 1$, the periodic oscillations necessarily lose stability before the global bifurcation at ε^*. Similar results are obtained in the case where the leading stable symmetry-breaking eigenvalue at B is complex, viz. $-\alpha_B + i\omega_B$, $\alpha_B > 0$, as suggested by the simulations. In this case stable periodic oscillations will persist down to ε^* if $\rho > 1$, but if $0 < \rho < 1$ complex dynamics of Shil'nikov type will be present. In fact the leading unstable eigenvalues α_A and λ_B are also expected to be complex, since in the partial differential equations the bifurcations at H_1 and H_2 are both Hopf bifurcations.

When λ_B is real a trajectory escaping from B describes an exponentially growing chevron state. This growth phase, including the states A and B, is clearly visible in the time series in Figure 1. When the growing chevron reaches the vicinity of A it becomes unstable to symmetry-breaking oscillations which take it back near B. This is the collapse phase of the repeated transient state (compare Figure 12d with Figure 1). The frequency of the decaying oscillations observed in the time series in Figures 12d is given by ω_B. This frequency will in general be of the same order as the blinking frequency associated with the branch of blinking states when these bifurcate from the small amplitude chevron B, but quite different from (and in general larger than) the blinking frequency of the *stable* blinking states beyond H_3. This observation explains the coincidence of the period of the blinking states and of the oscillations during the collapse phase of the repeated transient noted by Kolodner [15]. Note also that since α_B decreases as ε decreases (it passes through zero at H_2, i.e., at $\varepsilon = \varepsilon_2$) the collapse becomes slower and slower, cf. Figure 1, although the collapse rate is still finite when the three-frequency states disappear in the global bifurcation at ε^* (since $\varepsilon_2 < \varepsilon^* < 0$) and the system makes a hysteretic transition back to the conduction state. The fact that α_B decreases with ε makes it likely that the Shil'nikov condition $0 < \rho < 1$ holds at ε^*, resulting in *chaotic* repeated transients prior to their disappearance. However, despite this suggestion we have only succeeded in locating such chaotic repeated transients with *stress-free* boundary conditions (Figure 13).

The model equations describe the qualitative behavior of the system successfully whenever $|\varepsilon| \lesssim \Gamma^{-2}$. Identical transitions may also be found

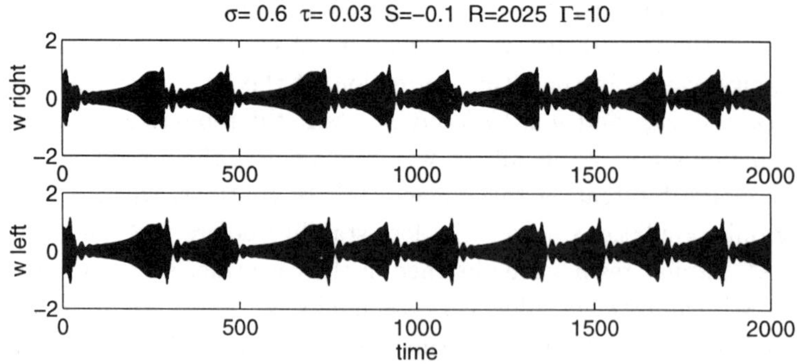

Fig. 13. A chaotic repeated transient in a ^{3}He-^{4}He mixture with stress-free and fixed temperature boundary conditions at $x = 0, \Gamma$ when $R = 2025$, $S = -0.1$. Courtesy O. Batiste.

in a degenerate Hopf bifurcation with broken O(2) symmetry in which the continuous part of O(2) is broken (by distant lateral walls), much as discussed for the nondegenerate case by Dangelmayr and Knobloch [13, 14] in their original explanation of the blinking state. However, for larger values of ε the model begins to break down. Among the new phenomena that occur once $|\varepsilon| \gtrsim \Gamma^{-2}$ are the *fish states* originally observed by Kolodner et al. in water-ethanol mixtures [23] and various types of dynamically localized travelling wave states also seen in experiments [24–26], all of which require a much larger number of modes for their description than retained in the model.

3. Natural Doubly Diffusive Convection

Natural doubly diffusive convection, i.e., convection driven by horizontal concentration or temperature gradients, provides another example of the phenomenon described in the preceding section, this time in three dimensions [8]. We assume that the wall at $x = 0$ is maintained at temperature $\Delta T > 0$ and concentration $\Delta C > 0$ above those imposed at $x = L$, and measure the relative importance of the temperature and concentration gradients by the buoyancy ratio $N = \rho_C \Delta C / \rho_T \Delta T$, where $\rho_T \equiv \partial\rho/\partial T < 0$, $\rho_C \equiv \partial\rho/\partial C > 0$. The dimensionless governing equations are

$$\frac{\partial \mathbf{u}}{\partial t} = -(\mathbf{u}\cdot\nabla)\,\mathbf{u} - \nabla p + \nabla^2\mathbf{u} + Gr(T + NC)\mathbf{e}_z, \quad \nabla.\mathbf{u} = 0, \tag{15}$$

$$\frac{\partial T}{\partial t} = -(\mathbf{u}\cdot\nabla)\,T + \frac{1}{\sigma}\nabla^2 T, \qquad \frac{\partial C}{\partial t} = -(\mathbf{u}\cdot\nabla)\,C + \frac{\tau}{\sigma}\nabla^2 C, \tag{16}$$

where $Gr \equiv g\rho_T \Delta T L^3/\nu^2$ is the Grashof number. The heat and mass fluxes vanish along the boundaries $y = 0, A_y$ and $z = 0, A_z$, and $\mathbf{u} \equiv (u, v, w) = 0$ on all walls.

When $N = -1$ the problem has the conduction solution $\mathbf{u} = \theta = \Sigma = 0$, where $T = 1-x+\theta(x, y, z)$, $C = 1-x+\Sigma(x, y, z)$. With the given boundary conditions the equations for $\mathbf{u}$, θ, Σ are invariant under the two operations

$$S_y : (x, y, z) \rightarrow (x, A_y - y, z), \qquad (u, v, w, \theta, \Sigma) \rightarrow (u, -v, w, \theta, \Sigma), \tag{17}$$

$$S_\Delta : (x, y, z) \rightarrow (1 - x, y, A_z - z), \quad (u, v, w, \theta, \Sigma) \rightarrow (-u, v, -w, -\theta, -\Sigma). \tag{18}$$

Thus S_y is a reflection in the plane $y = A_y/2$ while S_Δ is a rotation by π about the line $x = 1/2$, $z = A_z/2$. It follows that the equations are also invariant under the operation $S_\Delta \circ S_y = S_y \circ S_\Delta \equiv S_C$, corresponding to a point symmetry with respect to the center of the container. Like the operations $\{R_x, \kappa\}$ of the previous section these symmetries constitute the symmetry group D_2. If S is a nontrivial element of D_2 and $\mathbf{e}$ is an eigenvector of the linearized problem, then $S\mathbf{e} = \pm\mathbf{e}$, i.e., the instability either respects or breaks the symmetry S. As a result each neutral stability curve is characterized by a particular symmetry. In particular, if $S\mathbf{e} = \mathbf{e}$ for all $S \in D_2$, the generic bifurcation from the conduction state is transcritical. In contrast, if one of the reflections in D_2 is broken, the bifurcation is a pitchfork. In Figure 14, obtained for $\sigma = 1$, $\tau = 1/11$ and $A_y = 1$, $A_z = 2.5$, these facts are indicated using the notation T_j for the jth transcritical bifurcation from the conduction state, and P_j for the jth pitchfork bifurcation. The diagram shows $W_M \equiv |w|_{\max}$, i.e., the maximum of the absolute value of the vertical velocity in the enclosure, as a function of Gr. Thus two distinct branches of solutions, unrelated by any of the problem symmetries, emerge from points labelled T_j. In contrast, the two solutions emerging from points P_j are related by the broken symmetry and so have identical values of W_M at fixed Gr; consequently only a single solution branch emerges from each P_j. The figure shows that the first instability, at $Gr_{P_1} = 997.5$, is a subcritical pitchfork bifurcation. The resulting unstable branch terminates in a pitchfork bifurcation at $Gr_{S_2} = 959.2$ on the subcritical branch emanating from the transcritical bifurcation at T_1 ($Gr_{T_1} = 1045$). After S_2 the latter branch undergoes a second pitchfork bifurcation at S_1 ($Gr_{S_1} = 770$), and then a saddle-node bifurcation at $Gr_{SN} = 679$, but remains unstable throughout. The supercritical branch emanating from T_1 is initially once unstable and undergoes two Hopf bifurcations before terminating on the conduction state at T_2. Thus over the entire range of parameter values explored the only stable steady solutions are those located on the branch created at S_1, associated with *longitudinal* rolls, between the two Hopf bifurcations H_1 ($Gr_{H_1} = 1267$) and H_2 ($Gr_{H_2} = 1568$). The "stability" gap $Gr_{P_1} < Gr < Gr_{H_1}$ is created when S_1 moves down past SN, as in Section 2 [8].

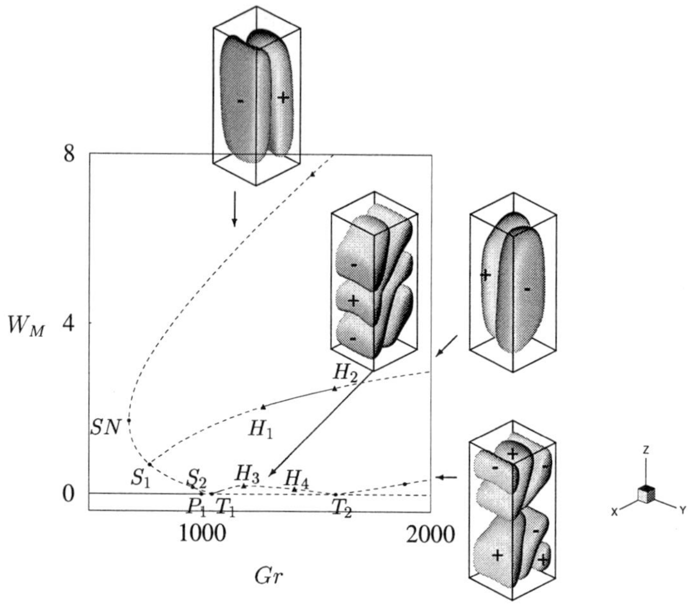

Fig. 14. Bifurcation diagram for the case $N = -1$, $A_y = 1$, $A_z = 2.5$, showing W_M as a function of the Grashof number Gr. Solid dots (triangles) indicate steady (Hopf) bifurcations. The insets show surfaces $w = \pm K$ for suitable choices of K at the locations indicated by the arrows. Resolution is $13 \times 13 \times 19$.

The bifurcation at Gr_{H_1} is supercritical and so produces stable small amplitude oscillations in $Gr < Gr_{H_1}$. In contrast the bifurcation at H_2 is subcritical and leads to large amplitude oscillations. These persist down to $Gr \approx 936.5$ and up to the largest values of Gr explored ($Gr > 3000$), coexisting with the small amplitude oscillations created at Gr_{H_1} in the interval $1085 < Gr < Gr_{H_1} \approx 1267$. In order to detect the symmetry of the corresponding flow we construct three indicators of symmetry breaking: for a generic point M inside the container we compute the quantities $I_y = w(M) - w(S_y(M))$, $I_\Delta = w(M) + w(S_\Delta(M))$ and $I_C = w(M) + w(S_C(M))$ as a function of time. If any of these indicators vanishes identically the flow has the corresponding symmetry. Using this technique we show that all the oscillations found for $A_z = 2.5$ are S_C-symmetric. Figure 15a shows the time evolution of I_y and $w(M)$ when $Gr = 936.5$, while Figure 15b shows a plot of I_y against $w(M)$. The figure shows that the oscillation is almost heteroclinic, spending a long time near two fixed points, labelled A and B. Figure 15b shows that these fixed points are S_y-symmetric, in addition to having the symmetry S_C of the oscillation. Thus both correspond to D_2-symmetric steady states. Both solutions lie on the subcritical steady state branch emanating from T_1 with the solution B located on the lower part of the branch and A on the upper part. At this value of Gr ($Gr_{S_1} < Gr < Gr_{S_2}$) the solution B is unstable with respect to D_2-symmetric perturbations but no other. A solution starting near B will therefore follow the unstable manifold of B which takes it to A. This solution is stable with respect to D_2-symmetric perturbations since it lies above SN but because it also lies above the point S_1 it is unstable with respect to perturbations breaking S_y (and hence S_Δ as well). This is confirmed in Figure 15b which shows that $I_y > 0$ once the solution escapes from A. The one-dimensional unstable manifold of A then takes the system back to B; during this phase the flow is dominated by longitudinal rolls. As Gr decreases to Gr^* the oscillation becomes more and more burst-like (Figure 16). Computation of the eigenvalue ratio ρ (Section 2) reveals that for our parameters $\rho < 1$ and hence that the branch of oscillations must lose stability in a saddle-node bifurcation prior to its termination in a heteroclinic bifurcation at $Gr = Gr^*$. It follows that the branch extends into $Gr < Gr^*$, and hence that $936 < Gr^* < 936.5$ [8].

Thus in this system the instability to longitudinal rolls plays the role of the onset of blinking in Section 2, but its outcome is the same: the first nontrivial state of the system is a long-period burst-like oscillation, and the origin of these oscillations is described by a similar model to that discussed in Section 2, viz. [8]

$$\dot{y} = (-\nu + cz)y - \delta y^3 \tag{19}$$

$$\dot{z} = (\mu + az - z^2)z - y^2 z. \tag{20}$$

Here z refers to the amplitude of the D_2-symmetric state, while y denotes the amplitude of the I_y contribution, that is, of the D_2-breaking part of the solution. Both variables are real. Figure 17a shows the various codimension one bifurcation lines in the (μ, ν) plane, while Figures 17b,c show the bi-

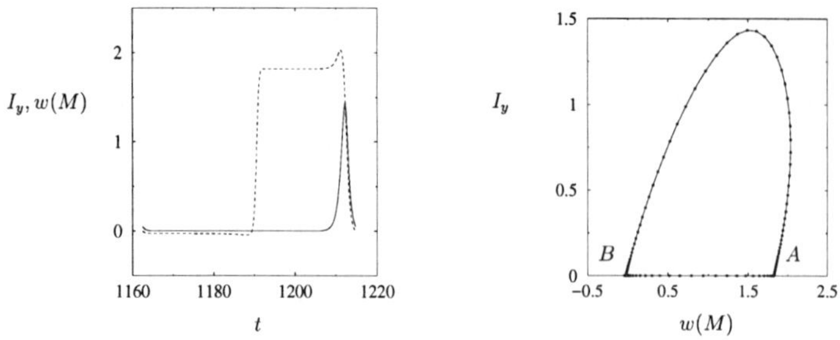

Fig. 15. Nonlinear oscillations at $Gr = 936.5$, when $A_y = 1$, $A_z = 2.5$. (a) Time evolution over one period of $I_y \equiv w(M) - w(S_y(M))$ (continuous line) and $w(M)$ (dashed line) at a generic point M. (b) Phase plane representation of the oscillation in (a). When $I_y = 0$ the solution is D_2-symmetric; when $I_y \neq 0$ it is only S_C-symmetric.

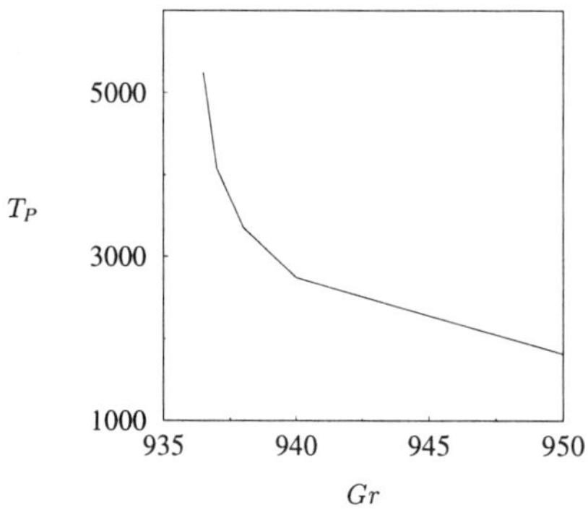

Fig. 16. The oscillation period T_P as a function of Gr for $A_y = 1$, $A_z = 2.5$, showing that $936 < Gr^* < 936.5$. Resolution is $13 \times 13 \times 19$.

furcation diagrams obtained by traversing the (μ, ν) plane along the two directions indicated by the heavy lines, obtained by integrating equations (19,20) for a particular choice of the coefficients. These lines correspond to increasing the Grashof number for fixed aspect ratio A_z (and fixed values of the other parameters as well), and capture the two fundamentally different bifurcation diagrams that characterize the present system. Figures 17d,e show the time series for $(y(t), z(t))$ for an oscillation near γ and the corresponding phase portrait. Observe that during the growth phase of the variable z the variable y vanishes, indicating that the growing state is

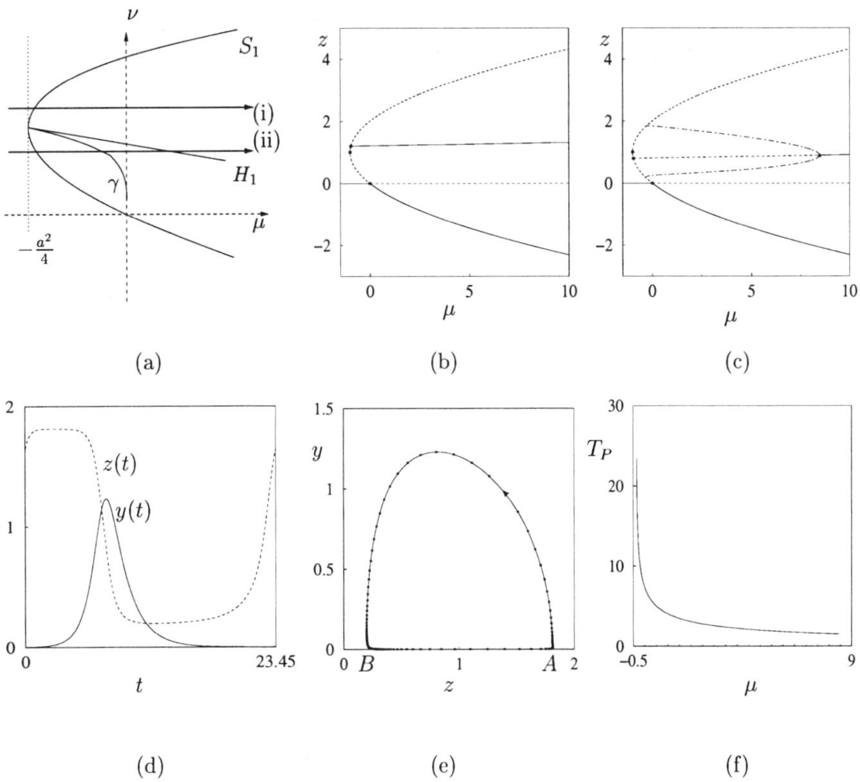

Fig. 17. Solutions of the model equations (19,20) for $a = 2$, $c = 1$, and $\delta = 0.01$. (a) Sketch of the parameter plane (μ, ν) showing the various codimension-one bifurcation surfaces. (b,c) Bifurcation diagrams constructed by traversing the (μ, ν) plane in (a) along the heavy lines labelled (i), (ii), respectively, corresponding to (i) $\nu = 1.2$, (ii) $\nu = 0.8$. In (c), the dot-dashed line represents the envelope (z_{max}, z_{min}) of the limit cycle $z(t)$ created at H_1 ($\mu = 8.45$) and labelled with a triangle. The global bifurcation γ occurs at $\mu \approx -0.35$. (d,e) The time series $(y(t), z(t))$ for $\mu = -0.34$ and the corresponding phase plane representation for comparison with Figure 15a,b, showing that the burst phase is initiated by the loss of symmetry at the large amplitude state A. (f) The period T_P of the time-dependent solution in (c) as a function of μ for comparison with Figure 16.

D_2-symmetric; y becomes nonzero only during the burst phase, indicating that the burst is triggered by a symmetry-breaking instability of the growing D_2-symmetric state when it reaches A. These figures bear an almost uncanny resemblance to Figures 15a,b obtained from the partial differential equations, and indicate that equations (19,20) do indeed capture the essence of the oscillations found numerically. Finally, Figure 17f shows the period $T_P(\mu)$ of the oscillations in the model for comparison with Figure 16. Theory predicts that as μ decreases to μ^* the oscillation period diverges as $\ln|\mu - \mu^*|$, and a similar rate of divergence is expected in the partial differential equations as well. However, the loss of stability of the oscillations near Gr^* is not captured by the model.

Since the longitudinal mode is only present in the three-dimensional formulation, the inclusion of the y direction has dramatic consequences for the behavior of the system: in two dimensions the branch S_1 is absent and the transverse rolls that appear at T_1 are stable beyond SN. Thus no stability gap is present and no dynamical behavior ensues.

4. Fast-slow Systems: Faraday Oscillations

Burst-like behavior can also be expected in parametrically excited surface gravity-capillary waves. A stunning example of this behavior was inadvertently discovered by Simonelli and Gollub [27]. These authors thought that they were studying Faraday waves in a vertically vibrated square container, and were surprised to find a burst-like state (and chaos) as the first nontrivial state of the system. However, the container proved to be slightly rectangular in horizontal cross-section, and Simonelli and Gollub were able to demonstrate that in a truly square container this phenomenon is absent.

Although the origin of this behavior is still not fully understood we describe below a closely related problem, Faraday waves in a slightly *elliptical* container in the nearly inviscid regime. In this regime a *streaming flow*, driven by time-averaged Reynolds stresses in oscillatory viscous boundary layers near rigid walls and the free surface [28–30], interacts with the waves producing it, leading to a description of the Faraday system in terms of amplitude equations coupled to a Navier-Stokes-like equation for the streaming flow with boundary conditions obtained by matching to the oscillatory boundary layers [30, 31]. These equations represent a new class of pattern-forming dynamical systems.

The new terms in the amplitude equations are formally of cubic order, indicating that the effects of the streaming flow cannot be neglected, even in the limit of vanishing viscosity. The resulting equations have thus far been studied only in the simplest cases: with single frequency forcing, periodic boundary conditions and one horizontal dimension. Moreover, when the spatial period is comparable to the wavelength selected by the parametric forcing the coupled amplitude-streaming flow equations simplify further: all solutions are attracted to standing waves of constant amplitude, and the streaming flow only advects the spatial phase of these waves; this phase in turn appears in the boundary conditions for the streaming flow, giving rise to a pair of simpler coupled phase-streaming flow equations. Analysis of these equations reveals that the coupling to the streaming flow is significant, and can lead to a number of *new* instabilities, which are absent if the streaming flow is neglected [32]. These include parity-breaking instabilities that lead to pattern drift, and oscillatory instabilities that produce oscillations in the spatial phase of the pattern. In an elliptical domain, however, these drifts are frustrated, and the streaming flow couples to both phase *and* amplitude of the surface gravity-capillary waves. The analysis that follows is based on the assumption that the *effective* Reynolds number of the streaming flow, suitably defined, remains small. This assumption permits us to project the Navier-Stokes-like equation for the streaming flow onto the dominant spatial eigenfunction, replacing it by a single ordinary differential equation for the evolution of the amplitude of this eigenfunction. The coupling of this viscous mode to the amplitude of the competing standing waves is retained as a free parameter. This procedure leads to the following system of scaled equations: [33, 34]

$$
\begin{aligned}
A'_\pm(\tau) &= -(1+i\Gamma)A_\pm + i\Lambda A_\mp + i(\alpha_1|A_\pm|^2 + \alpha_2|A_\mp|^2)A_\pm \\
&\quad + i\mu\bar{A}_\mp \mp i\gamma v_1 A_\pm, \qquad (21)\\
v'_1(\tau) &= \varepsilon(-v_1 + |A_-|^2 - |A_+|^2), \qquad (22)
\end{aligned}
$$

where $A_\pm$ are the (complex) amplitudes of clockwise and counterclockwise waves, and v_1 is the (real) amplitude of the first purely azimuthal viscous mode. In these equations time has been expressed in units of the viscous decay time of surface gravity-capillary waves. Once this time is known Γ, α_1, α_2 can be computed from inviscid theory for a cylinder [35]. The remaining coefficients Λ and μ are proportional to the ellipticity and acceleration amplitude of the container. When $\varepsilon \ll 1$ the viscous mode decays more slowly than the waves, and is therefore a *slow* mode. Such slow modes must be retained in any theory.

In the following we rewrite these equations in the form

$$X' = -(1+i(\Gamma+\Lambda))X + i((\alpha_1+\alpha_2)|X|^2 + 2\alpha_1|Y|^2)X - i(\alpha_1-\alpha_2)\overline{X}Y^2 + i\mu\overline{X} - 2\gamma v Y, \quad (23)$$

$$Y' = -(1+i(\Gamma-\Lambda))Y + i((\alpha_1+\alpha_2)|Y|^2 + 2\alpha_1|X|^2)Y - i(\alpha_1-\alpha_2)\overline{Y}X^2 + i\mu\overline{Y} + 2\gamma v X, \quad (24)$$

$$v' = \varepsilon(-v + i(\overline{X}Y - X\overline{Y})), \quad (25)$$

where

$$X = i(A_+ - A_-)/2, \quad Y = (A_+ + A_-)/2, \quad v = -v_1/2. \quad (26)$$

This form of the equations makes it clear not only that the (axisymmetric part of the) streaming flow vanishes if either $X=0$ or $Y=0$, i.e., for pure standing waves, but also that both modes must be present in order to drive such a flow, i.e., all instabilities of standing waves within Eqs. (23)-(25) will be due to *mode interaction*, at least when $\Lambda \neq 0$.

Equations (23)-(25) are equivariant with respect to the group D_2 generated by the two reflections

$$R_1 : (X,Y,v) \to (-X,Y,-v), \qquad R_2 : (X,Y,v) \to (X,-Y,-v). \quad (27)$$

As a result as μ increases the flat state loses stability to a pure mode. These are given by $P_+ \equiv (0,Y,0) = (0, R_+e^{i\varphi_+}, 0)$ and $P_- \equiv (X,0,0) = (R_-e^{i\varphi_-},0,0)$, with P_+ invariant under R_1 and P_- under R_2. Linearization about the pure modes identifies the stability properties of these states. These fall into one of two classes: the eigenvector may respect the symmetry of the solution, or it may break it. A steady state bifurcation of the former type corresponds to saddle-node bifurcations (SN), while a steady state bifurcation of the latter type is a pitchfork that produces a pair of *mixed modes* (M), i.e., steady states that do not have any symmetry. In the following we refer to this bifurcation as a symmetry-breaking bifurcation (SB). This bifurcation is present only when $\Gamma\Lambda \neq 0$ and produces *steady* solutions of the form (X,Y,v), $XYv \neq 0$, in contrast to the parity-breaking bifurcation that occurs when $\Lambda = 0$ and produces a *drifting* pure mode. The pure modes $P_\pm$ may also experience Hopf bifurcation. Only one type, a symmetry-breaking Hopf bifurcation, is possible and occurs on P_+ if $\gamma\Lambda < 0$, and P_- if $\gamma\Lambda > 0$. Note that this instability involves the excitation of streaming flow ($\gamma \neq 0$) and requires a finite ellipticity of the container ($\Lambda \neq 0$). The oscillations that result resemble trapped direction-reversing waves, oscillating about the major or minor axes, but exhibit no net drift.

Such states correspond to two-frequency Faraday oscillations. The mixed modes can also lose stability at a saddle-node or Hopf bifurcation; in contrast to the pure modes the latter is possible even when $\gamma = 0$, but then occurs only if P_+ and P_- bifurcate in *opposite* directions. However, if $\gamma \neq 0$, a Hopf bifurcation may occur along the mixed mode branch even when it connects two supercritical pure modes.

We now examine a situation that is reminiscent of that considered in Sections 2 and 3, and take $\Gamma = 0.15$, $\Lambda = 0.18$, $\alpha_1 = 0.48$, $\alpha_2 = -0.58$, $\varepsilon = 0.001$. Figures 18a,b shows the resulting bifurcation diagrams for $\gamma = -0.35$ and $\gamma = 0$, respectively. In both cases the pure mode branches are identical: P_+ comes in first and bifurcates slightly subcritically, closely followed by a supercritical bifurcation to P_- (off-scale). However, the secondary instabilities are markedly different. Figure 18b shows that in the absence of the coupling to the streaming flow the bifurcation diagram is identical to that discussed in Sections 2 and 3. A stability gap is present between the threshold at $\mu_0 = 1.00045$ and the Hopf bifurcation H on the mixed mode branch M. The gap is filled with stable and unstable periodic states. Since this situation is by now well understood we focus in the following on the new phenomena involving the streaming flow ($\gamma \neq 0$). Figure 18a shows that in this case the SB bifurcation to the mixed modes is *replaced* by a Hopf bifurcation producing R_1-symmetric oscillations. The resulting oscillatory branch follows the unstable mixed modes for $\gamma = 0$, at least initially, suggesting that the oscillations consist of an extended phase during which the system resembles an unstable mixed mode, i.e., that the oscillations created in the Hopf bifurcation on P_+ are *relaxation oscillations*. The coupling to the streaming flow has another effect as well: it introduces a new symmetry-breaking bifurcation SB *above* the saddle-node bifurcation. The resulting mixed modes are steady solutions of the five-dimensional system (23)-(25) and hence differ from the corresponding states in Figure 18b. Since the corresponding bifurcation on the P_- branch is supercritical the mixed mode branch is twice unstable near P_+ but is stable near P_-. It follows that there must be a tertiary Hopf bifurcation on the mixed mode branch, and numerically we find that this bifurcation occurs at $\mu_H \approx 1.034$. Thus the coupling to the streaming flow enlarges substantially the stability gap in comparison with the case $\gamma = 0$.

The close proximity of three secondary bifurcations on the P_+ branch makes this case particularly interesting. The saddle-node/Hopf interaction is responsible for the properties of the oscillations created at the Hopf bifurcation, and its unfolding corresponding to Figure 18a confirms not only

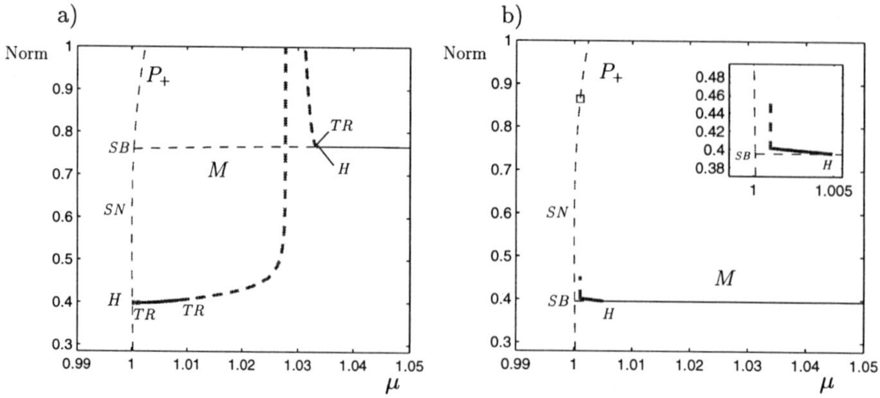

Fig. 18. Bifurcation diagrams for steady states and periodic orbits, in terms of the Euclidean norm $||(X, Y, v)||$ and the L_2 norm $||(X, Y, v)||_{L_2}$, respectively, as a function of μ, for $\Gamma = 0.15$, $\Lambda = 0.18$, $\alpha_1 = 0.48$, $\alpha_2 = -0.58$, $\varepsilon = 0.001$ when (a) $\gamma = -0.35$ and (b) $\gamma = 0$. Thin solid (dashed) lines correspond to stable (unstable) steady states. Thick solid (dashed) lines correspond to branches of stable (unstable) periodic orbits generated in Hopf bifurcations on P_+ and M, indicated by the letter H. Saddle-node, symmetry-breaking pitchfork and torus bifurcations are indicated by SN, SB and TR, respectively. In (b) the symbol □ indicates the homoclinic global bifurcation where the branch of periodic orbits, created at H, terminates. Note the rapid growth in the oscillation amplitude (inset) as this point is approached. Courtesy M. Higuera.

the presence of a branch of oscillatory solutions that bifurcates towards larger μ but also shows that this branch must be initially unstable before acquiring stability in a tertiary torus bifurcation [36, 37]. This theoretical prediction is confirmed in Figure 18a where the torus bifurcation is labelled TR ($\mu \approx 1.00136$). The figure shows, however, that the interval in which the periodic oscillations (thick line) are stable is quite narrow, and that these oscillations lose stability at larger μ at a second torus bifurcation ($\mu = 1.0083$). We presume that this loss of stability is a consequence of the proximity of these parameter values to the second codimension-two interaction: the saddle-node/pitchfork bifurcation. Indeed, Figure 18a shows that the periodic oscillations eventually become stable again (at a third torus bifurcation) before transferring stability at $\mu = \mu_H$ to the mixed modes M created at SB as already discussed. This Hopf bifurcation is in turn found in the appropriate unfolding of the saddle-node/pitchfork interaction [36, 37]. The asymmetric oscillations created at μ_H grow in amplitude with decreasing μ and at $\mu = \mu_g \approx 1.03$ glue pairwise and form the R_1-symmetric oscillations produced in the Hopf bifurcation on P_+. This gluing bifurcation takes place off-scale in Figure 18a, and eigenvalue computations show that it is of standard 'figure eight' type. Thus no chaos is associated with this global bifurcation. The thin regions of chaos associated with the termination of the torus branches [20,38,39] are usually too narrow to be of interest. However, there is much else that takes place in the stability gap, as we now describe.

In Figure 19a,b we show a two-frequency attractor obtained at $\mu = 1.0005$, that is, immediately after the trivial state becomes unstable ($\mu_0 = 1.00045$), with the mean flow included (Figure 18a). Figure 20a shows the corresponding time series. This attractor appears to describe oscillations *about* the small amplitude R_1-symmetric oscillation created at the Hopf bifurcation on the P_+ branch. The absence of exact R_1 symmetry suggests that this attractor is related to the symmetry-breaking torus bifurcation at $\mu \approx 1.00136$. This bifurcation appears to be supercritical, producing stable two-frequency oscillations in $\mu \lesssim 1.00136$. The R_1-symmetric oscillations undergo a second torus bifurcation at $\mu \approx 1.0083$ but this time the bifurcation is subcritical, i.e., the resulting two-frequency oscillations are present in $\mu < 1.0083$ but are now unstable. However, Figure 21 shows that this observation only scratches the surface of the complexity that is present in this parameter regime. In addition to the R_1-symmetric oscillations just mentioned (Figure 18 and lowest curve in Figure 21) there is a large number of *additional* branches of R_1-symmetric oscillations, four of which are included in the figure. These are labelled (a,b,c,d) and are stable near minimum period (solid lines) and unstable elsewhere (dashed lines). Figure 22 shows the projections of these stable oscillations in four cases, one from each branch. It is clear that the oscillations corresponding to the different branches differ in the number of revolutions about the two mixed modes $M^{\pm}$, and that multiple stable oscillations are present simultaneously. However, in all cases the oscillation period is of order $1/\varepsilon$ as expected of relaxation oscillations. The torus bifurcations at which these oscillations acquire stability with increasing μ may be responsible for the presence of chaotic oscillations in this regime, such as that shown in Figures 19b and 20b, but the details of this process are beyond the scope of this report.

The behavior reported in Figure 22 is a consequence of the two distinct slow time scales present in the problem, i.e., of the small value of the parameter ε. To see this we write Eqs. (23)-(25) in the form

$$\mathbf{X}' = \mathbf{F}_1(\mathbf{X}, \mathbf{Y}, v; \mu), \quad \mathbf{Y}' = \mathbf{F}_2(\mathbf{X}, \mathbf{Y}, v; \mu), \quad v' = \varepsilon G(\mathbf{X}, \mathbf{Y}, v), \tag{28}$$

where $\mathbf{X} = (\mathrm{Re}(X), \mathrm{Im}(X))$, $\mathbf{Y} = (\mathrm{Re}(Y), \mathrm{Im}(Y))$, and examine the case

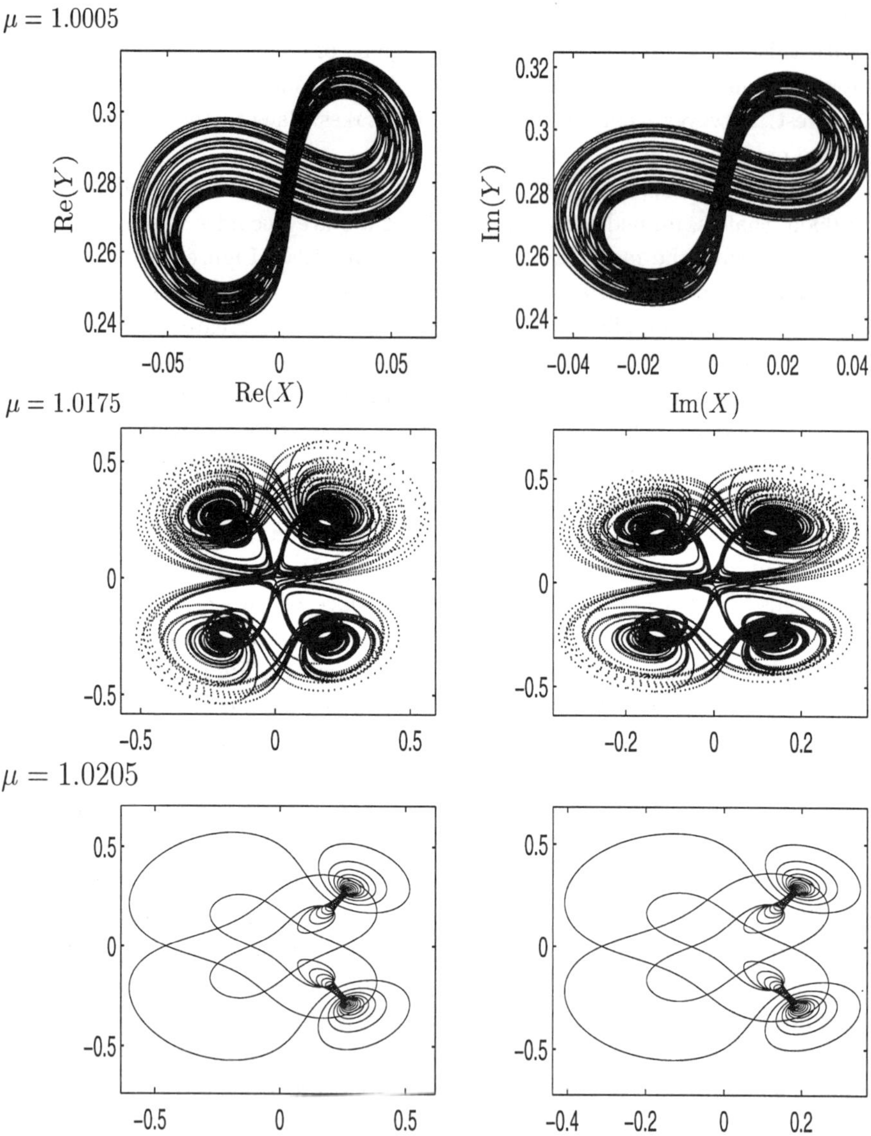

Fig. 19. Stable attractors of Eqs. (23-25) for $\Gamma = 0.15$, $\Lambda = 0.18$, $\alpha_1 = 0.48$, $\alpha_2 = -0.58$, $\gamma = -0.35$, $\varepsilon = 0.001$. (a) An asymmetric torus at $\mu = 1.0005$. (b) A chaotic attractor at $\mu = 1.0175$. (c) A R_2-symmetric periodic orbit at $\mu = 1.0205$. Courtesy M. Higuera.

$\varepsilon = 0$. In this case v becomes a parameter, and Eqs. (28) become

$$\mathbf{X}' = \mathbf{F}_1(\mathbf{X}, \mathbf{Y}; v, \mu), \quad \mathbf{Y}' = \mathbf{F}_2(\mathbf{X}, \mathbf{Y}; v, \mu). \tag{29}$$

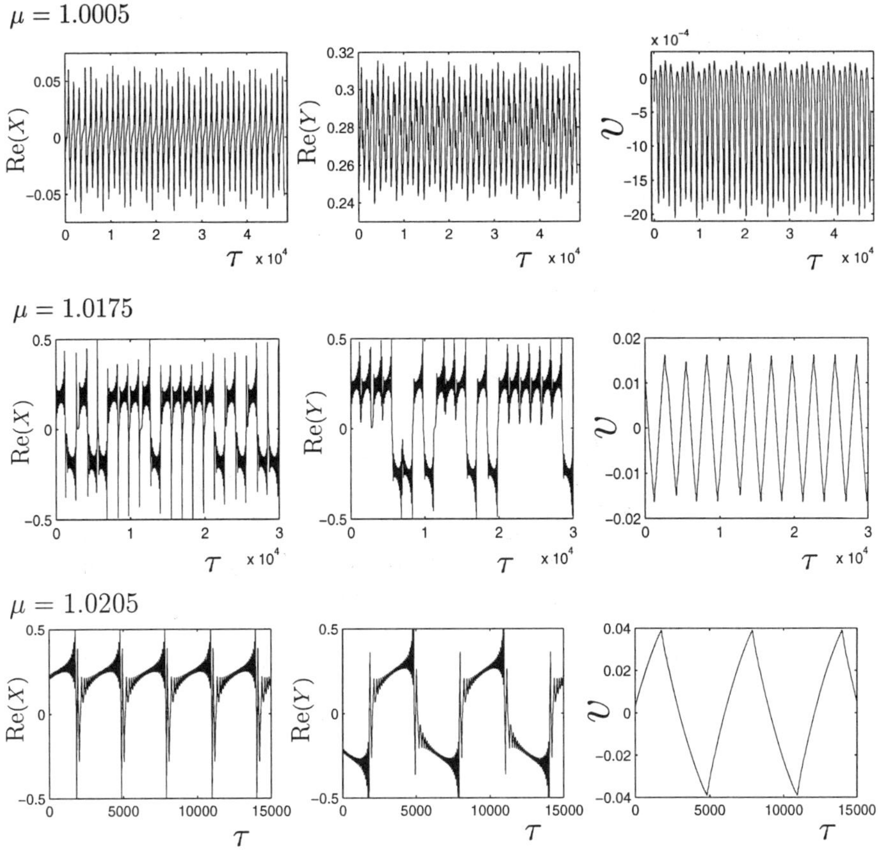

Fig. 20. Time series corresponding to the attractors in Figure 19. Courtesy M. Higuera.

In the following we refer to this system as the *fast* system even though the timescale for the evolution of $\mathbf{X}$, $\mathbf{Y}$ is long if $|\mu - \mu_0| \ll 1$. The nullcline $\Sigma : \mathbf{F}_1(\mathbf{X}, \mathbf{Y}, v; \mu) = \mathbf{F}_2(\mathbf{X}, \mathbf{Y}, v; \mu) = 0$ consists of fixed points of this system, parameterized by v. In the present case Σ consists of symmetry-related stable ($S^{\pm}$) and unstable ($U^{\pm}$) steady states. Of these the $S^{\pm}$ are created in pitchfork bifurcations from the trivial state as $|v|$ decreases. The $S^{\pm}$ states lose stability via subcritical Hopf bifurcations at $h^{\pm}$ and turn into $U^{\pm}$ (see Figure 23). The branches p of periodic oscillations created in these Hopf bifurcations are always unstable and terminate in a 'figure eight' gluing bifurcation involving an R_1-symmetric fixed point U_0. No chaos is associated with this bifurcation, and its presence does not appear to be

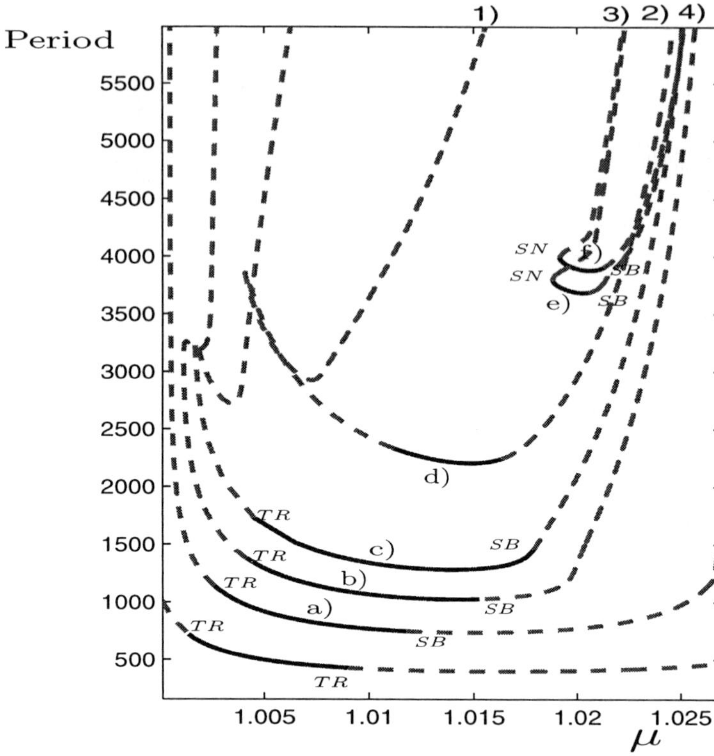

Fig. 21. The period of several branches of periodic oscillations as a function of μ when $\varepsilon = 0.001$. The branches (a,b,c,d) are R_1-symmetric, while (e,f) are R_2-symmetric. Solid (dashed) lines indicate stable (unstable) solutions. Saddle-node, symmetry-breaking pitchfork and torus bifurcations are indicated by SN, SB and TR, respectively. The lowest branch corresponds to that shown in Figure 18a. Courtesy M. Higuera.

relevant for the dynamics shown in Figure 22. Rather, the observed behavior appears to be a consequence of the fact that all four eigenvalues of $S^{\pm}$ are *complex*. When $0 < \varepsilon \ll 1$ these states all couple to the slow evolution of the variable v, and the manifolds of steady states and periodic orbits become part of the *slow manifold* of the system (28). In the following we speak of the solutions as drifting along this manifold (the slow phase); this drift proceeds until the system is forced away from the slow manifold, heralding the onset of the fast phase of the oscillation that takes it back to the slow manifold. Thus when $0 < \varepsilon \ll 1$ the solutions of Eqs. (28) drift slowly along the attracting slow manifolds $M_S^{\pm}$ associated with $S^{\pm}$; this drift is in the direction of increasing v for M_S^- and decreasing v for M_S^+. However,

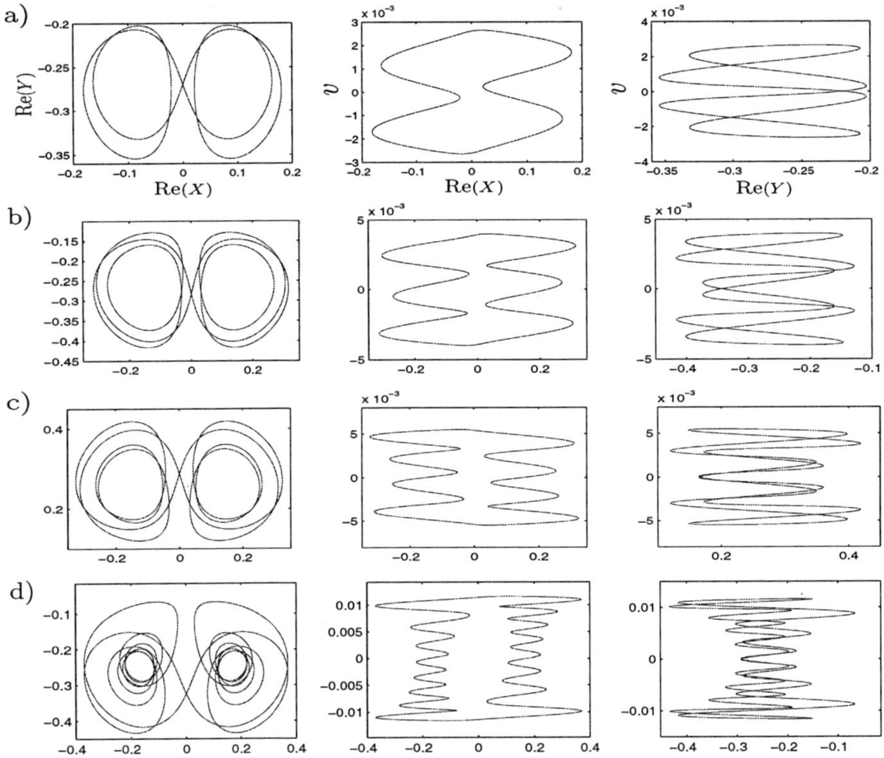

Fig. 22. Examples of stable R_1-symmetric oscillations on the branches (a,b,c,d) in Figure 21 at $\mu = 1.01185$. Courtesy M. Higuera.

when ε is sufficiently small the loss of stability of the fast motion is *delayed*, and the system drifts an $\mathcal{O}(1)$ distance in v before leaving the vicinity of the unstable manifold $M_U^{\pm}$ [40, 41]. This behavior can be seen clearly in Figure 24, computed for $\varepsilon = 0.0001$: the trajectory moves slowly along M_S^-, passes through a Hopf bifurcation (in the fast system) at h^-, and continues along the unstable manifold M_U^- for a time of order $1/\varepsilon$. The trajectory then spirals away from M_U^- and towards M_S^+. Once it is sufficiently close to M_S^+ the same process repeats, but this time with v decreasing. As shown in Figure 25 the drift along the slow manifolds $M_S^{\pm}$ and $M_U^{\pm}$ is faster at larger values of ε. With less time to approach $M_S^{\pm}$ the oscillations appear less and less "damped" and the orbits begin to resemble those in Figure 22.

The above point of view also permits us to understand the origin of the prominent cusp-like feature that develops on the higher branches of oscil-

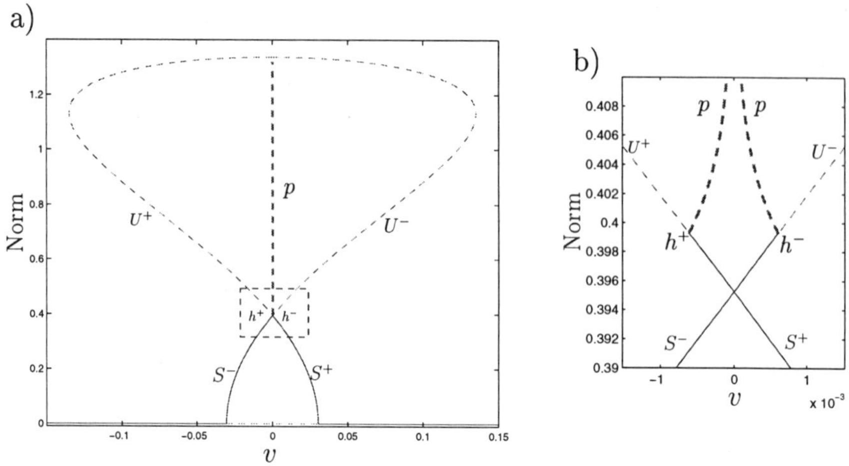

Fig. 23. (a) Bifurcation diagram for the fast system (29) showing the norm $||(X, Y)||$ as a function of v when $\mu = 1.01185$ and $\varepsilon = 0$. Thick dashed lines labelled p indicate branches of periodic orbits generated in Hopf bifurcations at $h^{\pm}$. Thin solid (dashed) lines indicate stable (unstable) steady states. The remaining parameters are as in Figure 19. (b) Detail of the bifurcation diagram in (a). Courtesy M. Higuera.

latory states as μ decreases (Figure 21). This structure is not due to an incipient global bifurcation. Instead what appears to be happening is the following. As μ decreases the real part λ of the unstable eigenvalue at the detachment point also decreases, implying a longer oscillation period. At the same time the Hopf frequency ω at h^- also decreases. For the transition from U^- to S^+ one must make at least half a turn around U^- before detachment, otherwise the trajectory ends up on the large amplitude U^- state (see Figure 23). This change in the type of oscillation becomes inevitable once ω/π falls below λ. Numerically we find that this condition is quite accurately satisfied at the tip of the cusp, while before the cusp is reached (i.e., for larger values of μ) $\lambda \lesssim \omega/\pi$. After the cusp the unstable two-dimensional manifold of U^- starts to veer away from the stable manifold of S^+ and begins to approach the three-dimensional stable manifold of the large amplitude U^- state. As this happens the period drops slightly but the oscillation amplitude starts to increase. Figure 26 shows the oscillations on either side of the (secondary) minimum in the period along branch (d) in Figure 21. Thereafter the period increases rapidly as the trajectory spends more and more time drifting along the slow manifold associated with the large amplitude unstable state U^- instead of the stable small amplitude

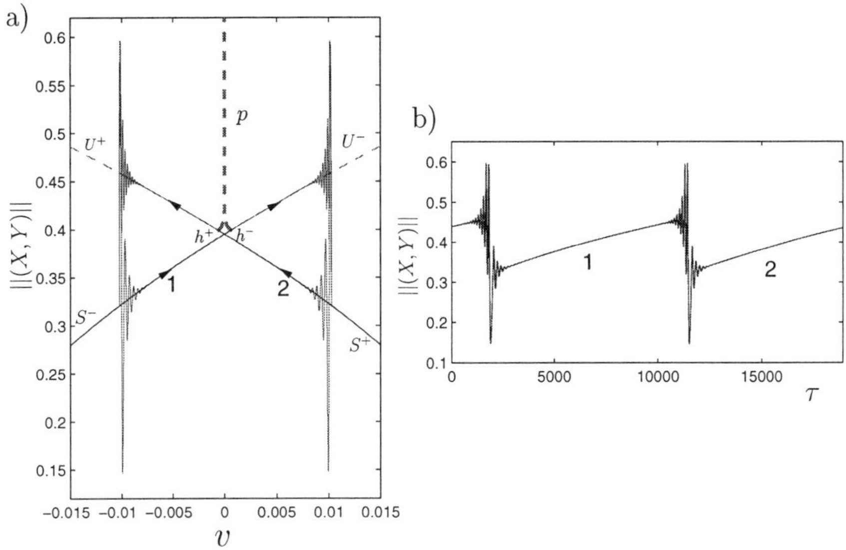

Fig. 24. (a) Stable relaxation oscillation computed from Eqs. (28) when $\mu = 1.01185$, $\varepsilon = 0.0001$, projected onto the $(v, ||(X,Y)||)$ plane for comparison with Figure 23. (b) The corresponding time series $||X(\tau), Y(\tau))||$. The remaining parameters are as in Figure 19. Courtesy M. Higuera.

state S^+, thereby acquiring an altogether different appearance. On the left each branch eventually terminates in a homoclinic connection to the pure mode P_+ contained in the slow manifold of U^- (see Figure 27, panel 1), while on the right it terminates in a homoclinic connection to the origin (see Figure 27, panel 2). Except for the complications arising from the presence of reflection symmetry much of the above phenomenology resembles that studied recently by Krauskopf and Wieczorek [42] and attributed to a nearby saddle-node Hopf bifurcation with no stable fixed points [43], cf. Figure 18a.

Figure 21 also shows two branches of R_2-symmetric oscillations labelled (e,f). These are also stable near minimum period; a stable R_2-symmetric oscillation on branch (e) is illustrated in Figures 19c and 20c. This type of solution can be described in an analogous manner to the R_1-symmetric orbits on branches (a,b,c,d) discussed above. The divergence of the oscillation period with both increasing and decreasing μ indicates that *all* of these branches appear and disappear through global bifurcations (see Figure 21). To identify these we show in Figures 27 and 28 high period unstable oscillations on the (d) and (f) branches, one near the initial appearance of each

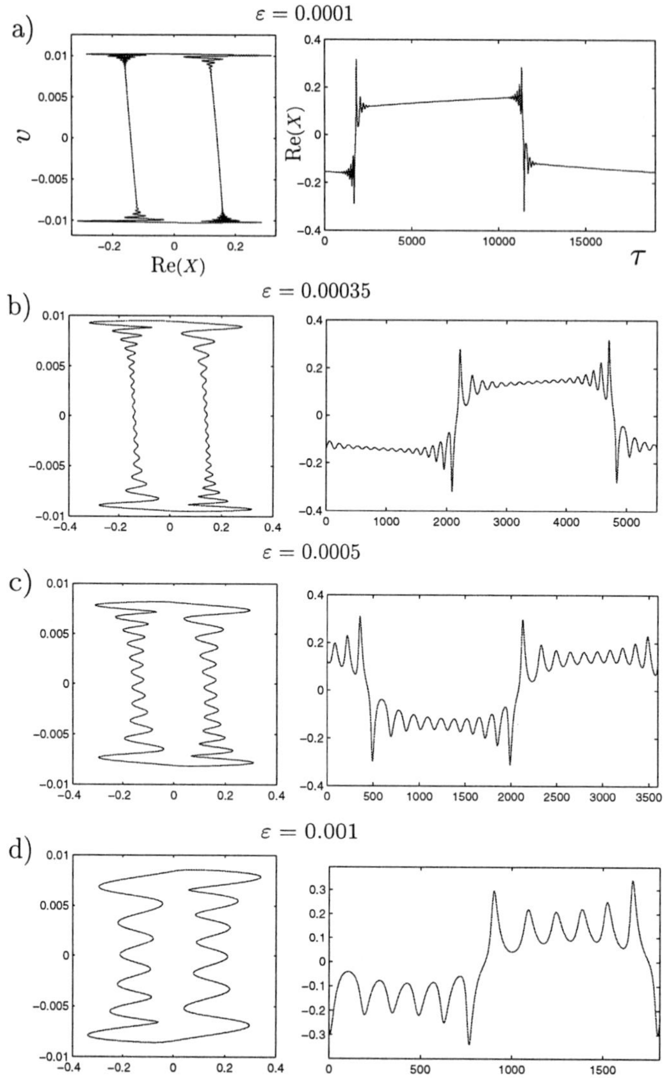

Fig. 25. Stable R_1-symmetric oscillations for $\mu = 1.01185$ and several different values of ε. The remaining parameters are as in Figure 19. In each case the time series (right column) show a single period only. Note that (d) corresponds to a branch omitted from Figure 21. Courtesy M. Higuera.

branch and one near its end. In contrast to the (d) branch the (f) branch originates and terminates in global bifurcations involving the origin and the steady states $\pm P_{\pm}$. However, despite the appearance of Figures 27 and

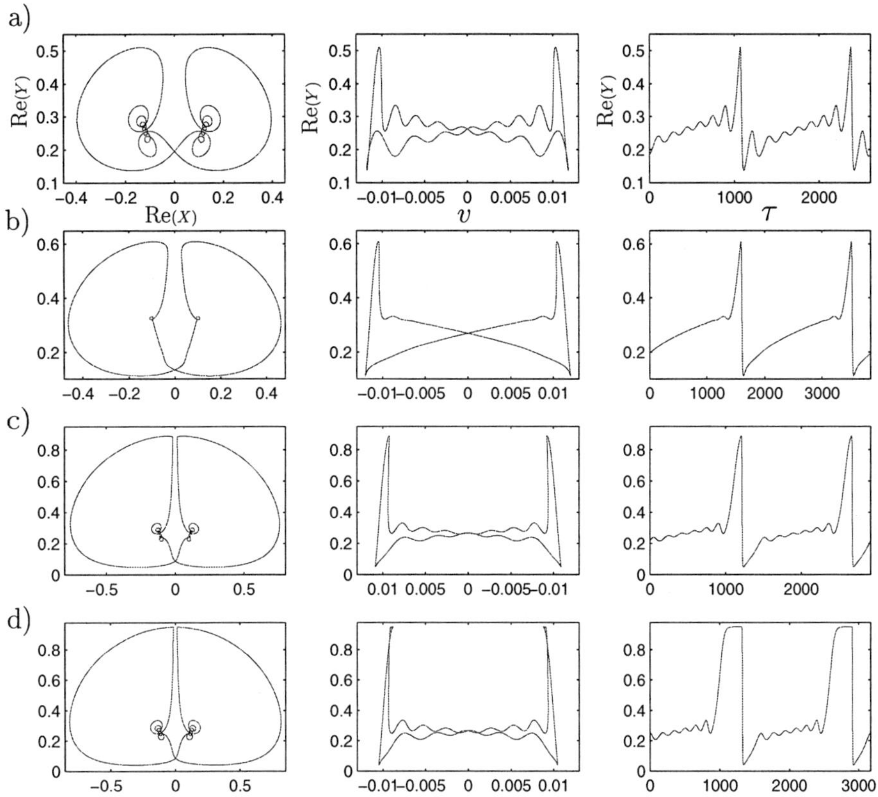

Fig. 26. Unstable periodic solutions in the cusp region on branch (d) of Figure 21, showing the continuous transition from relaxation oscillations involving the states S^+ to oscillations involving the large amplitude states U^- (see Figure 23). (a) $\mu = 1.008389$, near minimum period, (b) $\mu = 1.004072$, near the tip of the cusp, (c) $\mu = 1.00733$, near the secondary minimum, (d) $\mu = 1.00844$, to the right of the secondary minimum. Courtesy M. Higuera.

28, no heteroclinic connection to a periodic orbit actually occurs. In fact the observed behavior appears to be organized by codimension-two points corresponding to connections between the steady states $P_\pm$ and the origin. The figures suggest, and computations confirm, that all the eigenvalues of the fixed points involved are real.

It is clear, therefore, that the system (23)-(25) exhibits multiple co-existing stable states, some periodic (see Figure 21), others quasiperiodic or chaotic (see Figure 19). The latter are readily located between the symmetry-breaking bifurcation on branch (c) and the saddle-node bifur-

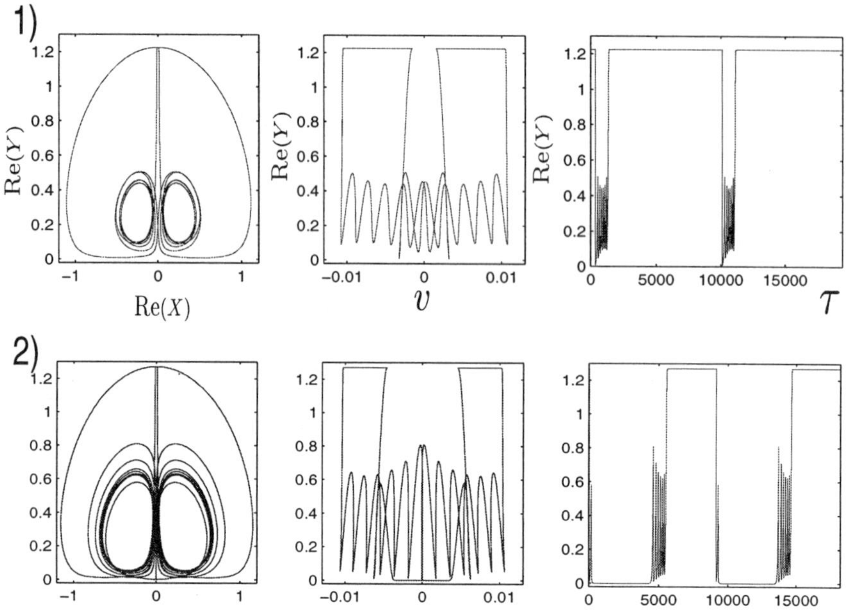

Fig. 27. Examples of large period unstable R_1-symmetric oscillations on branch (d) at the locations indicated in Figure 21. 1) $\mu = 1.0230$, 2) $\mu = 1.0258$. Courtesy M. Higuera.

cation on branch (e), and exist on either side of $\mu = 1.017$. Since we do not follow unstable tori we cannot identify the transitions that might lead from the various two-frequency states to the observed stable chaotic oscillations. It is likely, however, that each of the periodic oscillations depicted in Figure 22 bifurcates into chaos, and hence that the example in Figure 19b is but one from a number of coexisting intervals of chaos [42]. In fact, almost all of the behavior described here can also be found in a simpler three-dimensional system derived from Eqs. (23)-(25) that permits additional analysis, as reported elsewhere [44].

5. Conclusions

In this review we have described a very general principle, the opening of a stability gap due to the interaction with a competing mode, that is often responsible for the presence of complex time-dependence (periodic, quasiperiodic or chaotic) right at onset of a primary instability. This is so even when the primary instability is a steady state one, and occurs because of the destabilization of the saturated state by the competing mode. We have discussed how this occurs in three quite distinct circumstances, and illus-

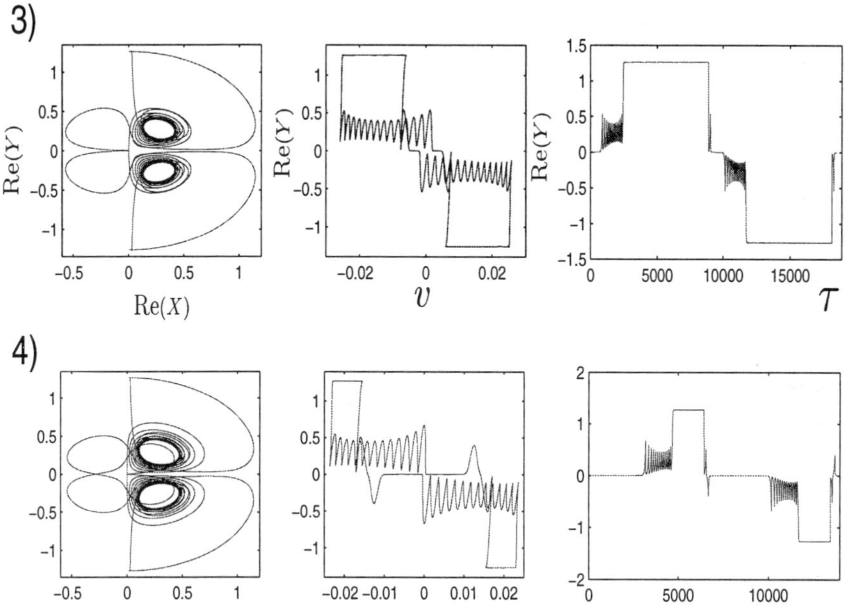

Fig. 28. Examples of large period unstable R_2-symmetric oscillations on branch (f) at the locations indicated in Figure 21. 3) $\mu = 1.02632$, 4) $\mu = 1.02680$. Courtesy M. Higuera.

trated the results with direct numerical simulations. From this comparison it is quite clear that the mechanism is robust and does occur in partial differential equations. Indeed the finite-dimensional models constructed on the basis of this insight reproduced essentially all the direct numerical simulation results, and in the case of binary fluid convection, the details of experimental observations as well. The reason for this success may be attributed to the fact that the models constructed in Sections 2 and 3 extend dramatically the standard codimension-two analysis used to identify the origin of complex dynamics in fluid dynamics. We expect, therefore, that a number of other systems of interest in fluid dynamics will be found where a similar mechanism may be identified.

The notion of a stability gap is undoubtedly a useful one. Indeed, in the Faraday system with $\Gamma = -0.5$, $\Lambda = 0.4$, $\alpha_1 = 0.4$, $\alpha_2 = -2.58$, $\gamma = -0.6$, $\varepsilon = 0.01$ such a gap is present for $\mu > \mu_{SB} \approx 2.8$ (Figure 29). As in Section 3 these are distinguished by their symmetries, and again take the form of *relaxation oscillations*. Figure 30 shows a dramatic example. This type of oscillation can again be interpreted in terms of a slow phase

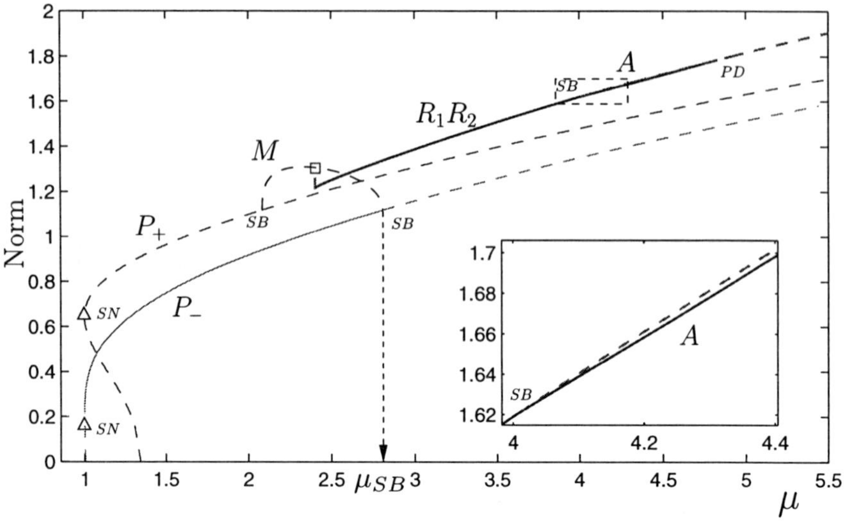

Fig. 29. Bifurcation diagram for steady states and periodic orbits, in terms of the Euclidean norm $||(X, Y, v)||$ and the L_2 norm $||(X, Y, v)||_{L_2}$, respectively, as a function of μ, when $\Gamma = -0.5$, $\Lambda = 0.4$, $\alpha_1 = 0.4$, $\alpha_2 = -2.58$, $\gamma = -0.6$, $\varepsilon = 0.01$. Thick solid (dashed) lines correspond to stable (unstable) periodic orbits; thin solid (dashed) lines correspond to stable (unstable) steady states. The arrow indicates the location of a subcritical bifurcation on P_- to unstable M. Courtesy M. Higuera.

(consisting of slow drift along the slow manifold, this time consisting of steady states and periodic orbits) interrupted by fast transitions between them (Figure 31). These and much more exotic sequences of bursts are discussed in greater detail in [34]. It is perhaps remarkable that the time series in Figure 30 appear indistinguishable from the type of (so-called parabolic) bursting behavior exhibited by neurons [45], even though in our case it involves visits to saddle-*foci*. The presence of relaxation oscillations is unusual in fluid mechanics, and ones that resemble bursting neurons doubly so, but here it is clearly a consequence of the weak damping of the streaming flow and its interaction with the gravity-capillary waves that drive it, in conjunction of course with the stability gap created by the destruction of the circle of standing waves when the ellipticity of the container becomes nonzero.

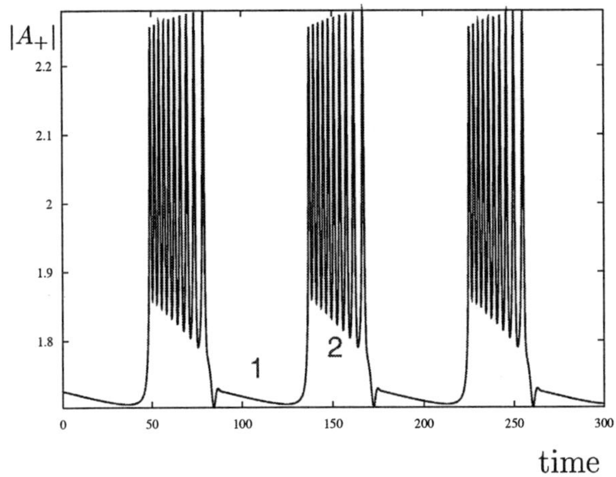

Fig. 30. Solution of Eqs. (21)-(22) with $\Gamma = -0.5$, $\Lambda = 0.4$, $\alpha_1 = 0.4$, $\alpha_2 = -2.58$, $\gamma = -0.6$, $\mu = 9.0$ and $\varepsilon = 0.01$, showing a sequence of bursts reminiscent of neuronal bursting. The labels 1 and 2 indicate the two slow phases, a drift along a branch of steady states and of periodic orbits of the fast system, respectively. Courtesy M. Higuera.

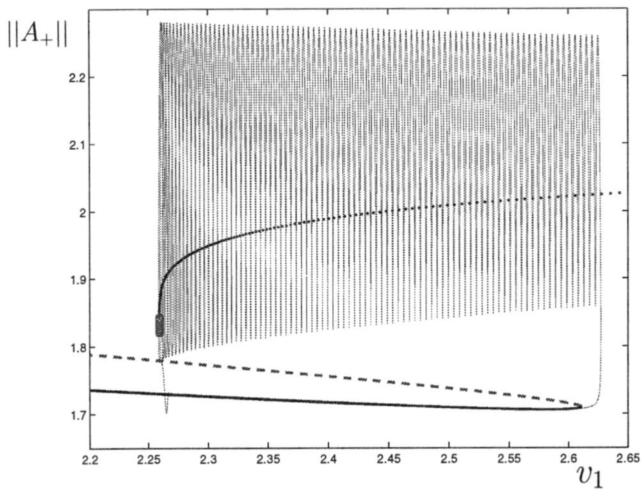

Fig. 31. The solution of Eqs. (21)-(22) with the parameters of Figure 30 but $\varepsilon = 0.0001$ ($||A_+|| \equiv (\text{Re}(A_+) + \text{Im}(A_+))^{1/2}$, thin line) projected onto the slow manifold ($\varepsilon = 0$). Courtesy M. Higuera.

Acknowledgement

The preparation of this article was supported by EPSRC under grant GR/R52879/01. The author is grateful to O Batiste, A Bergeon, M Higuera, C Martel, I Mercader, J Moehlis, M Net, J Porter and J M Vega for collaboration in the research reported here.

References

[1] E. Knobloch and J. Moehlis, in *Pattern Formation in Continuous and Coupled Systems*, M. Golubitsky, D. Luss and S. H. Strogatz (eds), Springer-Verlag: New York, pp. 157–174 (1999).
[2] E. Knobloch and J. Moehlis, in *Nonlinear Instability, Chaos and Turbulence*, vol. 2, L. Debnath and D. N. Riahi (eds), WIT Press: Southampton, pp. 237–287 (2000).
[3] P. Holmes, J. L. Lumley and G. Berkooz, *Turbulence, Coherent Structures, Dynamical Systems and Symmetry*, Cambridge Univ. Press: Cambridge (1996).
[4] T. S. Sullivan and G. Ahlers, *Phys. Rev.* ***A*** **38**, 3143 (1988).
[5] S. Rüdiger and E. Knobloch, Fluid Dyn. Research **33**, 477 (2003).
[6] J. Moehlis and E. Knobloch, Phys. Rev. Lett. **80**, 5329 (1998).
[7] O. Batiste, M. Net, I. Mercader and E. Knobloch, Phys. Rev. Lett. **86**, 2309 (2001).
[8] A. Bergeon and E. Knobloch, *Phys. Fluids* **14**, 3233 (2002).
[9] E. Knobloch, M. R. E. Proctor and N. O. Weiss, in *Turbulence in Fluid Flows: A Dynamical Systems Approach.* IMA Volumes in Mathematics and its Applications **55**, G. R. Sell, C. Foias and R. Temam (eds), Springer-Verlag: New York, pp. 59–72 (1993).
[10] D. R. Caldwell, *J. Fluid Mech.* **74**, 129 (1976).
[11] O. Batiste, E. Knobloch, I. Mercader and M. Net, *Phys. Rev.* ***E*** **65**, 016303 (2001).
[12] O. Batiste and E. Knobloch, in *Perspectives and Problems in Nonlinear Science*, E. Kaplan, J. Marsden and K. Sreenivasan (eds), Springer-Verlag: Berlin, pp. 91–144 (2003); O. Batiste and E. Knobloch, preprint.
[13] G. Dangelmayr and E. Knobloch, in *The Physics of Structure Formation: Theory and Simulation*, W. Güttinger and G. Dangelmayr (eds), Springer-Verlag: Berlin, pp. 387–393 (1987).
[14] G. Dangelmayr and E. Knobloch, *Nonlinearity* **4**, 399 (1991).
[15] P. Kolodner, *Phys. Rev.* ***E*** **47**, 1038 (1993).
[16] E. Kaplan, E. Kuznetsov and V. Steinberg, *Phys. Rev.* ***E*** **50**, 3712 (1994).
[17] A. E. Deane, E. Knobloch and J. Toomre, *Phys. Rev.* ***A*** **37**, 1817 (1988).
[18] C. Martel and J. M. Vega, *Nonlinearity* **11**, 105 (1998).
[19] O. Batiste, I. Mercader, M. Net and E. Knobloch, *Phys. Rev.* ***E*** **59**, 6730 (1999).
[20] J. Guckenheimer, in *Dynamical Systems and Turbulence.* Lecture Notes in

Mathematics **898**, D. A. Rand and L. S. Young (eds), Springer-Verlag: New York, pp. 99–142 (1981).
[21] E. Knobloch and D. R. Moore, *Phys. Rev.* ***A* 42**, 4693 (1990).
[22] V. Kirk and E. Knobloch, *Int. J. Bif. Chaos*, in press.
[23] P. Kolodner, C. M. Surko and H. Williams, *Physica* ***D*** **37**, 319 (1989).
[24] R. Heinrichs, G. Ahlers and D. S. Cannell, *Phys. Rev.* ***A* 35**, 2761 (1987).
[25] E. Moses, J. Fineberg and V. Steinberg, *Phys. Rev.* ***A* 35**, 2757 (1987).
[26] P. Kolodner, *Phys. Rev. A* **43**, 2827 (1991).
[27] F. Simonelli and J. P. Gollub, *J. Fluid Mech.* **199**, 349 (1989).
[28] H. Schlichting, *Phys. Z.* **33**, 327 (1932).
[29] M. S. Longuet-Higgins, *Phil. Trans. R. Soc.* ***A* 245**, 535 (1953).
[30] J. M. Vega, E. Knobloch and C. Martel, *Physica* ***D*** **154**, 313 (2001).
[31] E. Knobloch and J. M. Vega, in *Geometry, Mechanics and Dynamics*, P. Newton, P. Holmes and A. Weinstein (eds), Springer-Verlag: New York, pp. 181–222 (2002).
[32] E. Martín, C. Martel and J. M. Vega, *J. Fluid Mech.* **467**, 57 (2002).
[33] M. Higuera, J. M. Vega and E. Knobloch, *J. Nonlin. Sci.* **12**, 505 (2002).
[34] M. Higuera, E. Knobloch and J. M. Vega, *Physica* ***D***, preprint.
[35] J. W. Miles, *J. Fluid Mech.* **149**, 1 (1984).
[36] J. Guckenheimer and P. Holmes, *Nonlinear Oscillations, Dynamical Systems, and Bifurcations of Vector Fields*, Springer-Verlag: New York (1986).
[37] R. W. Wittenberg and P. Holmes, *Physica* ***D*** **100**, 1 (1997).
[38] V. Kirk, *Phys. Lett.* ***A* 154**, 243 (1991).
[39] V. Kirk, *Physica* ***D*** **66**, 267 (1993).
[40] A. I. Neistadt, *Diff. Uravneniya* **23**, 2060 (1987); **24**, 226 (1988).
[41] V. I. Arnol'd, V. S. Afrajmovich, Yu. S. Il'yashenko and L. P. Shil'nikov, in *Encyclopaedia of Mathematical Sciences*, Vol. 5, V. I. Arnol'd (ed), Springer-Verlag: New York, pp. 1–205 (1994).
[42] B. Krauskopf and S. Wieczorek, *Physica* ***D*** **173**, 97 (2002).
[43] M. K. S. Yeung and S. H. Strogatz, *Phys. Rev.* ***E*** **58**, 4421 (1998). See also Corrigendum **61**, 2154 (2000).
[44] M. Higuera and E. Knobloch, in preparation.
[45] X. J. Wang and J. Rinzel, in *Brain Theory and Neural Networks*, M. A. Arbib (ed), MIT Press: Cambridge, MA (1995).

CHAPTER 14

BIOLOGICAL LATTICE GAS MODELS

Mark S. Alber*, Maria Kiskowski*, Yi Jiang†, and Stuart Newman+

* *Department of Mathematics, 255 Hurley Building, University of Notre Dame, Notre Dame, IN 46556-4618, USA*

† *Theoretical Division, Los Alamos National Laboratory Los Alamos, NM 87545, USA*

\+ *Department of Cell Biology & Anatomy, Basic Science Building New York Medical College, Valhalla, NY 10595, USA*

Modelling pattern formation and morphogenesis are fundamental problems in biology. One useful approach is lattice gas cellular automata (LGCA) model. This paper reviews several stochastic lattice gas models for pattern formation in myxobacteria fruiting body morphogenesis and vertebrate limb skeletogenesis.

The fruiting body formation in myxobacteria is a complex morphological process that requires the organized, collective effort of tens of thousands of cells. It provides new insight into collective microbial behavior since myxobacteria morphogenic pattern formation is governed by cell-cell interactions rather than chemotaxis. We describe LGCA models for the aggregation stage of the fruiting body formation.

Limb bud precartilage mesenchymal cells in micromass culture undergo chondrogenic pattern formation, which results in the formation of regularly-spaced "islands" of cartilage analogous to the cartilage primordia of the developing limb skeleton. An LGCA model, based on reaction-diffusion coupling and cell-matrix adhesion, is described for this process.

1. Introduction

Two main approaches have characterized mathematical modelling of pattern formation (generation of specific arrangements of cells or other biological agents) and morphogenesis (generation of 3D forms from such agents) in developmental biology. In the first approach, insight into basic mechanisms is sought by studying models based on simplified biological assumptions about cell behavior. The second approach involves attempts to build more comprehensive models and compare results of numerical simulations and biological experiments in detail. In this paper we will demonstrate both approaches by describing the modelling of aggregation in myxobacteria and the formation of chondrogenic patterns in limb cell cultures.

Mathematical and computational models of biological systems are by necessity extreme simplifications of the real biological systems. A model is validated by its ability to reproduce the relevant behavior and make predictions that may be empirically confirmed. Many examples of mathematical modelling applied to biological problems can be found in [21, 26].

Models of biological problems fall into two categories: continuous models that use families of differential or integro-differential equations to describe "fields" of interaction, and discrete models in which space, time or state may be discrete. Models may be deterministic or stochastic.

In biological applications, continuous models have been used to describe oceanic microbial cycles [4], microbial growth dynamics [52], the spread of species through an ecosystem [55], biofilm formation [59], and periodic fungal or microbial structures [25]. This is by no means a complete list; many other examples of the use of PDEs in biological applications can be found in [30, 36, 49]. Equations in these models often describe fields of concentration or force and long-range interactions. For example, cells are often modelled as a density field and long-range chemotaxis is modelled as cell motion in response to a chemical field gradient.

Discrete models describe individual (microscopic) behaviors. They are often applied to micro-scale events where a small number of elements can have a large (and stochastic) impact on a system. For example, while many periodic growth patterns can be modelled using continuous methods, such patterns which depend sensitively on substrate concentration are best modelled with discrete methods [61] including cellular automata (CA). The CA models have been reviewed in [3, 20].

In lattice gas models, particles move freely on a spatial grid and undergo state changes when they collide. These state changes can be stochastic or deterministic. Lattice gas models have been used for fibroblast aggregation, ant trail organization and topographical neural maps [20]. In classical lattice gases, there is an exclusion rule in which no two particles with the same orientation may occupy a node. Biological lattice gas models often relax this exclusion principle.

This paper reviews several stochastic lattice gas cellular automata (LGCA) models for biological pattern formation. Section 2 provides a general definition for LGCA and discusses the suitability of these models for describing biological pattern formation related to morphogenesis. LGCA are especially suited for modelling biological problems due to their versatility and the ease with which they may be designed and forward-evaluated. Their disadvantages are that they may require significant computer resources and sometimes are difficult to analyze.

Developmental morphogenesis is the molding of living tissues during development, regeneration, wound healing, and disease. It is a complex phenomenon involving gene regulation, changes in cell shape, cell-cell interactions, and cell division, growth and migration. Representing cell shape realistically is an important problem in modelling morphogenesis. An original method of cell representation consistent with lattice gas models is described in Section 3. In this method, particles may interact over a biologically-relevant geometric region but may be moved without violating the exclusion principle. This "template-based" approach is an efficient and generally applicable method for representing cells of any size, shape and orientation.

Our general approach is demonstrated in Sections 4 and 5 by modelling the aggregation stage of fruiting body formation in myxobacteria and pattern formation of precartilage cells in chick limb cell cultures. Each section includes a discussion of the choice and justification of the local rules for state evolution as well as an analysis of the novel properties of the models and comparison with experimental data.

2. Lattice Gas Models

With the availability of new computational techniques in combination with large scale computing facilities, *in silico* experiments are becoming more and more an important option, in additional to *in vivo* and *in vitro* experiments, to study pattern formation in biological systems. Lattice gas cellular automata (LGCA) have become a rather popular and extremely flexible modeling tool in this context [3, 16, 20, 40].

Cellular automata (CA) consist of discrete agents or particles, which occupy some or all sites of a regular lattice. These particles have one or more internal state variables (which may be discrete or continuous) and a set of rules describing the evolution of their state and position. Both the movement and change of state of particles depend on the current state of the particle and those of neighboring particles. Again, these rules may either be discrete or continuous (in the form of ordinary differential equations (ODEs)), deterministic or probabilistic. Updating can be synchronous or stochastic.

In 1973 Hardy, de Passis and Pomeau [24] introduced models to describe the molecular dynamics of a classical lattice gas (hence "Hardy, Passis and

Pomeau"(HPP) models). They designed these models to study ergodicity-related problems and to describe ideal fluids and gases in terms of abstract particles. Their model involved particles of only one type which moved on a square lattice and had four velocity states. Later models extended the HPP in various ways and became known as lattice gas cellular automata (LGCA). LGCA proved well suited to problems treating large numbers of uniformly interacting particles.

LGCA employ a regular, finite lattice and include a finite set of particle states, an interaction neighborhood and local rules which determine the particles' movements and transitions between states [16]. LGCA differ from traditional CA by assuming particle motion and an exclusion principle. The connectivity of the lattice fixes the number of allowed velocities for each particle. For example, a nearest-neighbor square lattice has four non-zero allowed velocities. The velocity specifies the direction and magnitude of movement, which may include zero velocity (rest). Particles follow the exclusion rule. Thus, a set of Boolean variables describes the occupation of each allowed particle state: occupied (1) or empty (0). Each lattice site can then contain from zero to five particles.

The transition rule of an LGCA has two steps. An interaction step updates the state of each particle at each lattice site. Particles may change velocity state, appear or disappear in any number of ways as long as they do not violate the exclusion principle. For example, the velocities of colliding particles may be deterministically updated, or the assignment may be random. In the transport step, cells move synchronously in the direction and by the distance specified by their velocity state. Synchronous transport prevents particle collisions which would violate the exclusion principle (other models define a collision resolution algorithm). LGCA models are specially constructed to allow parallel synchronous movement and updating of a large number of particles [16].

LGCA can model a wide range of phenomena including the diffusion of ideal gases and fluids [32], reaction-diffusion processes [10] and population dynamics [51]. For details about CA models in physics see [11] and specifically for lattice-gas models see [7,62]. In their biological applications LGCA treat cells as point-like objects with an internal state but no spatial structure. Because living cells function as "agents" interacting according to set rules, LGCA particularly suit modelling collective cellular behaviors [2,12,13,18,19].

To summarize, LGCA are inherently simple; their discrete nature makes them straightforward to implement by computer and they lend themselves to "agent-based" approaches which reflect the intrinsic individuality of cells. Local rules can be developed from a microscopic level phenomenon understanding which is direct and intuitive. LGCA provide an opportunity to study interactions and behaviors which are difficult to formulate as continuum equations.

General limitations of LGCA are: the difficulty of going from qualitative to quantitative simulations since backward evolution is difficult; the artificial constraints of lattice discretization; and the difficulty in interpreting the simulation outcomes. The patterns that may emerge from even simple rules can be so rich that determining whether the model has genuinely captured the relevant biological mechanisms is difficult. Also, in many cases there is no faster way to predict the outcome of a simulation than to run the simulation [61]. Another major problem is detecting and avoiding artificial behavior [16], (for example, 'checkerboard' patterns) which result from the overly simple geometry or local rules.

3. Representing Cell Shape

In classical LGCA each particle (cell) is dimensionless and is represented as a single occupied node on a lattice. Cells without dimension are untenable for a sophisticated model of morphogenesis. Biological cells may be very elongated and cell polarity determines movement and interaction. Also, a realistic model of cell overlap and cell stacking is needed since interaction may occur only at specific regions of highly elongated cells and cell density is a critical parameter throughout morphogenesis.

One way to resolve the problem of stacking is to introduce a semi-three-dimensional lattice where a third z-coordinate gives the vertical position of each cell when it is stacked upon other cells [8]. In a CA model for streaming and aggregation in myxobacteria, rod-shaped cells are modelled to occupy many nodes and have variable shapes [57]. These two models are not classical lattice gases since they do not incorporate synchronous transport along channels. Below we describe a novel way of representing cells which facilitates modelling variable cell shape, cell stacking and incomplete cell overlap while preserving the advantages of classical lattice gases; namely, synchronous transport and binary representation of cells within channels.

In a template-based model [1–3] cells are represented as (1) a single node which corresponds to the position of the cell's center (or "center of mass") in the xy plane, (2) the choice of occupied channel at the cell's position designating the cell's orientation and (3) a local neighborhood defining the physical size and shape of the cell with associated interaction neighborhoods. The interaction neighborhoods depend on the dynamics of the model and need not exactly overlap the cell shape. In the models for rippling and aggregation, the size and shape of the cell are defined as a $3 \times \ell$ rectangle, where ℓ is cell length. The cell width is greater than 1 to account for interaction in the vertical y direction. As ℓ increases, the cell shape becomes more elongated. A cell length of $\ell = 30$ corresponds to the 1×10 proportions of rippling *Myxococcus xanthus* cells [37]. Representing a cell as an oriented point with an associated cell shape is computationally efficient, yet approximates continuum dynamics more closely than assuming point-like cells. The cell stacking problem is also resolved, since overlapping cell shapes correspond to cells stacked on top of each other. This cell representation conveniently extends to changing cell dimensions and the more complex interactions of fruiting body formation.

4. A Model for Myxobacteria Aggregation

Myxobacteria are one of the prime model systems for studying cell-cell interaction and cell organization of a uniform cell type preceding differentiation. Myxobacteria are social bacteria which swarm, feed and develop cooperatively [34]. When starved, myxobacteria self-organize into a three-dimensional fruiting body structure. Fruiting body formation is a complex multi-step process of alignment, rippling, streaming and aggregation that culminates in the differentiation of highly elongated, motile cells into round, non-motile spores. (See Figure 4, from [41] with permission.)

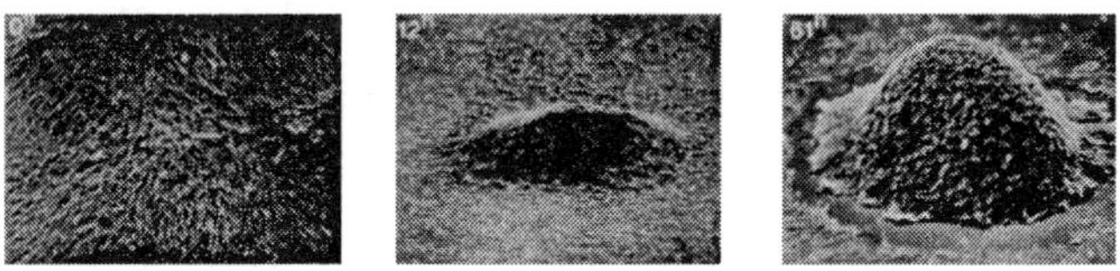

Fig. 1. Snapshots during the fruiting body formation of *M. xanthus* at 0h, 12h, and 61h.

During aggregation, myxobacteria cells are elongated with a 2:1 to 14:1 ratio (cells are typically 2 to 12 by 0.7 to 1.2 μm, see [53]), and they move on surfaces by gliding along their long axis [9]. Several hours after the onset of aggregation, cells begin transmitting a membrane-associated signaling protein called *C-signal* via their cell poles [37]. Repeated C-signaling between cells raises the level of C-signal by two signal amplification loops in the *act* pathway. Increasing levels of C-signal induce aggregation [56] and sporulation.

4.1. *Computational Model*

Models for bacteria aggregation have been based on attractive chemotaxis, e.g. *E. Coli* [58] and *Bacillus subtilis* [5,48]. Successful models for the fruiting body formation of the eukaryotic slime mold *Dictyostelium discoideum* also include chemotaxis [5,29,45–47]. Chemotaxis is a long range cell interaction. Initialization of chemotactic signals plays an important role in the initial position of aggregates and subsequent signaling biases cell motion towards developing aggregates [58]. In myxobacteria, however, aggregates appear to form without the aid of chemotactic cues; rather, myxobacteria fruiting body development is organized by short-range cell-cell contact-mediated interactions [17].

Computational models based on cell collisions, a non-chemotactic short-range interaction, were first applied to explain myxobacteria rippling patterns [2, 8, 27, 43]. A recent paper [28] has extended an earlier continuous model for rippling to include myxobacteria aggregation. Our model [1] is complementary to the continuum model, and focuses on a two-stage aggregate formation via streams.

Identical rod-shaped cells are modelled as 3×21 rectangles and assume a cell size of 1×7 μm. The model keeps track of the C-signal exchange neighborhood of each occupied node which defines the possible locations of head-to-end overlaps between C-signaling cells. The total C-signaling neighborhood for each cell is fourteen nodes; seven at each cell pole separated by one half a cell length from the cell center. Cells follow the following local rules:

(1) Cells move on a hexagonal lattice with periodic boundary conditions imposed at all four edges. Unit velocities are allowed in each of the six directions.
(2) Cells are initially randomly distributed with density δ, where δ is the total cell area divided by total lattice area.
(3) At each time-step, cells may turn 60 degrees clock-wise, counter-clockwise or stay in their current direction, with a preference for directions that maximize the overlap of C-signaling nodes at the head of a cell with the C-signaling nodes at the tails of other cells.
(4) During the transport step, all cells move synchronously one node in the

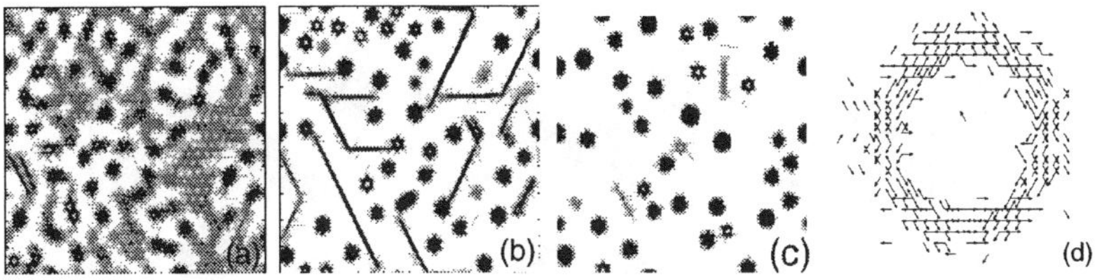

Fig. 2. Aggregation stages on a 500 × 500 lattice, which corresponds to an area of 2.8 cm^2. Local cell density after (a) 200, (b) 900, and (c) 25,000 timesteps. Average cell density is 10. The number of simulated cells is 39,507. The darker shade of gray corresponds to higher cell density. (d) Directions of cell centers within a typical annular aggregate.

direction of their velocity by updating the positions of their centers.

4.2. *Simulation Results*

Cells aggregate in distinctive stages in our simulations. During the first stage, cells turn from low density areas towards areas of slightly higher cell density. Initially randomly distributed cells condense into small stationary aggregates (Figure 2 (a)). These aggregate centers grow and absorb immediately surrounding cells. Next, some adjacent stationary aggregates merge and form long, thin streams which extend and shrink dynamically on their own and in response to interactions with other aggregates (Figure 2 (b)). In each stream cells move head-to-tail in either direction along the stream. These streams are transient and eventually disappear at later stages of the simulation, leaving behind a new set of larger, denser stationary aggregates which are stable over time (Figure 2 (c)). Cells in a typical aggregate form an annulus of aligned cells tangent to a hollow center (Figure 2(d)).

Figure 3 shows the details of stream formation from two interacting aggregates. Initial aggregates crowd as they grow. When the distance between aggregates is less than one cell length, they begin exchanging cells, and the cells reorganize into a stream. In contrast to stationary aggregates, cells travel long distances in streams.

A stream is bi-directional, with cells flowing equally in both directions along the stream. Given the end-to-end contacts required for C-signaling, an infinitely long stream of cells flowing in two directions is obviously a stable arrangement. However, there are a fixed number of cells within simulation streams, and thus streams are of finite length. Cells at the end of streams do not C-signal in the open space, hence will diffuse without any preferred direction. Although randomly diffusing cells often find their way back into

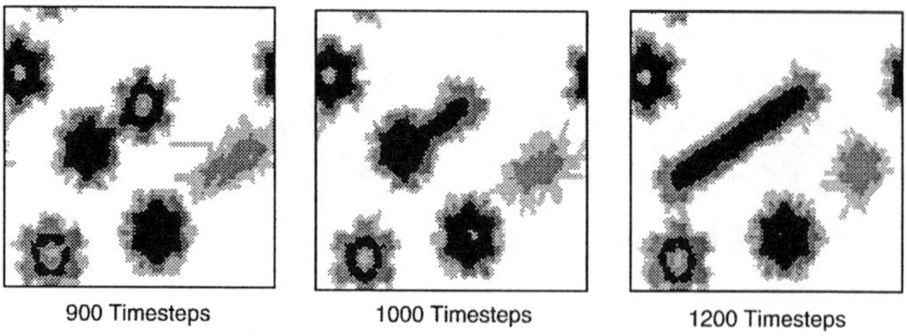

Fig. 3. The stream formation from two adjacent aggregates. Panels left to right correspond to 900, 1000, and 1200 timesteps, respectively. Lattice size is 128×128.

the stream, some cells escape away from the stream. Over time, the stream shortens as it gradually loses cells. A stream will lose cells more quickly if there is an aggregate near the end of the stream to absorb cells diffusing at the ends of the stream. The eventual fate of a stream is to become a small stable aggregate.

In simulation, depending on the aggregate size, aggregates form one of six distinctive types, shown in Figure 4 ordered by increasing size. Very small, typically early, aggregates in our simulation have the characteristics of early developing *Myxococcus xanthus* aggregates. Early *Myxococcus xanthus* aggregates are asymmetric and have been referred to as "traffic jams", since it is assumed that cell motility is hindered by many cells trying to move in antagonistic directions [35]. Likewise in simulations, the directions of cells in initial aggregates of type I are disordered and cells are analogously "jammed": tracking of cells in these aggregates has shown that cells rarely travel more than one quarter of a a cell length before turning several times and entirely reversing their direction. Also, simulation aggregates round up as more cells are added to the aggregate (compare aggregates of type I with aggregates of type II-VI in Figure 4), just as asymmetric aggregates in experiment grow and gain circular symmetry from 8 to 24 hours [41] (Figure4).

Larger aggregates in our simulation have the unique structure of mature myxobacteria aggregates for several myxobacteria species. In *Myxococcus xanthus*, the basal region of the fruiting body is a shell of densely packed cells which orbit in two directions, both clock-wise and counter-clockwise, around an inner region only one-third as dense [31, 54]. A magnified picture of the cell centers of aggregate type II (Figure 4) in our simulation show that cells are arranged in dense, concentric layer tangent to a relatively low-density inner region. Cell tracking shows that cells orbit either

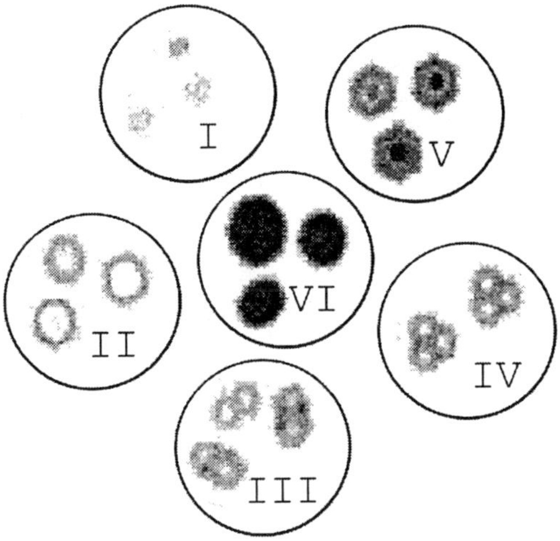

Fig. 4. Six aggregate types, I-VI, ordered by increasing aggregate size, aggregates identified within two simulations over 25,000 timesteps.

clockwise or anti-clockwise along the periphery of the orbit. The fruiting bodies of myxobacteria often occur in fused clusters called sporangioles (for example, *S. erecta* (Reichenbach, 1993)). Intermediate-sized aggregates in our simulation form in clusters of two or three closed orbits (III and IV in Figure 4). The largest simulation aggregates (Type V and VI) have no hollow center. Presumably, modelling in three dimensions would be required to resolve their three-dimensional structure.

The area-density phase diagram, Figure 5, enables a prediction of the final aggregate shape based on the number of cells within the stream. This diagram was obtained by plotting the areas and densities of every stationary aggregate which appeared over the course of two simulations. These aggregates fall within a narrow range in the area-density phase diagram, illustrating that for an aggregate of a given cell number, its area and density is prescribed within a narrow attractor region. In [1] the stability of this attractor region is analyzed with respect to two kinds of noise: 1) external noise, including those in the initial random distribution of cells and our perturbations to the system; 2) internal noise, which originate from the stochastic nature of the cell's turning process. Results from both kinds of perturbations suggest that this attractor region is a stable attractor in the

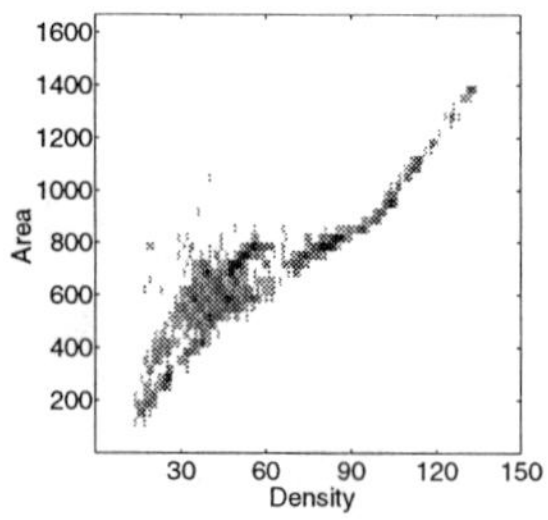

Fig. 5. Area-density phase diagram for 186 stationary aggregates identified within two simulations over 25000 timesteps.

area-density phase space of aggregates.

The area-density phase diagram not only prescribes the region of stable aggregates, it also helps ones understanding of the formation of streams. When two stationary aggregates interact, the area of interacting cells increases at the moment of interaction while the density remains approximately the same. Thus, the newly formed aggregate lies off the attractor region. Large aggregates with high cell density and area will fuse and quickly form a new stationary aggregate. Smaller aggregates have a lower cell density and lower cell C-signaling levels, so when small aggregates fuse, they have a longer transient stage and are more likely to form a stream.

5. A Model for Chondrogenic Patterning

Chondrogenesis is the development of cartilage, which is a tough flexible tissue from which the skeleton of the limb and vertebral column is initially established during embryonic development. The cartilage skeleton of the limb develops as a series of spot-like and bar-like condensations of the precartilage cells. In these condensations the cells (also called "mesenchymal" cells), which are initially dispersed in a gel-like medium called the extracellular matrix (ECM), move closer to one another, making transient contact, which triggers their differentiation into cartilage. In species with bony skeletons, such as birds and mammals, the cartilage skeleton serves as a structural template which is later replaced by bone. In low volume, high density ("micromass") cultures of precartilage mesenchymal cells, depending on the culture medium and the limb-type origin of the cells (wing buds vs leg buds), patterns of condensations form which may be spot-like, stripe-like, or fully confluent (sheet-like).

This section presents a stochastic lattice gas model for the formation of patterns of mesenchymal condensations in micromass cultures. This *in vitro* system provides a simplified, experimentally tractable model for skeletal patterning in the vertebrate limb. Its quasi-2D geometry suits computational modelling particularly well (see [39, 40] for details). For the computational model to be manageable one must select key processes from the hundreds of cell-cell and cell-gene product interactions in the limb. Our choice for the "core" set of patterning interactions described below comes from experiments performed on the limb-forming cells of several species.

Chondrogenenesis in micromass culture can be described as follows. Over 36-72 hours in a controlled experiment, a homogeneously distributed population of undifferentiated limb bud mesenchymal cells cluster into dense islands, or "condensations", of aggregated cells [42]. The roughly equally spaced patches of approximately uniform size are reminiscent of the patterns produced by the classical Turing reaction-diffusion mechanism. The condensations develop concurrently with increases in extracellular concentrations of a cell-secreted protein, fibronectin, a non-diffusing extracellular matrix macromolecule which binds adhesively to cell surface molecules, including receptors known as integrins, which can transduce signals intracellularly. The limb cells also produce the diffusible protein TGF–β, which positively regulates its own production as well as that of fibronectin [50]. In what follows, we describe a model for the production of fibronectin and subsequent limb bud patch formation using an LGCA-based reaction–diffusion process having TGF-β as the activator but with an unknown inhibitor.

5.1. *Computational Model*

Our computational model contains three components.

First, we model cells as occupied nodes of a square lattice (i.e., a rectangular grid) whose default behavior is random walk diffusion (analogous to Brownian motion). We describe the diffusion of cells and other diffusing agents on the lattice, below. We assume all cells are identical.

Second, we simulate on the lattice a cell-driven process that depends on the interaction between two molecular species: a diffusible activator, A, which we identify with TGF-β in the developmental model, and an inhibitor, B, which we identify with the laterally-acting inhibitory activity in the developmental model. Increased levels of activator stimulate production of activator and inhibitor while inhibitor decreases the effective activator concentration. For the purpose of the computational model we make the assumption that B is a diffusible molecule, with a faster diffusion rate than

A. Since cells produce these molecular species, only nodes of the lattice that contain cells produce morphogens.

Third, when cells encounter threshold levels of activator, they respond by producing a secreted, but otherwise immobile, molecular species to which cells attach. We term this substrate adhesion molecule (SAM) and identify it with fibronectin in the developmental model.

Our model defines a "morphogenic domain" on a square $n \times n$ lattice by an $n \times n$ matrix of 0s and 1s. A '0' indicates a node outside of, and a '1' indicates a node belonging to, the morphogenic domain. The domain, all portions of which need not be connected, and which can have holes, can freely change at each time step and could be calculated by an auxiliary program. The only restriction on the domain is that it is a union of overlapping rectangles of at least two lattice points in height and width. In the current simulations, the morphogenic domain is the entire $n \times n$ lattice. The components in the morphogenic domain of the lattice include cells, activator molecules, inhibitor molecules and SAM molecules. We store the concentration of each of these components as an $n \times n$ matrix of integers, where the matrix element (i,j) corresponds to the concentration of the various components at the node (i,j). Multiple cells and molecules of each type may occupy a node. Boundary conditions for the morphogenic domain of the lattice are reflective, so that particles (cells or molecules) cannot diffuse beyond the domain boundary.

We initially distribute a fixed number of cells uniformly on a disc-shaped region centered in the morphogenic domain of the lattice. We set initial densities of activator, inhibitor and SAM to zero. The total number of cells remains constant throughout the simulation and cells secrete activator, inhibitor, and SAM molecules. Activator and inhibitor molecules diffuse through the morphogenic domain of the lattice at every time-step, while SAM diffuses only during the time-step in which it is secreted.

Simulation results that depict cells, represent SAM-attached cells as black pixels and unattached cells as gray pixels. This representation corresponds to the appearance of cells in micromass cultures visualized by Hoffman Modulation Contrast microscopy, where the rounded cells in condensations appear darker than the flatter cells outside the condensations.

5.2. *Simulation Results*

In this section we demonstrate comparison between simulation and experimental results by describing chemical peaks and cell clustering under standard "leg" conditions.

In living leg-cell cultures the initial cell distribution is homogeneous, but, by the second day of growth, cells begin to form tightly packed focal condensations. Spacing between condensations is irregular, with a measurable average distance between centers (i.e., the average peak interval). The average condensation size generally increases in wing-cell cultures as the condensations expand and often merge [15, 42], whereas it stays fixed in leg-cell cultures where most condensations remain discrete [14,15]. The terminal pattern in the leg cultures occurs around 3 days and in the wing cultures around 4 days. The condensations differentiate fully into cartilage nodules by day 6.

We initially explored the behavior of the CA model under conditions that simulate those of a typical limb-cell micromass culture. The initial micromass diameter *in vitro* is 3 mm. Although living cells in standard *in vitro* experiments are plated at greater than confluent density (see. e.g., [15]), a layer of ECM rapidly separates the cells. Our simulations assumed a matrix to cell area ratio of 60:40, a cell diameter of 15 mm, and a "culture" spot diameter of 120 cells. Thus we model cells with average density 1 on a disk 120 nodes in diameter (see Figure 6).

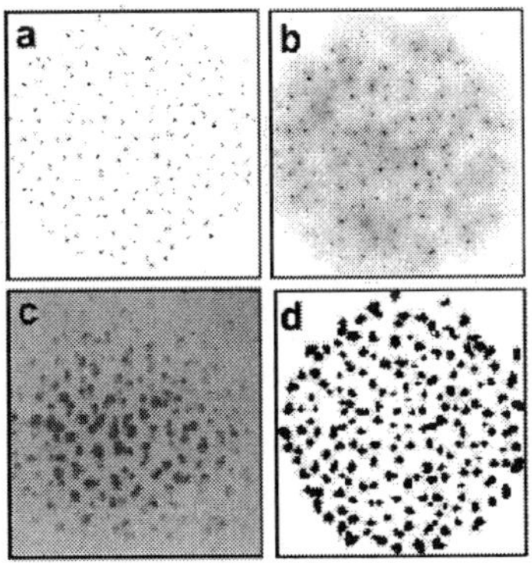

Fig. 6. Simulation results for scaled concentrations of key variables compared to the leg condensation pattern. (a) Morphogen A concentration, and (b) morphogen B concentration, after 4000 time-steps for control leg parameters; (c) Leg condensations visualized by Hoffman Contrast Modulation optics after 72 hours. Actual diameter of the circular culture is 3 mm ("low magnification view"); (d) Scaled SAM density after 4000 time-steps for control leg parameters. Simulations correspond to a low magnification view.

As cells diffuse, one or more cells, or no cells, may occupy a node, so the production of activator and inhibitor varies over the lattice. Peaks of morphogen A and B begin to appear early in the simulation. Morphogen A peaks are only 1 or 2 nodes in size and levels drop from over 1000 units in a peak to zero units in immediately surrounding nodes (Figure 6a). Morphogen B peaks are larger in size and much more diffuse (Figure 6b). For comparison, we show *in vitro* condensations at the scale of a full micromass culture for the comparable experimental stage (Figure 6c). When the level of morphogen A in the simulation reaches a threshold, the cells begin to deposit SAM (Figure 6d). Cells stick to these SAM deposits and local cell density increases. Activator, SAM and cell concentration peaks are all co-local (compare Figure 7 d, e and f).

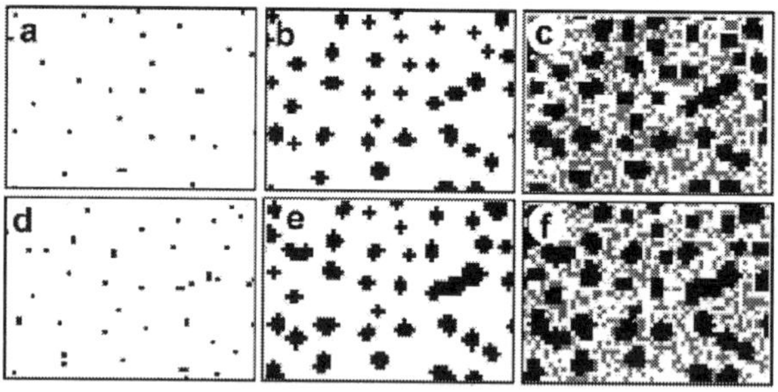

Fig. 7. Simulation dynamics of activator concentrations, SAM accumulation, and cell density over 1000 to 4000 time-steps for control parameters. (a) Morphogen A concentrations greater than threshold (1000 units), (b) SAM locations and (c) smoothed free (gray) and stuck (black) cell locations after 1000 time-steps. (d) Morphogen A concentrations greater than threshold (1000 units), (e) SAM locations and (f) smoothed free (gray) and stuck (black) cell locations after 4000 time-steps. Simulation images correspond to a high magnification view.)

Comparison of the development of condensations in experiments and simulations indicated that the period spanned by computational time steps 1000-4000 corresponded to 50-72 hours in culture. In particular, the cell density distribution after 1000 time-steps for the optimized parameters qualitatively resembles precartilage condensations after 50 hours. During the next 3000 times-steps the islands' areas grow but their number remains unchanged, as in experiments after 72 hours.

Simulated morphogen peaks, SAM deposits and cell clusters develop in both time and space (Figure 7). The number of SAM deposits does not increase between 1000 and 4000 time-steps, indicating that almost all activator peaks form within 1000 time-steps. Activator peaks remain colocal with SAM deposits. SAM clusters grow, however, and occasionally fuse over 4000 time-steps (compare Figure 7 b and e).

The model was shown [39] to successfully determine a non-trivial parameter set that reproduces the number, size, and distribution of condensations of the standard culture, and demonstrate the ability of this parameter set to produce qualitatively accurate simulations of cultures under diluted, TGF-β-stimulated, reduced-inhibitor and fibronectin-transfected conditions. Therefore, the model captures important aspects of development.

Indeed, LGCA modelling as presented in [39], far from being a retrospective summary of existing experiments, is actually a parallel means of experimentation on systems with partially characterized relevant variables and parameters have been (e.g., chondrogenic patterning *in vitro*). It is an efficient and cost-effective tool for homing in on the range of potential manipulations that can provide decisive tests of *in vitro* and *in vivo* experimental models.

Acknowledgments

This work was supported by NSF Grant No. IBN-0083653 and by DOE under contract W-7405-ENG-36. We acknowledge support from the Center for Applied Mathematics and the Interdisciplinary Center for the Study of Biocomplexity at the University of Notre Dame.

References

[1] M.S. Alber, Y. Jiang and M.A. Kiskowski, *Phys. Rev. Lett.* in press.
[2] M.S. Alber, Y. Jiang, and M.A. Kiskowski, *Physica D* **191**, 343 (2004).
[3] M.S. Alber, M.A. Kiskowski, J.A. Glazier and Y. Jiang, in J. Rosenthal and D.S. Gilliam (Eds.), *Mathematical Systems Theory in Biology, Communication, and Finance*, Springer-Verlag, New York, 2002, p. 1.
[4] A.P. Belov, J.D. Giles, *Hydrobiologia* **349**, 87 (1997).
[5] E. Ben-Jacob, I. Cohen, and H. Levine, *Advances in Physics* **49**, 395 (2000).
[6] D. Beysens, G. Forgacs, and J. A. Glazier, *Proc. Natl. Acad. Sci. USA* **97**, 9467 (2000).
[7] J. Boon., D. Dab, R. Kapral, and A. Lawniczak, *Phys. Reps.* **273**, 55 (1996).
[8] U. Borner, A. Deutsch, H. Reichenbach, and M. Bar, *Phys. Rev. Lett.* **89**, 78101 (2002).
[9] R.P. Buchard, *Annu. Rev. Microbiol.* **35**, 497 (1981).

[10] S. Chen, S. P. Dawson, G. D. Doolen, D. R. Janecky, and A. Lawniczak, *Computers & Chemical Engineering* **19**, 617 (1995).
[11] B. Chopard and M. Droz, *Cellular automata modeling of physical systems*, Cambridge University Press, NY, 1998.
[12] A. Deutsch, *J. Biosc.* **24**, 115 (1999).
[13] A. Deutsch, *Mathematical Biosciences* **156**, 255 (1999).
[14] S.A. Downie, and S.A. Newman, *Dev. Biol.* **162**, 195 (1994).
[15] S.A. Downie, and S.A. Newman, *Dev. Biol.* **172**, 519 (1995).
[16] S. Dormann, *Pattern Formation in Cellular Automation Models*, Dissertation, Angewandte Systemwissenschaft FB Mathematik/Informatik, Universität Osnabrück, 2000.
[17] M. Dworkin, *Microbiol. Rev.* **60**, 70 (1996).
[18] L. Edelstein-Keshet, G.B. Ermentrout, Differentiation **45** (1990), 147.
[19] L. Edelstein-Keshet, G.B. Ermentrout, *J. Math. Biol.* **29**, 33 (1990).
[20] G.B. Ermentrout and L. Edelstein-Keshet, *J. Theor. Biol.* **160**, 97 (1993).
[21] C. Fall, E. Marland, J. Wagner, and J. Tyson (Eds.), *Computational Cell Biology: An Introductory Text on Computer Modeling in Molecular and Cell Biology*, Springer-Verlag, New York, 2002
[22] G. Forgacs, R. Foty, Y. Shafrir, and M. Steinberg, *Biophys. J.* **74**, 2227 (1998).
[23] D. Godt and U. Tepass, *Nature* **395**, 387 (1998).
[24] J. Hardy, O. de Pazzis, and Y. Pomeau, *Phys. Rev. A* **13**, 1949 (1976).
[25] F.C. Hoppensteadt and W.P. Jager, *A hysteresis model for bacterial growth patterns: modelling of patterns in time and space*, Springer-Verlag, Berlin, **55** (1984), p. 123.
[26] F.C. Hoppensteadt, C.S. Peskin, *Modeling and simulation in medicine and the life sciences*, 2nd edition, Springer, Berlin, 2001.
[27] O. Igoshin, A. Mogilner, D. Kaiser, and G. Oster,*Proc. Natl. Acad. Sci. USA* **98**, 14913 (2001).
[28] O. Igoshin, R. Welch, D. Kaiser, G. Oster, *Proc. Natl. Acad. Sci. USA* **101**, 4256 (2004).
[29] Y. Jiang, H. Levine, and J.A. Glazier, *Biophys. J.* **75**, 2615 (1998).
[30] D.S. Jones and B. D. Sleeandhoopesman, *Differential Equations and Mathematical Biology*, CRC Press, Boca Raton, USA, 2003.
[31] B. Julien, D. Kaiser and A. Garza, 2000. Spatial control of cell differentiation in *Myxococcus xanthus*. *Proc. Natl. Acad. Sci. USA*. 97:9098–9103.
[32] L. P. Kadanoff, G. R. McNamara, and G. Zanetti, *Phys. Rev. A* **40**, 4527 (1989).
[33] D. Kaiser, private communication.
[34] D. Kaiser and L. Kroos, in *Myxobacteria II*, M. Dworkin and D. Kaiser (Eds.), Am. Soc. Microbiol., Washington DC, 1993, p. 257.
[35] D. Kaiser and R. Welch, *J. Bacteriol.* **186**, 919 (2004).
[36] J. P. Keener, J. Sneyd, *Mathematical Physiology*, Springer, New York, 1998.
[37] S.K. Kim and D. Kaiser, *Science* **249**, 926 (1990).
[38] S. Kim and D. Kaiser, *J. Bacteriol.* **173**, 1722, (1991).

[39] Kiskowski, M.A., M.S. Alber, Thomas, G.L., Glazier, J.A., Bronstein, N., Pu, J., and Newman, S.A., *Dev. Biol.* **271**, 372 (2004).

[40] M.A. Kiskowski, *Discrete Stochastic Models for Morphological Pattern Formation in Biology*, Ph.D. Thesis, University of Notre Dame (2004).

[41] J. Kuner and D. Kaiser, *J. Bacteriol.* **151**, 458 (1982).

[42] Leonard, C.M., Fuld, H., Frenz, D.A., Downie, S.A., Massagu, J., and Newman, S.A., *Dev. Biol.* **145**, 99 (1991).

[43] F. Lutscher and A. Stevens, *J. Nonlinear Sciences* **12**, 619 (2002).

[44] C. Ma, Y. Zhou, P.A. Beachy, and K. Moses, *Cell* **75**, 927 (1993).

[45] A. Marée, *From pattern formation to morphogenesis: Multicellular coordination in Dictyostelium discoideum*, Ph.D. Thesis., Utrecht Univ. (2000).

[46] A.F.M. Marée and P. Hogeweg, *Proc. Natl. Acad. Sci. USA* **98**, 3879 (2001).

[47] J. Martiel and A. Goldbeter, *Biophys. J.* **52**, 807 (1987); T. H *o*fer, J.A. Sherratt, and P.K. Maini, *Proc. R. Soc. London B* **259**, 249 (1995).

[48] M. Matsushita and H. Fujikawa, *Physica A* **168**, 498 (1990); I. Golding, Y. Kozlovsky, I. Cohen, and E. Ben-Jacob, *Physica A* **260**, 510 (1998); A. Komoto, K. Hanaki, S. Maenosono, J.Y. Wakano, Y. Yamaguchi and K. Yamamoto, *J. Theor. Biol.* **225**, 91 (2003).

[49] J. Murray, *Mathematical Biology*, Springer-Verlag, New York, 2003.

[50] S. Newman,*Bioessays* **18**, 171 (1996).

[51] C. Ofria, C. Adami, T. C. Collier, and G. K. Hsu, *Lect. Notes Artif. Intell.* **1674**, 129 (1999).

[52] N.S. Panikov, *Microbial Growth Kinetics*, Chapman and Hall, London, 1994.

[53] H. Reichenbach, in *Myxobacteria II*, M. Dworkin and D. Kaiser (Eds.), American Soc. Microbio., Washington DC, 1993.

[54] B. Sager, and D. Kaiser, *Proc. Natl. Acad. Sci. USA* **90**, 3690 (1993).

[55] E.M. Scott, E.A.S. Rattray, J.I. Prosser. K. Killham, L.A. Glover, J.M. Lynch, M.J. Bazin, *Soil Biol Biochem.* **27**, 1307 (1995).

[56] L.J. Shimkets, R.E. Gill and D. Kaiser, *Proc. Natl. Acad. Sci. USA* **80**, 1406, (1983); S.K. Kim and D. Kaiser, *J. Bacteriol.* **173**, 1722 (1991); S. Li, B. Lee and L.J. Shimkets, *Genes Dev.* **6**, 401 (1992).

[57] A. Stevens, *SIAM J. Appl. Math.* **61**, 172 (2000).

[58] L. Tsimring, H. Levine, I. Aranson, E. Ben-Jacob, I. Cohen, O. Shochet, and W.N. Reynolds, *Phys. Rev. Lett.* **75**, 1859 (1995).

[59] O. Wanner, *Biofouling* **10**, 31 (1996).

[60] J. Wartiovaara, M. Karkinen-Jääskelänen, E. Lehtonen, S. Nordling, and L. Saxén, *Progress in Differentiation Research*, North-Holland Publishing Company, Amsterdam, 245 (1976).

[61] J. Wimpenny in C.R. Bell, M. Brylinsky, and P. Johnson-Green (Eds.), *Microbial Biosystems: New Frontiers Proceedings of the 8th International symposium on Microbial Ecology*, Atlantic Canada Society for Microbiol Ecology, Halifax, Canada (1999).

[62] D. Wolf-Gladrow, *Lattice-gas cellular automata and lattice Boltzmann models - an introduction*, Springer-Verlag, Berlin, Lect. Notes in Math., **1725** (2000).

CHAPTER 15

A COMPARISON OF OPTIMAL LOW DIMENSIONAL PROJECTIONS OF A HURRICANE SIMULATION

Amanda Fox *, Michael Kirby *, John Persing †, and Michael Montgomery †

* *Department of Mathematics, Colorado State University, Fort Collins, CO 80523-1874, USA*

† *Department of Atmospheric Science, Colorado State University, Fort Collins, CO 80523-1371, USA*

We introduce the Signal Fraction Analysis (SFA) methodology for understanding large data sets associated with the simulation of a multi-scale physical system, i.e., hurricanes. We compare the results of this approach with the well-known Karhunen-Loève (KL) procedure which has been widely applied for characterizing coherent structures in a variety of physical systems. The SFA method separates the simulation data into subspaces with independent scales while the KL approach separates into components based on maximizing the capture of statistical variance. It appears that this study represents the first application of these methods to the hurricane problem.

1. Introduction

As coastal populations continue to grow and hurricane track forecasts improve, intensity predictions for landfalling hurricanes will become an increasing priority for the United States. At the present time hurricane intensity forecasts leave much to be desired [2]. More than seventy percent of the tropical cyclone damage and destruction in the United States is caused by the small percentage of intense hurricanes with sustained tangential winds greater than 50 m s^{-1} [11]. Recently, with winds over 140mph, the category 4 Hurricane Charley caused an estimated $15 billion in damages along the Florida coastline: the numerical simulations by the National Hurricane Center actually predicted significantly more damage.

A hurricane comprises a multitude of interacting scales, spanning individual convective cells to the larger synoptic-scale. Observations using ground-based and airborne Doppler radar provide a helpful but still incomplete picture of intensity variations (see [14], and references therein). At the heart of the hurricane intensity problem is the interaction of a rotating convective region that is of a very small scale ($\sim$ 50 km) with broader scale influences such as winds, moisture, temperature, and surface heterogeneities (including land/sea differences) that influence the storm from scales as large as $\sim$ 1000 km. This large dynamic range of interactions presents an enormous challenge for the accurate prediction of hurricane intensity.

To date, current methodologies used to forecast intensity are limited to heuristic approaches (climatology), multiple regression limited to synoptic-scale meteorological variables (SHIPS), and direct numerical simulations at coarse resolution [10] extracting only minimum surface pressure and maximum tangential velocity. While these methods have increased our understanding of hurricane intensity to some degree they leave many fundamental questions unanswered.

Recent advances in high performance computing are permitting the development of high (spatio-temporal) resolution numerical simulations of hurricanes and their surrounding environment, e.g., full-physics models such as the Pennsylvania State University/National Center for Atmospheric Research, Mesoscale Model 5 (MM5). These simulations offer detailed information that should furnish new insights into the nature of the interactions among different scales and how these interactions govern the intensity of the hurricane. Because of the massive data sets generated by these models it seems apparent that a mixture of physical and mathematical pattern analysis techniques will be required to identify the key processes that control hurricane maximum intensity in an idealized tropical environment.

In this paper we propose a preliminary study of the complexity of coherent structures exhibited by a 2D axisymmetric hurricane simulation using the Rotunno and Emanuel [15] model. In Section 2 we present the basic ideas behind the Karhunen-Loève and Signal Fraction approaches for characterizing large data sets. In Section 3 we present the results and in Section 4 we discuss our conclusions and future work.

2. Reduction Methodology

We consider two methods and compare their results. The Karhunen-Loève (KL) procedure (and the related methods of principal component analysis, singular value decomposition, empirical orthogonal functions) is by now well-documented, see, e.g., [8, 13] and references therein. A suite of papers illustrating the application of the KL methodology [16–18] still provide an excellent introduction to ideas of particular interest in fluid dynamics. See also [3] for more details related to this investigation. Also of interest in the context of hurricanes is the Empirical Normal Modes (ENM) study of Chen et. al. [1]. The second, less well-known method, is referred to as the maximum noise fraction approach (MNF) [5,20], or signal fraction analysis (SFA) [9]. The KL technique is based on optimizing the entropy in the data while the SFA approach serves to *separate* the data according to a quotient optimality criterion. KL captures variance in the data while SFA can split a signal from noise, or prominent coherent structures from each other [6]. See also [21] for a related study.

The representation of the simulation data in terms of an optimal expansion may be written

$$u(x,t) = \sum_{i=1}^{N} a_i(t)\varphi^{(i)}(x)$$

where the ith time-varying coefficient $a_i(t)$ is found by the projection of a data snapshot onto the basis function $\varphi^{(i)}(x)$.[a] We now review the two methods we use to compute the expansion set.

2.1. *Karhunen-Loeve Decomposition*

Now using vector notation, given an ensemble of vectors $\{x^{(\mu)}\}_{\mu=1}^{P}$, with each $x^{(\mu)} \in \mathbb{R}^N$, we seek a set of basis vectors $\{\varphi^{(j)}\}_{j=1}^{N}$ such that the error of the truncated expansion is minimized in the mean-square sense. The solution to this problem is provided by the eigenvectors of the equation

$$C\varphi = \lambda\varphi$$

where

$$C = XX^T$$

and the columns of X are the data points; see, e.g., [8] for details of this derivation. Of course this $N \times N$ problem is prohibitively large in our problem (in this investigation $N = 56000$) so we make use of the observation that writing

$$\varphi = X\psi$$

[a]Of course the actual calculations are all finite dimensional and the expansion is in terms of vectors.

results in the $P \times P$ (smaller) eigenvector problem

$$X^T X \psi = \lambda \psi$$

These alternative formulations produce the left and right singular vectors in the singular value decomposition. This approach was developed under the name of the *snapshot* method [16–18] and has a broad range of interesting applications, including face recognition [7, 19]. The basis is ordered based on the eigenvalues which capture the statistical variance, or energy. The total energy captured by the data is denoted

$$E_N = \sum_{i=1}^{N} \lambda_i$$

where λ_i is the i-th eigenvalue.

2.2. *Signal Fraction Analysis*

Now consider the situation where the observed data actually consists of the sum of two signals, i.e.,

$$X = A + B$$

If A and B share distinct subspaces then the Signal Fraction Analysis algorithm provides a basis that can separate them. Additionally, under some side assumptions, it is possible to show that this same basis can be used to solve the mixing problem, i.e., solve for S in the equation

$$X = MS$$

where M is an unknown mixing matrix and X is the observed signal. This is known as the Blind Signal Separation (BSS) problem. We refer the reader to [6] for details of how SFA may be used to solve the BSS problem.

We proceed as above by writing the SFA basis vectors in their data dependent format

$$\varphi = X\psi$$

where of course the φ and ψ are now different. If we use the decomposition of X we can further decompose φ as

$$\varphi = A\psi + B\psi$$

where $\varphi_A = A\psi$ and $\varphi_B = B\psi$ represent the dependency of the new basis vector on the individual data matrices A and B, respectively. A useful separation criterion would then be to compute φ such that

$$D(\varphi) = \frac{\varphi_A^T \varphi_A}{\varphi_B^T \varphi_B}$$

is a maximum (or minimum). Or, in terms of ψ,

$$D(\psi) = \frac{\psi^T A^T A \psi}{\psi^T B^T B \psi}$$

We can interpret the quantities in this ratio $\alpha^2 = \psi^T A^T A \psi$ and $\beta^2 = \psi^T B^T B \psi$ as the average alignment of the rows of A and B with ψ, respectively. Clearly, this criterion achieves the separation of the data by optimizing one at the expense of the other via the quotient.

The resulting solution to this optimization problem can be shown to be a generalized singular value decomposition, i.e.,

$$\beta A^T A \psi = \alpha B^T B \psi$$

Further details and references on the GSVD are available in [4]. Now the orthonormal basis vectors $\varphi^{(i)} = X \psi^{(i)}$ possess a natural ordering based on the signal fraction $\lambda_i = \alpha_i / \beta_i$.

In our particular application we have taken A to be the data set X and B to be the derivatives of the data set X by applying differencing in the vertical direction. This is tantamount to generating a basis φ that separates the original signal from its detail. This follows as the differencing we applied is equivalent to projecting the data onto the finest scale Haar wavelet subspace. Here we see the rational for the term signal fraction analysis which emphasizes the flexibility with which the quantities A and B may be chosen.

3. Results

The model used here is the 2D axisymmetric hurricane simulation due to Rotunno and Emanuel [15]. This model simulates seven state variables (pressure, liquid water content, water vapor content, potential temperature, radial, tangential and vertical winds) over a physical domain of 1500 km by 25 km defined on a grid of 100 by 80 points. Thus, each snapshot of the state of the hurricane consists of 56000 points. The simulation was run for eight days with sampling every 5 minutes, yielding 2305 snapshots. The variables are then each concatenated (in alphabetical order) to create a single vector of size 56,000 x 1 for each of the 2305 output times resulting in a data matrix D of size 56000 x 2305. The result of averaging the snapshots over time is shown in Figure 1. This mean is removed from the data in subsequent calculations. In a data set such as this one, there are many options for data scaling.[b] In this study, we scale each variable in every snapshot to have unit variance.

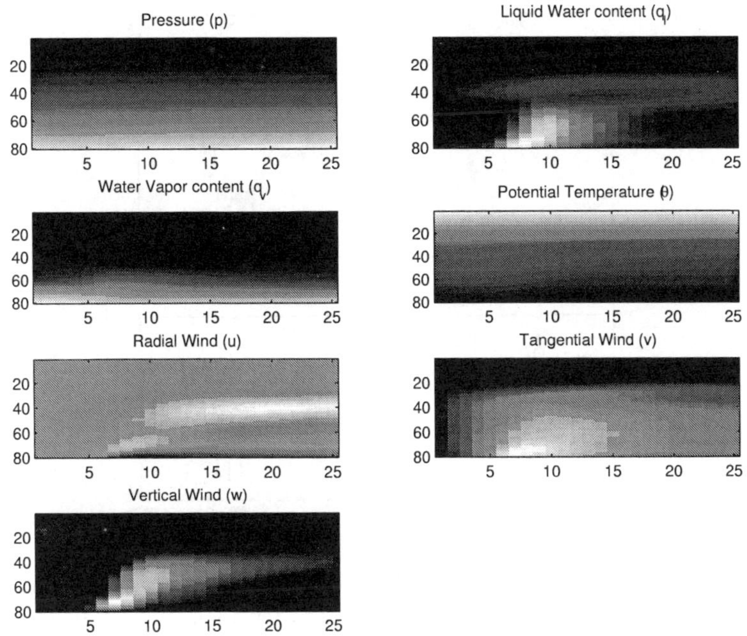

Fig. 1. The mean

3.1. *KL Results*

In this section we present the results of applying the KL procedure to the hurricane simulation data described above. We present and interpret the coherent structures represented by the eigenvectors of the ensemble averaged covariance matrix and examine the manner in which some of these modes, or *eigenhurricanes* are modulated.

[b]It is well-known that data scaling may drastically change the basis vectors in the KL algorithm, however the SFA algorithm is scale invariant; see [3] for additional details.

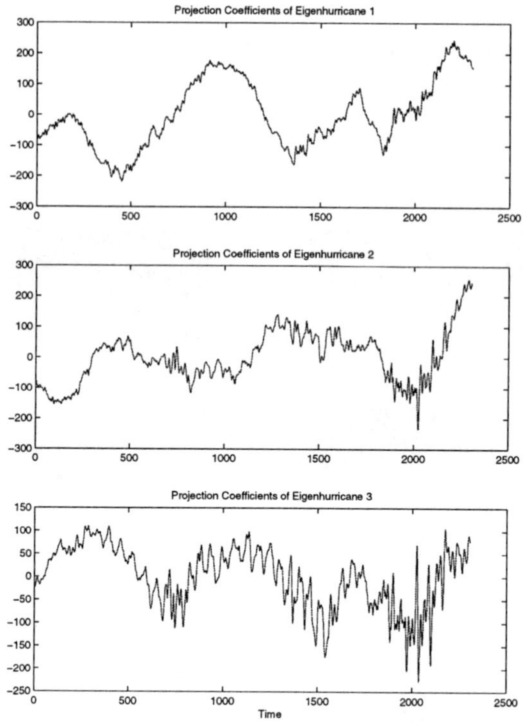

Fig. 2. The time projection coefficients of the first three eigenhurricanes from the KL algorithm onto the data, i.e., $a_i(t), i = 1, \ldots, 3$. The time axis refers to the model time output. In other words axis time label 500 refers to 500 5-minute intervals, or model hour 41.58.

KL Projection Coefficients

Figures 2 and 3 are representations of the time varying projections of the data onto specific KL eigenhurricanes. These are analyzed in more detail in the following sections. However, there are a few points that are worth making here:

(1) The projections on the first few eigenhurricanes (Figure 2) exhibit oscil latory behavior with a period of roughly 2 to 3 days. Overlapping with this longer oscillation, is a shorter scale of a few hours, or a convective time-scale.

(2) The projections on the last few eigenhurricanes (Figure 3) exhibit mainly very short, low amplitude variability. It is interesting to note however that this short time, low amplitude variability does seem to

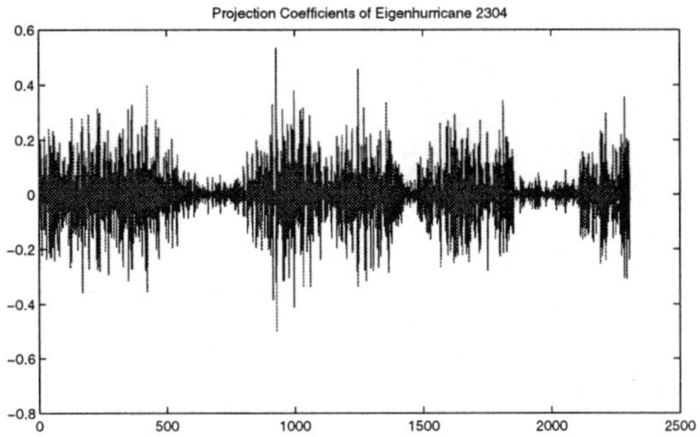

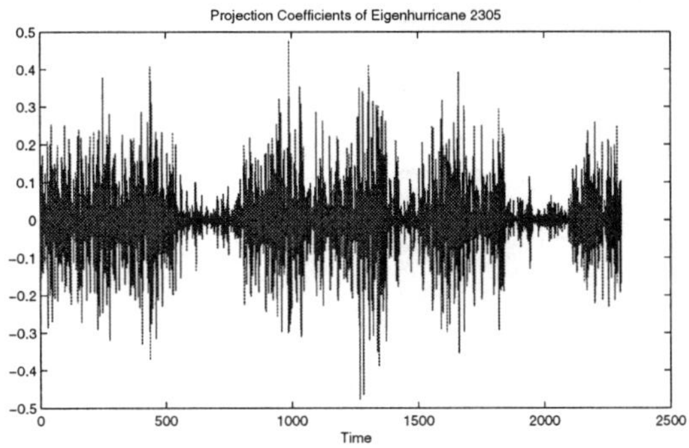

Fig. 3. The low-energy time projection coefficients $a_{2304}(t)$ and $a_{2305}(t)$. The time axis is the same as in Figure 2.

oscillate in a pulsing time scale that could be around the three-day scale seen in the early projections.

Note that it takes approximately 250 eigenvalue/eigenvector pairs to represent 90 percent of the energy in the system (see [3] for additional figures). Hence the structure in the time varying modes associated with a very low energy eigenspace is noteworthy; whether this is a real phenomenon or an artifact of the numerical simulation is unknown.

KL Coherent Structures

We focus our attention on interpretting the first 3 modes. Additional modes are described in more detail in [3].

First Mode. The time varying characteristics of the projection on mode 1 (Figure 2) shows at a glance variability on two scales, the inertial scale

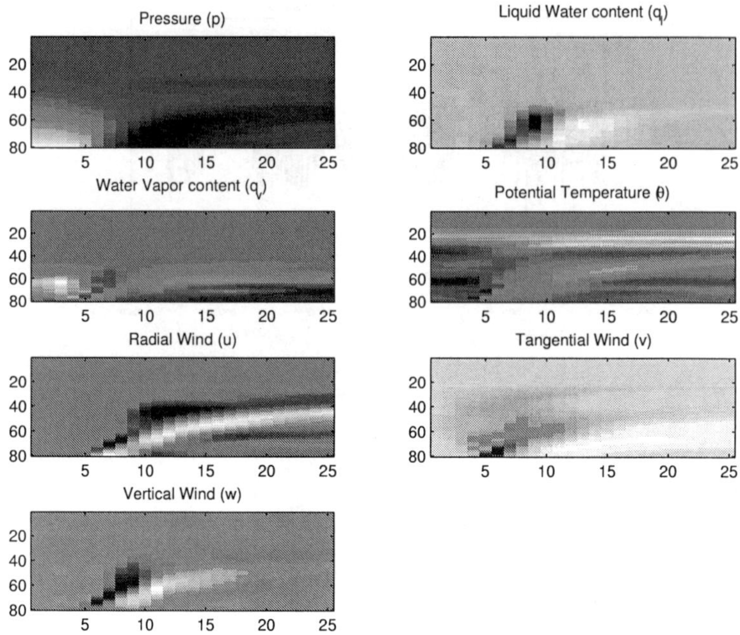

Fig. 4. The 1st eigenhurricane from the KL algorithm

(order of a day) and the convective time scale (order 30 minutes). This mode can explain 27 mb excursions in the storm central pressure, which is approximately that observed in this period [12]. The radius of maximum winds varies roughly in phase with the magnitude of the projection with changes of about 7 km.

In general, the signal (Figure 4) characterizes gross changes in the vortex. The θ field shows a general cooling while the pressure gradient weakens. An outward migration of the eyewall is also seen in the q_v and w fields.

Consistent with this is a spin-down of v as locations find themselves in the now-expanded eye, a calm region. Because the eyewall is a slanted ascent, u and w are in phase in the eyewall. Exterior to the eyewall, from the u field, we see that the outflow level is lower in the far environment, probably because of enhanced stability ($\partial\theta/\partial z$), but it is difficult at this point to extract cause and effect here. The higher pressures (p) and cooler temperatures (θ) in the eye are consistent with a weakening of the vortex. It is not clear why the eye should moisten and the low-level environment should dry out with such an event.

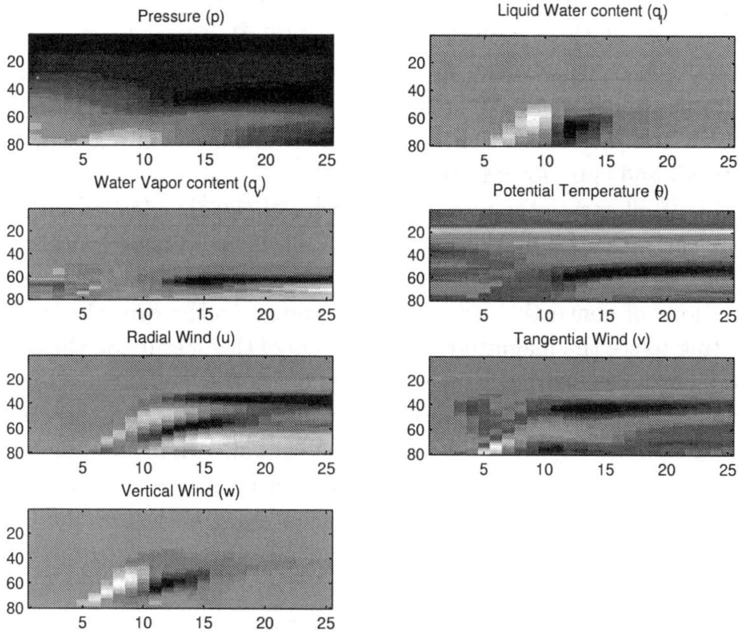

Fig. 5. The 2nd eigenhurricane from the KL algorithm

Second Mode. We see the second mode (shown in Figure 5) is again associated with outward migration of the eyewall. In contrast with mode 1, the outflow (u) emerges from the storm at its typical altitude, so rather than raising/lowering of the outflow, the mass-flux away from the storm is enhanced. The low-level source of mass is not the friction layer, but the middle-levels just above the friction layer. Conservation of angular momentum provides for spin-up (v) where this mid-level air encounters the eyewall, but the radius of maximum winds spins down, so the storm weakens slightly. By looking at the radial flow field u, mode 2 emphasizes an interaction with the environment and mode 1 emphasizes more an interaction with the eye. In contrast with mode 1, the signature of outward migration of the eyewall (negative of mode 2) shows a deepening of the central pressure deficit.

Third Mode. In contrast to modes 1 and 2, the magnitudes of the projections of mode 3 (see [3] for additional figures)) show variability on the convective time scale, and are less indicative of the coarse variability of the entire vortex.

The liquid water mixing ratio (q_l) shows a strong mid-level positive

signal within the eyewall, suggesting the structure of updraft enhancements in the middle of their evolution. Pressure (p) shows a structure that is a natural consequence of localized enhancement of convection with low-pressure in the lee of the updraft enhancement (w). The tangential wind field (v) is suggestive of a tightening of the radial gradient with spin up of the eyewall and spin down of the eye.
Mode 13. To illustrate the physical structure of eigenhurricanes with intermediate variance, we consider KL mode 13. From the q_l and vertical motion (w) fields (see [3] for additional figures), this mode seems associated with modulations of convective activity just exterior to the eyewall. It will be interesting to see the magnitude of the effects of this feature on the eyewall. Tangential winds (v) are enhanced directly below this updraft due to vortex stretching, and the radius of maximum winds spins down slightly at low-level and significantly at mid-levels of the eyewall. Under the assumption of a weakening of the eyewall convection, spin up just interior to the eyewall is not unexpected as the angular momentum is allowed to flux inward from the eyewall to the calmer eye. Again pressure (p) perturbations are a natural consequence of the convective action. Variations in the u, and θ and q_v seem to be consequential to this interpretation.
Two Modes with Least Variance. Here we consider the eigenhurricanes associated with modes 2304 and 2305 to illustrate that although the energy is relatively small, interesting structure is still captured at small scales. Note that these eigenhurricanes and the associated time projections are somewhat similar to the results produced by the SFA algorithm. This seems to indicate that the structures of least variance are those associated with the convective scales and their time variance is also very similar to the time projections of virtually all the SFA modes.

3.2. *SFA Results*

As discussed in the theory section, SFA may be used as a tool to separate signals that are mixed by some unknown mixing process. Thus we may speculate that the observed variables may afford an interesting separation that could serve to enhance our understanding of the mixing process in hurricane evolution. While we do not pretend to answer this question here in this preliminary paper, the application does give insight into the general methodology and illustrates the nature of the information one must analyze to attempt to answer this question.

SFA Projection Coefficients

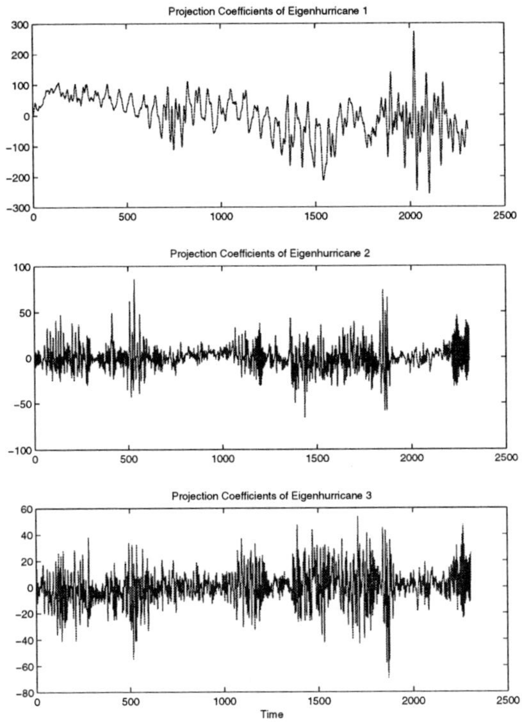

Fig. 6. The time projection coefficients of the three largest scale SFA modes. The time axis refers to the model time output. In other words axis time label 500 refers to 500 5-minute intervals, or model hour 41.58.

Here we proceed as with the KL algorithm but now the amplitude modulation is with respect to the basis vectors generated by the SFA procedure.

Figures 6 and 7 are representations of the time varying projections of the data onto specific SFA spatial modes. These are analyzed in more detail in the following sections. However, there are a few points that are worth making here:

(1) Unlike the KL algorithm, only the first set of SFA time projection coefficients (Figure 6) exhibit any sort of a slowly varying, sinusoidal signal.
(2) The other SFA projection coefficients favor more the sort of behavior seen in the KL low variance mode projections (Figure 3), i.e., highly

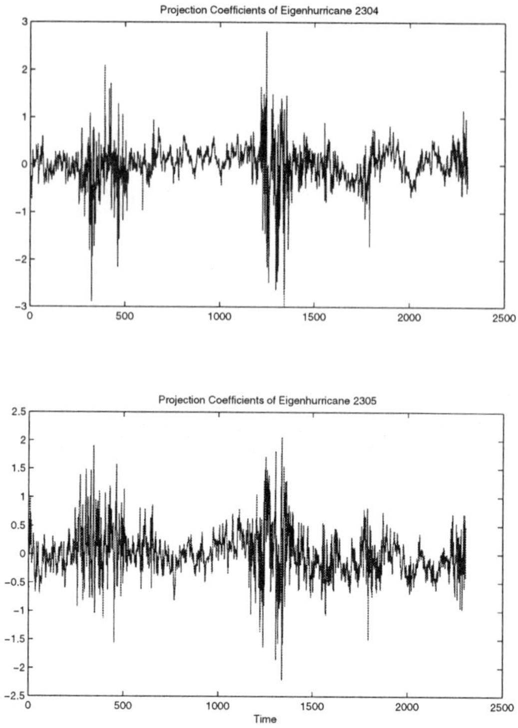

Fig. 7. The time projection coefficients of the smallest scale SFA modes (2304 and 2305). The time axis is the same as in Figure 6.

oscillatory behavior on a very fast time-scale (*sub-convective*).

Coherent Structures

First Mode. This mode does not lend itself to ready physical interpretation. Rather this mode is a measure of the maximum signal fraction which in turn relays the features that are most dominant (in a signal-to- noise ratio sense).

Three features show strongly in the projection upon the mode that might aid interpretation (Figure 6). First, there is a significant amount of variability on the convective time scale (about 30 minutes). Second, this variability shortens in period towards the end of the time series. Third, there is a downward trend in the projection coefficients.

Different variables of the hurricane show differing relationships to each of these characteristics, and that these features might be related to each other raises surprising questions without immediate answers. Note that there is

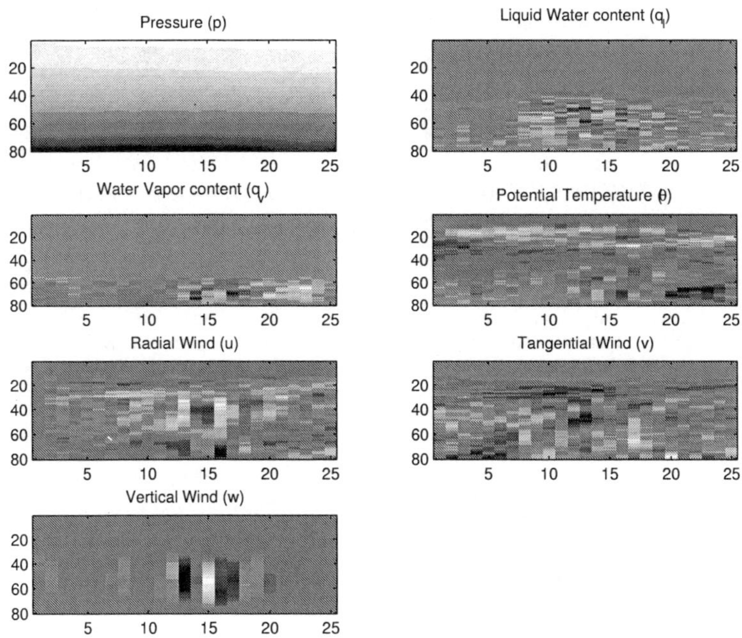

Fig. 8. The 1st signal eigenhurricane from the MNF algorithm

an "event" of some kind near the end of the time series of maximum tangential winds (intensity) (not shown) that may be associated with changes in behavior near time slice 2000. This weakening of the storm of about 10 m s^{-1} for a period of about a day is the most extreme excursion of storm intensity from the quasi-steady state noted in the simulation ([12]).

First, despite appearances, there is very little structure in the q_l and q_v fields for this mode (Figure 8). This might be expected when there are long-term trends in these fields with time if the mode also has a trend with time in its projection on the data. Pressure (p) is similar with the weakening of the central pressure deficit (a gross measure of vortex strength) with time being associated with drying of the entire atmosphere (q_l, q_v). The vertical motion field (w) shows perhaps a modulation in locations of updrafts exterior to the eyewall, not within the eyewall itself. Aspects of the u- and v-fields are most easily interpreted in relation to the w-field which might seem representative of the convective time scale variability.

Second Mode. The projection of the data upon this mode (Figure 6) shows variation on the *sub-convective* time scale; i.e. a process faster than the

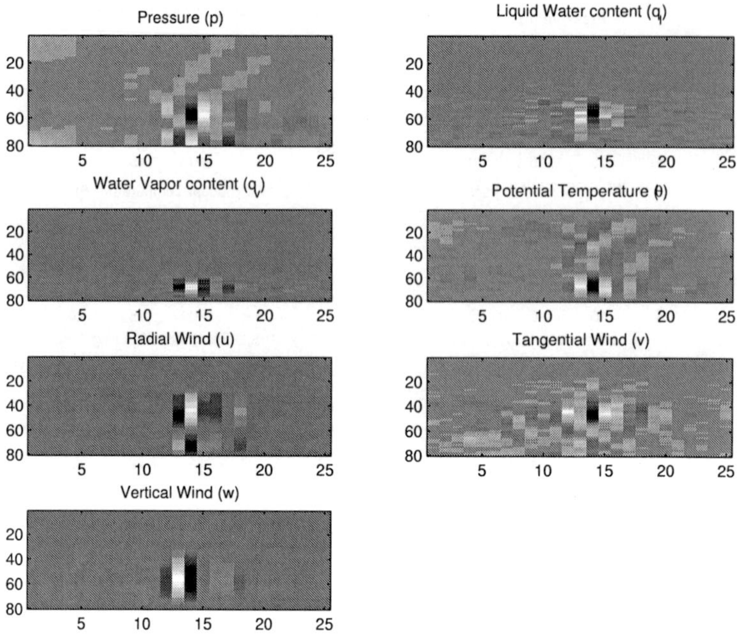

Fig. 9. The 2nd signal eigenhurricane from the MNF algorithm

convective time scale. Individual fields from this mode (Figure 9) can be analyzed. From the w-field, this mode is associated with convective activity exterior to the eyewall. With negative sign, the convection occurs in one of two neighboring locations.

Since updrafts tend to be spatially constrained to the smallest resolvable scales (outside of the eyewall), it is not surprising that convection at one location should be negatively correlated with convection at a neighboring location. Many features in other fields are entirely consistent with this interpretation, such as spin up (v) with the updraft, a secondary circulation (u) for mass conservation, with trailing negative (leading positive) pressure (p) perturbations relative to the updraft core, and local heating (θ) due to latent heat release. The details of the q_v and q_l fields are consistent with the w structure, but note that uniform change is the dominant response of these fields to this mode, which is somewhat surprising.

Third Mode. There are many similarities between mode 3 (not shown) and mode 2 (Figure 9), even with the uniform behavior of the moisture fields. However, mode 3 differs in that the updraft has a larger horizontal scale.

Finest Scale Modes. The projection onto these modes, as described in [3], was not particularly revealing. The structures are associate with the finest time scales and appeared to be non-physical. Clear understanding of these modes and the effect of the differencing procedure is the subject of future work.

4. Conclusions and Future Work

In this preliminary investigation we were concerned with comparing a well-known methodology (albeit surprisingly sparsely applied to the hurricane problem) to a relatively new approach, at least in the context of understanding large data sets associated with physical systems. A number of interesting features were determined in this comparison between the Karhunen-Loéve (KL) and Signal Fraction Analysis (SFA) Algorithms but there is much work to be done to fully understand the relative utility of these methods.

As expected, the methods emphasized different structures in their primary modes. In this study the KL modes yielded more to physical interpretation while the SFA modes appeared to be dominated by the convective time scale behavior. The numerical differentiation technique employed here in the SFA emphasizes vertical structure. Different approaches need to be examined to determine the impact on the interpretability of the SFA modes.

While the application of SFA did not immediately produce any increased understanding of the hurricane mixing problem, this study has generated a number of fundamental questions and directions for future work: Are these KL structures generic to hurricanes? If not, can such structures be found via other optimization criteria? Can we employ alternative methods to the differencing to split the SFA basis functions according to scale? How would other methods, e.g., Empirical Normal Modes (ENM), Canonical Correlation Analysis (CCA) or Partial Least Squares (PLS), compare in analysis of this data set? In particular, it would be interesting to apply these methods to analyze transient properties present in the data set as well as to analyze simulations from the spin-up stage. Can these methods be used to detect non-physical noise artifacts arising in the discretization process in the numerical simulation?

Acknowledgments

This work was supported by the National Science Foundation under grants ATM-0101781 and DMS 9973303.

References

[1] Y. Chen, G. Brunet, and M.K. Yau. *Journal of the Atmospheric Sciences*, 60:1239–1256, 2003.

[2] M. DeMaria and J. Kaplan. *Wea. Forecasting*, 14:326–337, 1999.

[3] Amanda Fox. A comparison of the Karhunen-Loeve (KL) and Maximum Noise Fraction (MNF) algorithms on axis-symmetric hurricane model data. Master's thesis, Colorado State University, Department of Mathematics, 2003.

[4] Gene H. Golub and Charles F. Van Loan. *Matrix Computations*. Johns Hopkins, Baltimore, third edition, 1996.

[5] Andrew A. Green, Mark Berman, Paul Switzer, and Maurice D. Craig. *IEEE Transactions on Geoscience and Remote Sensing*, 26(1):65–74, January 1988.

[6] Douglas R. Hundley, Michael J. Kirby, and Markus Anderle. *Signal Processing*, 82(10):1505–1508, 2002.

[7] M. Kirby and L. Sirovich. *IEEE Trans. Pattern Anal. Mach. Intell.*, 12(1):103–108, 1990.

[8] Michael Kirby. *Geometric Data Analysis: An Empirical Approach to Dimensionality Reduction and the Study of Patterns*. Wiley, 2001.

[9] Michael Kirby and Chuck Anderson. In Ehud Kaplan, Jerry Marsden, and Katepalli R. Sreenivasan, editors, *in press*, Springer Applied Mathematical Sciences Series. Springer-Verlag, 2003.

[10] Y. Kurihara, R. E. Tuleya, and M. A. Bender. *Mon. Wea. Rev.*, 121:2030–2045, 1998.

[11] C. W. Landsea. *Mon. Wea. Rev.*, 121:1703–1713, 1993.

[12] J. Persing and M.T. Montgomery. *Journal of the Atmospheric Sciences*, 60:2349–2371, 2003.

[13] R.W. Preisendorfer. Number 17 in Developments in Atmospheric Science. Elsevier Science Publishing Co., 1988.

[14] P. D. Reasor, M. T. Montgomery, F. D. Marks, and J. F. Gamache. *Mon. Wea Rev.*, 128:1653–1680, 2000.

[15] R. Rotunno and K.A. Emanuel. *Journal of the Atmospheric Sciences*, 44(3):542–561, 1987.

[16] L. Sirovich. *Quart. of Appl. Math.*, XLV(3):561–571, 1987.

[17] L. Sirovich. *Quart. of Appl. Math.*, XLV(3):573–582, 1987.

[18] L. Sirovich. *Quart. of Appl. Math.*, XLV(3):583–590, 1987.

[19] L. Sirovich and M. Kirby. *J. of the Optical Society of America A*, 4:529–524, 1987.

[20] Paul Switzer. In L. Billard, editor, *Computer Science and Statistics*, pages 13–16. Elsevier Science Publishers, 1985.

[21] T. Yokoo, B.W. Knight, and L. Sirovich. *NeuroImage*, 14:1309–1326, 2001.

CHAPTER 16

LINEAR AND NONLINEAR NUSSELT NUMBER MEASUREMENTS DURING ELECTROCONVECTION OF A LIQUID CRYSTAL

Jim T. Gleeson

Department of Physics, Kent State University,
Kent, Ohio 44242, USA

Energy transport measurements, as quantified by the Nusselt number, are shown to be a detailed, quantitative and accurate technique for characterizing electroconvection of nematic liquid crystals. This method is not only a all-purpose tool for identifying bifurcations and other transitions, but also offers a powerful discriminator for testing competing models describing the underlying mechanisms of electroconvection. We present experimental results on a variety of different electroconvection experiments. We show how these measurements can yield direct insight into the nonlinearities present in electroconvective flow, and discuss challenges for future experiments.

1. Background

Electroconvection (EC) in nematic liquid crystals is an excellent model system, both experimental and theoretical, for detailed, quantitative study of pattern formation, spatio-temporal chaos and turbulence in fluid systems displaced from equilibrium [1, 2]. In this system, under the appropriate condition, a thin layer of nematic liquid crystal, under the influence of an applied electric field, spontaneously is set into motion. The principal mechanism for this motion is the Carr-Helfrich instability [3,4]. Basically, because of the liquid crystal's anisotropic electrical conductivity, an applied electric field in one direction can induce a current in a perpendicular direction. This current in turn leads to a non-zero charge density, which results in a body force on the fluid, setting it into motion. Some of the reasons that make this a particularly attractive system for experimental and theoretical study are the relative ease with which all relevant variables can be controlled, the extremely wide and rich variety of nonequilibrium phenomena that are exhibited, and, the particularly convenient length scales. The latter stems from the ability to study pattern formation with large enough aspect ratio that lateral boundary conditions can be ignored. It also allows the possibility of creating fully developed turbulence within a volume comparable to that of a postage stamp.

EC as a model system for studying pattern formation and fluid dynamics possesses some differences with both mechanically driven flow and the better known Rayleigh-Bénard convection (RBC); it is useful to point out the important ones here. In mechanically driven flow (*e.g.* stirred, pumped or sheared flow), only the flow itself is the active field, and the system is fully described by the Navier-Stokes equation (with attendant boundary conditions). Normally, the control parameter is the Reynold's number. In RBC, in addition to Navier-Stokes and the flow, one must consider the heat equation for the temperature field, and the two fields (and the two equations) are coupled. The main effect of these additions is twofold: first, there are now two control parameters, the Rayleigh number and the Prandtl number. Second, the flow velocity (as characterized by the Reynolds number) is no longer the control parameter, but rather becomes the response to the new control parameters. The other response is the rate of heat transport as characterized by the Nusselt number.

EC is perhaps most similar to RBC, in that the motion is not mechanically driven, but rather is induced because of the body force coupling the flow to other active fields. However, the anisotropic nature of liquid crystals also introduces significant additional factors not found in RBC. Specifically, the direction of anisotropy is characterized by a unit vector field (the *director*) [5]. This field couples to the flow field via the Ericksen-Leslie equations for flow of a uniaxially anisotropic fluid. Moreover, because EC is driven by sustaining an electric potential difference across the liquid crystal layer, one must also include Gauss' Law for the induced potential, as well as constitutive relations for the electric current and charge density. Thus, formulating a mathematical description of EC poses a formidable problem. This problem is compounded when one considers the mechanism of charge transport in liquid crystals. The simplest model, called the standard model, (SM) assumes Ohm's law to apply. A more realistic, yet far more complicated model, assumes the charge is carried by ionic carriers which originate from weakly dissociating neutral species. This is the weak electrolyte model (WEM) [6].

A schematic diagram of EC is shown in Figure 1. A thin layer of nematic liquid crystal (NLC) is confined between two parallel plane conducting electrodes, separated by distance d. These electrodes are constructed by applying a transparent conductive coating onto glass plates. This permits study of the optical properties of the NLC, which reflect the structure of the director field. The lateral dimensions of this layer are typically $10^2 - -10^3$ times larger than (d), making it possible to ignore the lateral boundary conditions. Furthermore, d is typically less than $100\mu m$; small enough that one may neglect effects of gravity. The plates are pre-treated to induce a desired direction for $\hat{n}$ where the NLC is bounded by the plates. The NLC is introduced between the plates, and in the absence of an electric field, n will be undistorted throughout the NLC, and its direction specified by the pre-treatment.

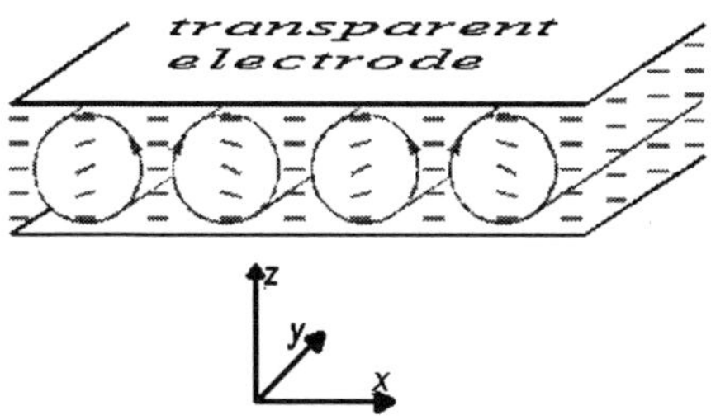

Fig. 1. Schematic description of EHC. The short segments represent the director field, and the arrows denote the direction of flow.

EC can result when a sufficiently large potential difference is sustained across the electrodes. In order to avoid electrolysis, an *ac* potential difference $\sqrt{2}V\cos(\omega t)$ is employed. The threshold amplitude, V_c, depends strongly on ω. As V is increased above V_c, the fluid spontaneously begins to move in the cylindrical fashion shown in Figure 1. The first instability results in a pattern formation state, in which a regular array of parallel, cylindrical rolls is observed. Depending on the experimental conditions and ω, one can either observe normal rolls (perpendicular to the undistorted $\hat{n}$), or oblique (zig-zag), with the rolls at angles $\pm\theta_{zz}$ to the undistorted $\hat{n}$. Furthermore, under appropriate conditions, the first instability may be a Hopf bifurcation to a travelling wave state. Figure 2 shows typical shadowgraph [7, 8] images of the normal roll and the oblique roll states.

As V continues to be increased, one observed a cascade of instabilities, ending up in the turbulent, dynamic scattering mode.

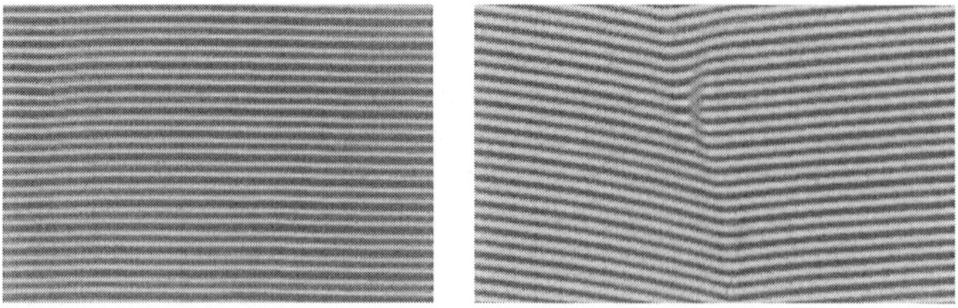

Fig. 2. Example shadowgraphs of the normal roll and oblique roll states of electroconvection in nematic liquid crystals.

1.1. *Nusselt Numbers*

In RBC, the rate of heat transport across the layer when the temperature difference is ΔT is characterized by the Nusselt number, $\mathcal{N}$, defined as $Q/(\kappa \Delta T d)$ where Q is the rate of heat transport and κ is the thermal conductivity of the fluid. When $\mathcal{N}$ exceeds unity, heat transport is greater than is possible by conduction alone; the excess transport is due to convection. Indeed, measurement of $\mathcal{N}$ led to the first accurate determination of the critical value of ΔT [9]. Such measurements remain a technique of choice for not only identifying instabilities or transitions in RBC, but also permit sensitive tests of competing theoretical models of turbulence during RBC at extremely large Rayleigh numbers [10,11]. We shall show the direct analogues of such measurements in EC in liquid crystals.

RBC is sustained by a temperature difference, and is a more efficient way to transport heat than conduction through the quiescent fluid. Electroconvection, which is sustained by a electric potential difference, is also a more efficient way to transport charge than conduction through the quiescent fluid. Motivated by this, our goal is to use measurements of charge transport, characterized by electrical Nusselt numbers, to more quantitatively and comprehensively study EC in liquid crystals. Furthermore, such measurements are particularly valuable for experimental tests of theories that rely on charge transport mechanisms like the SM and the WEM. Lastly, because one needs only to measure the electrical current, Nusselt number measurements in EC are quite straightforward (at least in principal). We note here the first attempts [12,13] to describe quantifying charge transport during electroconvection.

Specifically, the response to an applied potential difference $\sqrt{2}V\cos(\omega t)$ is a current $I(t)$. The linear portion of this response is $I_L(t) = I_r\cos(\omega t) +$

$I_i \sin(\omega t)$ In the absence of convection (within the SM), $I_r^0 = \sqrt{2} g_\perp V$ and $I_i^0 = \sqrt{2}\omega C_\perp V$. $g_\perp$ and $C_\perp$ are (respectively) the conductance and the capacitance of the quiescent LC layer when $\hat{n}$ is perpendicular to the applied field. A potential difference across a LC with finite conductance results in energy dissipation; the time averaged power dissipated is $< P > = \sqrt{2} I_r V$. With these, we can define *two* reduced Nusselt numbers for EC, corresponding to the in-phase and out-of-phase current:

$$\mathcal{N}_r \equiv I_r/I_r^0 - 1; \tag{1}$$

$$\mathcal{N}_i \equiv I_i/I_i^0 - 1; \tag{2}$$

The former is exactly analogous to the familiar Nusselt number from RBC. The latter quantifies not energy dissipation but rather how a capacitive load under an *ac* potential can momentarily store energy; it has no analogue with RBC. For convenience, both of these are expressed as reduced quantities so that they are zero (rather than unity) in the quiescent state.

2. Experiments

In principle, experimental determination of Nusselt numbers in EC is straightforward; at any given potential difference, one measures the current through the LC. The specific details are found in the literature [14,15]. Some important features of our experiments include a) making measurements at constant temperature in order to ensure not only that the material parameters remain constant during the experiment, but also to minimize the effects of Joule heating in the LC; b) using etched electrodes so that only the LC is within the area in which the electric field is applied, and c) employing a FET input current-to-voltage converting preamplifier to enable us to measure currents as small as 10^{-9}A. An example of such a measurement is shown in Figure 3. This figure represents the raw current vs voltage data. The "knee" represents the onset of convection. Reducing data such as this to express it in terms of the reduced Nusselt numbers is shown in Figure 4. This figure reveals a number of features. First, the reduced Nusselt number increases linearly with ε, which confirms the supercritical nature of the bifurcation from the quiescent to the convective state. Second, this linear regime ends at a second knee at higher ε. This indicates a secondary bifurcation. Independent optical measurements confirm that at this point, the zig-zag pattern appears. Another bifurcation to the defect chaos state,

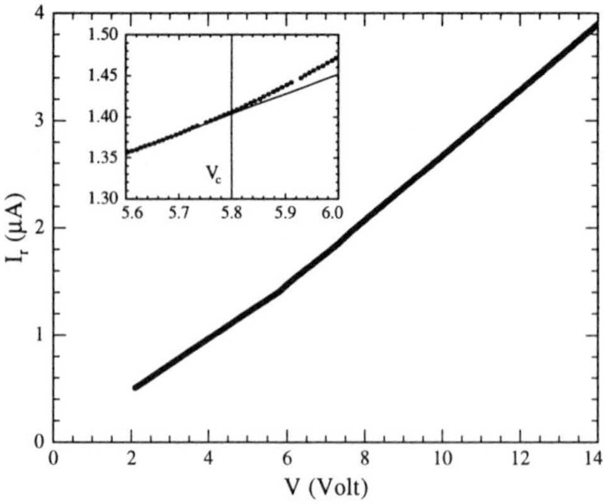

Fig. 3. Example current vs voltage data for electroconvection in the nematic liquid crystal methoxy butylidene butyl aniline. The inset shows a blowup of the voltage range around V_c.

in which dislocations in the roll pattern are spontaneously created and annihilated, is seen at the second knee. Results such as these reveal the power of Nusselt number measurements to not only identify bifurcations, but also to quantitatively determine their onset.

Figure 5 shows how Nusselt numbers at onset as measured compared with calculations of the same quantity within the framework of the standard model [14]. These quantities vary strongly with ω (scaled by the charge relaxation time $C_\perp/g_\perp$). The experiments generally follow the trends predicted, especially the change in sign of the out-of-phase Nusselt numbers. Thus, the SM correctly predicts the Nusselt numbers within 50% or so, but, as also can be seen, systematic discrepancies remain. While the agreement shown here is far from perfect, it should be borne in mind that this is not a fit; *no* parameters were adjusted to bring the experiments into agreement with the predictions. Possible explanations for the remaining discrepancies are discussed subsequently. Because electroconvection is sustained by an *ac* electric potential, charge transport in this system has a substantial number of engaging, additional features compared to heat transport in RBC. The out-of-phase Nusselt number shown in Figure 4 is one such example. Another more intriguing aspect is the direct measurement of nonlinear charge transport. Specifically, when the liquid crystal is excited by a potential difference alternating at frequency ω, the response can be a current with Fourier components at $n\omega$ where n is odd. That is, we can directly measure

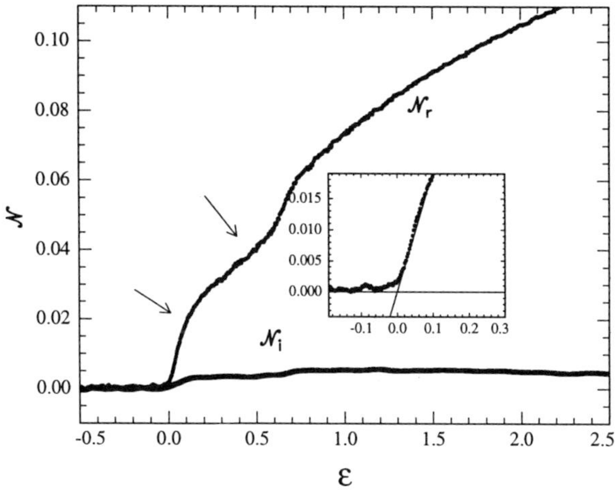

Fig. 4. Real and imaginary Nusselt numbers vs ε. The inset is a blow up of the real part showing the linear increase with ε at onset. The arrows indicate the knees which are the signatures of secondary bifurcations.

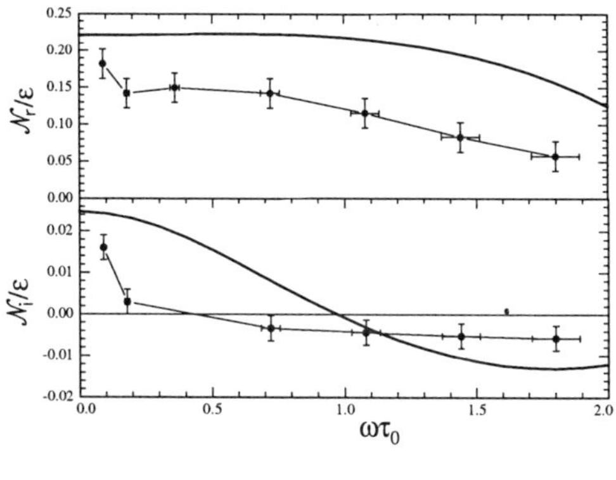

Fig. 5. Slope of $\mathcal{N}$ vs ε vs frequency. The solid line is the result of a comprehensive standard model calculation. [14].

a response that is a nonlinear function of the excitation. Our technique of probing the current using a digital-signal-processing lock-in amplifier makes such measurements routine. Figure 6 shows one such example. This shows the 3ω out-of-phase component of the current vs voltage. Note that this component is almost three orders of magnitude smaller than the fundamental component. Again, there is a pronounced "knee" in the curve right at the onset of convection, and the current increases linearly after that.

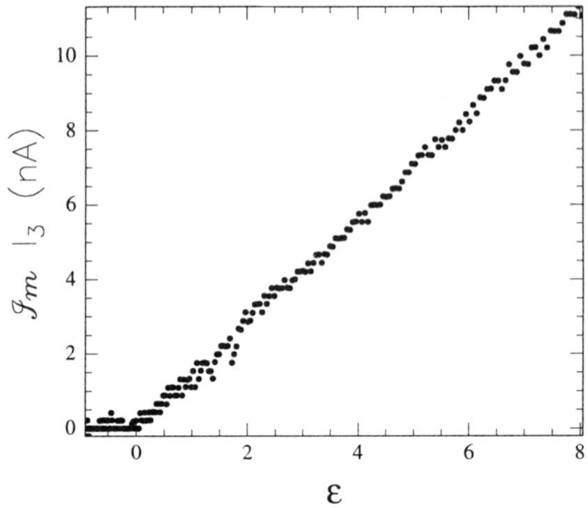

Fig. 6. 3ω Fourier component of current vs applied voltage. The component is zero up until V_c.

Nonlinear measurements such as seen above represent a powerful test for competing theoretical models of electroconvection. One reason for this is that the standard model predicts that such components will be strictly zero in the quiescent state. Indeed, that is the case in the example shown above. However, at different values of frequency, we observe a non-zero 3ω component even in the quiescent state. Thus, the SM is incapable of describing such a nonlinear response; more sophisticated models of charge transport will be required.

In RBC, when the Rayleigh number is far above onset, the flow becomes turbulent. In this regime, there has been a large body of work both calculating and measuring the Nusselt number. Much effort has been focused on a simple power-law scaling relationship:

$$\mathcal{N} \propto Ra^{\alpha} \quad . \tag{3}$$

Specific focus has been on determining whether or not such a relationship exists, and, the value of α. The existence of such a simple relationship is attractive, because it implies that even though the turbulent flow itself is unbelievably complicated and unpredictable, there exist underlying unifying and simplifying principles. Does EC exhibit similar phenomena?

We have measured the (in-phase) Nusselt number over an extremely large range of applied potential difference to examine this issue. Fig 7 shows an example of our results. This figure reveals a number of notable features. Firstly, power-law scaling indeed exists, albeit over a limited range. Secondly, the exponent found is similar to what is observed in RBC. While these two similarities with RBC are welcome, it is the differences that are especially intriguing. In particular, as noted above, the scaling regime does not span an extended range. Looking more closely, the manner in which this range terminates is wholly unexpected. As seen in the upper left portion of Figure 7, $\mathcal{N}$ begins to deviate from the power-law scaling because as V^2/V_C^2 continues to increase, $\mathcal{N}$ saturates, and then decreases. We are unaware of another example of a diminution of the excess power dissipated by a fluid flow system as the stress is increased. The other notable difference is the values found for α. Note that in RBC this exponent is never far from 0.3. In stark contrast, in EC, as $\omega C_\perp/g_\perp$ is varied from 1.25 to 20, α changes from being greater than 0.4 to less than 0.1; from both much larger to much smaller than ever observed for RBC.

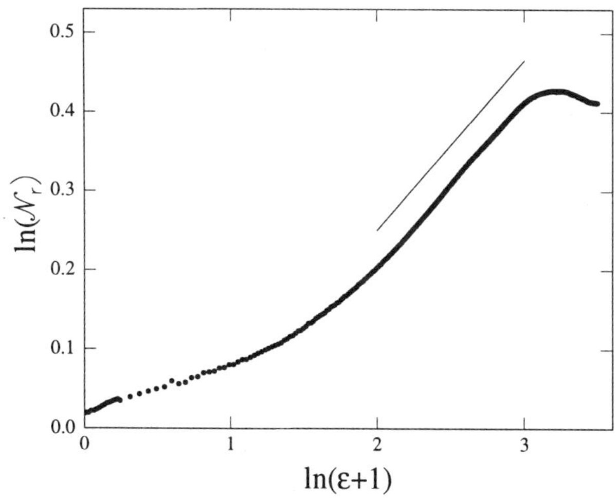

Fig. 7. Log-log plot of Nusselt number vs V^2/V_c^2 for EC in MBBA. This data corresponds to a dimensionless frequency $\omega C_\perp/g_\perp = 0.25$. The line has slope 0.215.

3. Conclusions

Nusselt number measurements represent a powerful technique for studying electroconvection in liquid crystals. Not only does this method provide a sensitive indicator of transitions and bifurcations, but it also permits detailed, quantitative comparison between theory and experiment. We have demonstrated that such comparison can be made using no fitting parameters whatsoever. Charge transport measurements can also directly probe the nonlinear aspects of electroconvection. This approach brings into sharp focus the shortcomings of the standard model of electroconvection. This is perhaps unsurprising given how the mechanism of charge transport is the heart of the difference between the SM and the WEM. Lastly, as in RBC, there is a evidence of a power law scaling relationship between the Nusselt number and the stress applied to the system.

Acknowledgments

This work was supported by the National Science Foundation (DMR-9988614). The benefit of invaluable contributions and discussions from E. Plaut, J. deBruyn, T. Toth-Katona, N. Gheorghiu and W. Pesch is gratefully acknowledged.

References

[1] M. C. Cross and P. C. Hohenberg, *Rev. Mod. Phys.*, **65**, 851 (1993).
[2] L. Kramer and W. Pesch in *Pattern Formation in Liquid Crystals*, Á. Buka and L. Kramer, eds, Springer, NY 1996.
[3] E.F. Carr, *Molec. Cryst. Liq. Cryst.*, **7**, 253 (1969).
[4] W. Helfrich, *J. Chem. Phys.*, **51**, 4092 (1969).
[5] P. G. de Gennes and J. Prost, *The Physics of Liquid Crystals, 2nd ed.* (Clarendon, Oxford, 1993).
[6] M. Dennin, M. Treiber, L. Kramer, G. Ahlers and D.S. Cannell, *Phys. Rev. Lett.*, **76**, 319 (1996).
[7] S. Rasenat, G. Hartung, B. Winkler, I. Rehberg, *Exp. Fluids*, **7**, 412 (1989).
[8] A. Joets and R. Ribotta, *J. Phys. I France*, **4**, 1013 (1994).
[9] Schmidt R. J., Milverton S. W., *Proc. Roy. Soc. (Lond.)*, **A152** (1935) 486.
[10] E.D. Siggia, *Annu. Rev. Fluid Mech.*, **26**, 137 (1994).
[11] X-Z Wu and A. Libchaber, *Phys. Rev.* **A**, **45** 842 (1992).
[12] T. Kai, S. Kai, K. Hirakawa, *J. Phys. Soc. Jpn*, **43**, 717 (1977).
[13] I. Rehberg, B.L. Winkler,M. de la Torre-Juarez, S. Rasenat and W. Schöpf, *Festkörperprobleme*, **29** (1989) 35.
[14] J.T. Gleeson, N. Gheorghiu and E. Plaut, *Euro. J. Phys. **B***, **26**, 515 (2002).
[15] J. T. Gleeson, *Phys. Rev. **E***, **63**, 026306 (2001).

CHAPTER 17

CHARACTERIZATIONS OF FAR FROM EQUILIBRIUM STRUCTURES USING THEIR CONTOURS

Girish Nathan and Gemunu H. Gunaratne

Department of Physics, University of Houston, Houston, TX 77204, USA

Most details of far from equilibrium structures depend on uncontrollable and intractable factors like the initial state and stochasticity. However, some facets of these patterns depend only on the model system and control parameters. A comprehensive statistical description of the structures requires the analysis of as many such characteristics as possible. We discuss two classes of such measures that can be used to describe labyrinthine patterns and growth interfaces.

1. Introduction

The aim of the paper is to discuss two groups of characteristics that can be used to describe statistical features of far from equilibrium structures and their evolution. These measures are functionals of a scalar field $u(\mathbf{x}, t)$ that represents the structure. Some examples of such scalar fields are

1 Concentration of I_3^- in a CIMA reaction [1,2].
2 The intensity of light reflected from a vertically vibrated layer of brass beads [3,4].
3 The height of an epitaxially growing surface [5].
4 The local concentration of a constituent of an alloy during spinodal decomposition [6,7].
5 The temperature of a convecting fluid [8].

The characteristics are constructed to be equivariant under all Euclidean motions of the structure, and attempt to quantify statistical features of the contours [9]. Although mean curvature of contour lines [10] and fractal dimension of contour loops [11] have been proposed for the purpose, extraction of these quantities from a noisy field is quite challenging. For example, curvature of contour lines of a field $u(\mathbf{x}, t)$ is given by $\kappa = (u_{xx}u_y^2 + u_y y u_x^2 - 2u_{xy}u_xu_y)/(u_x^2 + u_y^2)^{3/2}$, and its calculation is extremely sensitive to noise and discretizations at locations where the denominator is small. The measures discussed in this paper are constructed from local densities that are related to the curvature of contours, but are less sensitive to noise.

2. Labyrinthine Patterns

By labyrinthine patterns, we refer to structures formed as a result of local bifurcations from a uniform state to one consisting of a linear combination of plane waves. Typical structures observed in experimental and model systems contain striped, square, triangular, and hexagonal planforms [12], although more complex super-lattice patterns have also been studied [13,14]. Due to the (assumed) Euclidean symmetry of the underlying physical or model system, and the simultaneous and independent formation of domains in spatially separated locations, these planforms are not synchronized over the entire domain. As they grow, these patches compete with each other to generate complex patterns such as those shown in Figure 1. The model system used to generate these patterns is an extension of the Swift-Hohenberg equation [12], which describes the dynamics of a scalar field $u(\mathbf{x}, t)$ via

$$\frac{\partial u}{\partial t} = D\Big(\varepsilon - (1 + k_0^{-2}\triangle)^2\Big)u - \gamma u^3 - \nu(\nabla u)^2 + \eta(\mathbf{x}, t). \qquad (1)$$

D sets the rate of diffusion, ε is the distance from pattern onset, ν is the strength of a non-variational term, and k_0 is the wave number of the underlying planform. η is a stochastic term such that $\langle\eta(\mathbf{x}, t)\eta(\mathbf{x}', t')\rangle = F\delta(\mathbf{x} - \mathbf{x}')\delta(t - t')$, F being the strength of the noise; $\eta \neq 0$ and $\eta = 0$ represent stochastic and deterministic spatio-temporal dynamics. For $\varepsilon > 0$, random initial states (whose amplitude is chosen to be much smaller than the saturated peak amplitude of the field) evolve to patterns under Eqn. (1).

2.1. *Disorder Function*

More ordered patterns (such as Figure 1(d)) contain large domains of stripes and a small amount of "disorder." The latter consists of regions with varying stripe orientations, locations with changing stripe width, those containing defects etc. Several functionals of $u(\mathbf{x}, t)$ have been used as measures to characterize the amount of irregularity in such structures. They include the structure factor [15–17], domain wall length [18], and stripe orientation correlation length [17]. Analysis of these measures show that relaxation of an initially random state consists of two stages. During the initial period, the

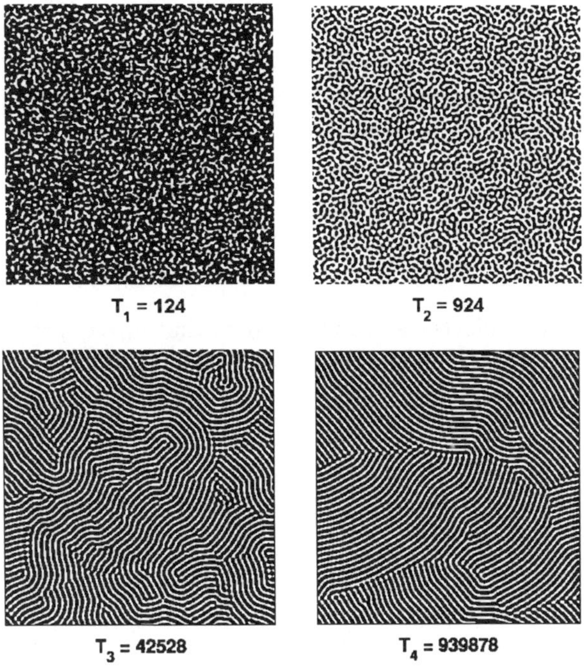

Fig. 1. Four snapshots of the spatio-temporal dynamics of a random initial state ($|u(\mathbf{x},t)| < 0.001$) under Eqn. 1 with $D = 0.01$, $\varepsilon = 0.1$, $k_0 = 1/3$, $\nu = 0$, $\gamma = 2$, and $F = 0$. Times $T_1 = 124$ and $T_2 = 924$ are in stage I, and $T_3 = 42,528$ and $T_4 = 939,878$ are in stage II. Each side of the domain is $96\pi/k_0$, and periodic boundary conditions are imposed on the model. The peak of the (radial projection) of the power spectrum in the spatio-temporal dynamics relaxes to k_0.

behavior of all measures is consistent with the growth of a lengthscale at a rate of $t^{0.5}$. During the second stage, however, the growth rate corresponding to each measure gives distinct values. Moreover, these rates depend on the model; e.g., they take different values when stochastic or non-variational terms are included [17,18]. Note that length scales associated with this description are not related to the width $(2\pi k_0^{-1})$ of the stripes.

The disorder function is a systematic way to study the behavior of multiple length scales that may be associated with the evolution of labyrinthine structures. Its definition is motivated by the observation that the Helmholtz operator $(\triangle + k_0^2)$ acting on a scalar field representing a perfect striped array vanishes, and hence any non-zero value of $|(\triangle + k_0^2)u|$ can be associated with a deviation from perfect stripes. (In these considerations, it is assumed that higher harmonics in $u(\mathbf{x},t)$ have been filtered out [19].) For example, for stripes with wave number k, $(\Delta + k_0^2)u \sim (k_0^2 - k^2)$; for an array of stripes with curvature κ, $(\Delta + k_0^2)u = k_0\kappa$ [9].

Once a field quantifying the "level of local disorder" is identified, one can define a set of corresponding global measures using the thermodynamic formalism [20]. For our case, they are defined by

$$\delta(\beta,t) = \left[\frac{(2-\beta)}{(\int d^2x)} \frac{\int d^2x |(\Delta + k^2)u(\mathbf{x},t)|^\beta}{\langle |u(\mathbf{x},t)| \rangle^\beta}\right] \Big/ k^{2\beta}. \tag{2}$$

The various normalizations are included so that $\delta(\beta,t)$ provides the ratio of the wave number corresponding to "moment" β and k_0.

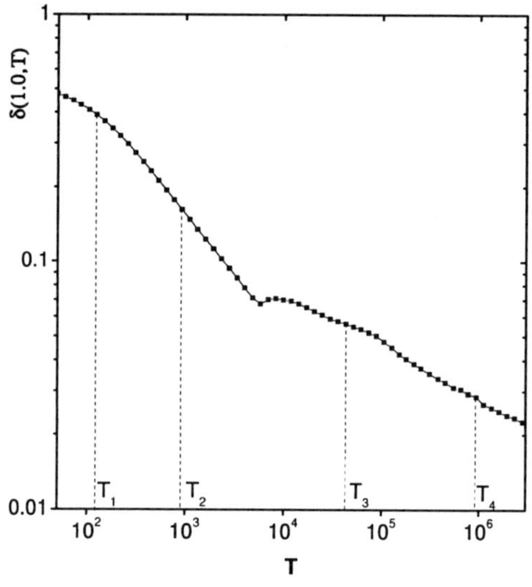

Fig. 2. The behavior of $\delta(1.0,t)$ for the spatio-temporal dynamics shown in Figure 1. The presence of two distinct stages of relaxation, each with a power law decay of $\delta(1.0,t)$ can clearly be observed.

2.2. *An Application: Stages of Relaxation*

It is clear from Figure 2 (and corresponding curves for other values of β) that there are two distinct stages of relaxation of deterministic ($\eta = 0$), variational ($\nu = 0$) systems and that the characteristics $\delta(\beta,t)$ exhibit power law decays in t. The rate of decay is

$$\sigma(\beta, t) = \frac{\log \delta(\beta, t)}{\log t}. \tag{3}$$

During the first stage, the mean value of $|u(\mathbf{x}, t)|$ is significantly smaller than its long-time value, and hence spatio-temporal dynamics can be assumed to be approximately linear. Under these conditions, it can be shown that $\sigma_I(\beta, t)$ takes on a universal value $\frac{1}{2}$, see [4] independent of β and t; i.e., all lengthscales associated with the pattern grow like $t^{1/2}$, as is already known to be the case for the structure factor [15, 16]. The top curve of Figure 3 shows that this conclusion is valid for patterns formed by Eqn. (1).

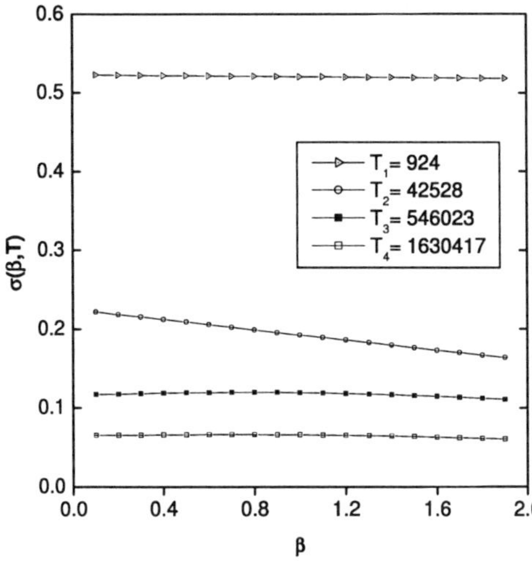

Fig. 3. The behavior of the function $\sigma(\beta, t)$. During the domain forming stage $\sigma(\beta, t) = 1/2$, while it is β dependent in the domain coarsening stage. As in the example shown, $\sigma(\beta, t)$ also varies with time during the second stage.

Beyond saturation of the field, the rate of relaxation becomes significantly slower. Furthermore, $\sigma(\beta, t)$ becomes β-dependent, see the curve corresponding to T_2 of Figure 3; i.e., distinct lengthscales associated with the structure grow at different rates. In addition, as seen from Figure 3, the rate for larger values of β (which emphasize more singular regions of the structure [9]) is smaller. These conclusions are consistent with previous assertions on measures such as structure factor and defect density [15, 18].

The rate of decay $\sigma_{II}(\beta, t)$ during the second stage increases when non-variational terms are included (i.e., $\nu \neq 0$) [21]. Once again, analogous results have been presented before; for example, the decay of the stripe orientation correlations for a variational model and a non-variational model were shown to be 0.24 and 0.51 respectively [17].

The addition of stochasticity ($\eta \neq 0$) also enhances $\sigma_{II}(\beta, t)$. Once again, analogous increases of decay rates for structure factor [15] and defect density [18] have been noted. The disorder function analysis can be used to deduce an additional feature of stochastic spatio-temporal dynamics; it is found that for sufficiently large times, there is a third stage in the dynamics where once again all lengthscales associated with $\delta(\beta, t)$ decay at a rate independent of β (see the two lower curves in Figure 3. However, unlike in stage I the value $\sigma_{III}(\beta, t)$ is not universal, and is furthermore time dependent. We have not been able to detect such a stage under deterministic spatio-temporal dynamics.

Disorder function analysis of patterns generated in a vertically vibrated layer of brass beads have confirmed the presence of the three stages of relaxations discussed [22].

3. Domain Growth

Structures formed during epitaxial growth [5] and spinodal decomposition [6, 7], or lamellar structures seen in reaction-diffusion systems [23] have characteristics that are different from labyrinthine patterns discussed in Section 2; for example, they do not relax to a state with a well-defined local planform. Statistical properties of patterns on a domain of size L can be quantified using the "surface roughness" $W_L(t)$ (standard deviation of $u(\mathbf{x}, t)$), correlation length [24], domain size [25] or the fractal dimension of contour loops [11].

Figure 4 shows several snapshots taken during the evolution of a random initial state under the paradigmatic Karder-Parisi-Zhang (KPZ) equation for surface growth [26]

$$\frac{\partial u(\mathbf{x}, t)}{\partial t} = \nu \Delta u + \lambda (\nabla u)^2 + \eta(\mathbf{x}, t). \tag{4}$$

It models the competition between surface diffusion and nonlinear growth in the presence of a stochastic term which represents random deposition. The surface roughness grows as $W_L(t) \sim t^b$ during an initial growth period which scales as L^z, and saturates to a final value $W_L(\infty) \sim L^\alpha$, see [24]. The growth exponent b, the dynamics exponent z, and the roughness exponent α is believed to only depend on a few generic features of the model. Evidence for this assertion is provided by comparisons of growth characteristics of the KPZ model and a second discrete model referred to as the Restricted Solid on Solid (RSOS) model [27]. The scaling indices for the two models are identical for sufficiently large values of λ. The need to determine limitations of universality of such processes motivates the introduction of the next set of statistical variables.

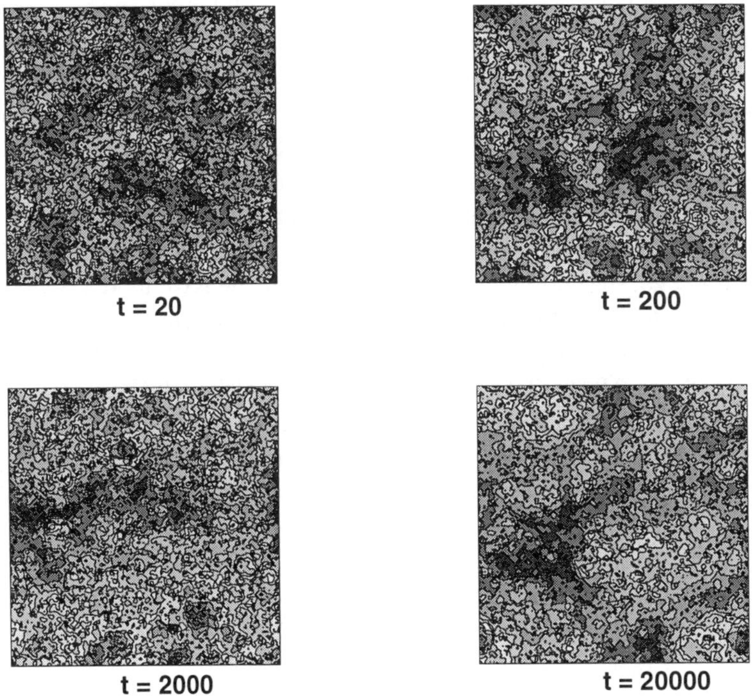

Fig. 4. Figure showing the dynamics of the Kardar-Parisi-Zhang model in time. Large height values are represented in red, and small height values in blue. The times at which the contours are plotted are t=20, t=200, t=2000, and t=20000. The value of the nonlinear coupling parameter is 15.0.

3.1. *New Measures to Study Surface Growth*

In the spirit of the definition of $\delta(\beta, t)$, we first search for a local field related to the "level of complexity" of the growth interface. For structures such as those in Figure 4 the determinant of the Hessian $H(\mathbf{x}, t) \equiv (u_{xx}u_{yy} - u_{xy}^2)$ provides one such density. Its value is typically proportional to the local curvature of contour lines of the field $u(\mathbf{x},t)$. The thermodynamic formalism motivates the introduction of the measures

$$\mu(\beta,t) = \left(\frac{\int d^2x |H(\mathbf{x},t)/Var(u)|^\beta}{\int d^2x}\right)^{1/4\beta}, \tag{5}$$

to analyze global properties of the far from equilibrium structures. Here $Var(u)$ denotes the surface roughness and the measures have been normalized so that for each β, $\mu(\beta,t)$ is the inverse of a lengthscale associated with the structure. Figure 5 shows the behavior of $\mu(\beta,t)$ for $\beta = 0.1, 1$, and 5. There are two distinct stages in the dynamics. It is found that the relaxation rate $\sigma_I(\beta,t)$ during stage I is independent of β, but depends on the model parameter λ. The second stage follows the saturation of $\mu(\beta,t)$ to a value $L^{\gamma(\beta)}$. For $\lambda = 25$, $\gamma(0.1) = -0.075 \pm 0.007$ and $\gamma(1.0) = -0.068 \pm 0.007$.

For $\lambda = 25$, the scaling indices b, z, and α for the KPZ model is identical to those of the RSOS model. It is interesting to compare the values of γ in the two cases. We find that for the RSOS model, $\gamma(0.1) = -0.095 \pm 0.004$ and $\gamma(1.0) = -0.083 \pm 0.007$. Thus, we find that although the growth exponent b, dynamical exponent z, and the roughness exponent α are identical for RSOS and the KZP ($\lambda = 25$) models, their saturation exponents $\gamma(\beta)$ take different values, providing one limitation of the assumed universality.

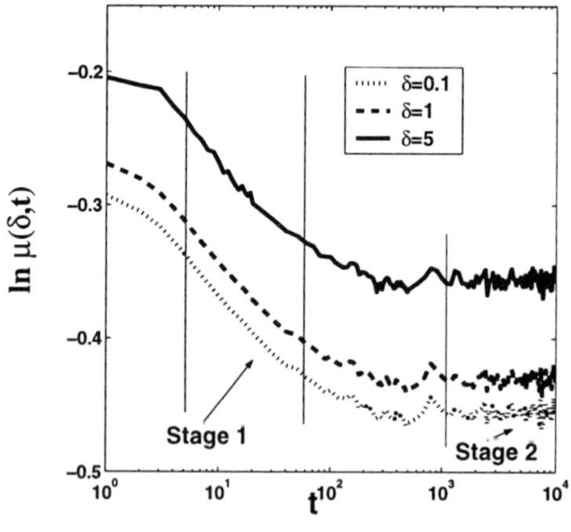

Fig. 5. Behavior of several moments of the measure $\mu(\beta,t)$ for the evolution of a random initial state under the KPZ equation.

4. Discussion

Most details of far from equilibrium structures depend on intractable factors like the nature of the initial state and stochastic terms. However, it is also clear that other facets depend only on control parameters of the system. A statistical description of patterns requires the analysis of such "configuration independent" characteristics. Further, the availability of as many such characteristics as possible will allow for a more comprehensive statistical description of a structure.

In this paper we have discussed two such classes. The form of the functionals was limited by the requirement for the measures to be equivariant under Euclidean motions of the underlying structure. We have shown the utility of these measures by identifying new features of pattern evolution under two sets of model dynamics.

Acknowledgments

The authors acknowledge discussions and collaborations with Dan Goldman, Dave Hoffman, Shaowen Hu, Donald Kouri, and Harry Swinney. This research is partially funded by a grant from the National Science Foundation.

References

[1] Q. Ouyang and H. L. Swinney, *Nature*, **352**, 610 (1991).
[2] J. Boissonade, V. Castets, E. Dulos, and P. de Kepper, *Phys. Rev. Lett*, **64**, 2953 (1990).
[3] F. Melo, P. Umbanhower, and H. L. Swinney, *Phys. Rev. Lett.*, **75**, 3838 (1995).
[4] D. I. Goldman, M. D. Shattuck, H. L. Swinney, and G. H. Gunaratne, *Physica* ***A***, **306**, 180 (2002).
[5] S. D. Sarma and S. V. Ghaisas, *Phys. Rev. Lett.*, **69**, 3762 (1992).
[6] J. S. Langer, *Solids Far From Equilibrium*, ed. C. Godreche (Cambridge University Press, 1992).
[7] A. J. Bray, *Adv. Phys.*, **43**, 357 (1994).
[8] M. S. Heutmaker and J. P. Gollub, *Phys. Rev.* ***A***, **35**, 242 (1987).
[9] G. H. Gunaratne, R. E. Jones, Q. Ouyang, and H. L. Swinney, *Phys. Rev. Lett.*, **75**, 3281 (1995).
[10] Y. Hu, R. Ecke, and G. Ahlers, *Phys. Rev.* ***E***, **51**, 3263 (1995).
[11] J. Kondev, C. L. Henley, and D. G. Salinas, *Phys. Rev.* ***E***, **61**, 104 (2000).
[12] M. C. Cross and P. C. Hohenberg, *Rev. Mod. Phys.*, **65(3)**, 851 (1993).
[13] A. Kudrolli, B. Pier, and J. P. Gollub, *Physica* ***D***, **123**, 99 (1998).
[14] M. Silber and C. M. Topaz, *Physica* ***D***, **172**, 1 (2002).

[15] H. R. Schober, E. Allroth, K. Schroeder, H. Müller-Krumbhaar, *Phys. Rev.* ***A***, **33**, 567 (1986).

[16] K. R. Elder, J. Viñals, and M. Grant, *Phys. Rev.* ***A***, **46**, 7618 (1990).

[17] M. C. Cross and D. I. Meiron, *Phys. Rev. Lett.*, **75**, 2152 (1995).

[18] Q. Hou, S. Sasa, and N. Goldenfeld, *Physica* ***A***, **239**, 219 (1997).

[19] D. K. Hoffman, G. H. Gunaratne, D. S. Zhang, and D. J. Kouri, *Chaos*, **10**, 240 (2000).

[20] D. Ruelle, "Statistical Mechanics, Thermodynamic Formalism," Addison-Wesley, Reading, Massachusetts, 1978; M. J. Feigenbaum, *J. Stat. Phys.*, **46**, 919 (1987).

[21] G. H. Gunaratne, A. Ratnaweera, and K. Tennekone, *Phys. Rev.* ***E***, **59**, 5058 (1999).

[22] S. Hu, D. I. Goldman, H. L. Swinney, D. J. Kouri, D. K. Hoffman, and G. H. Gunaratne, "Stages of Relaxation of Patterns and the Role of Stochasticity on the Final Stage," to appear in *Nonlinearity.*

[23] K. Lee and H. L. Swinney, *Phys. Rev.* ***E***, **51**, 1899 (1995).

[24] A.-L. Barabasi and H. E. Stanley, *Fractal Concepts in Surface Growth* (Cambridge University Press, 1995).

[25] A. Chakrabarti, R. Toral, and J. D. Gunton, *Phys. Rev.* ***B***, **39**, 4386 (1989).

[26] M. Kardar, G. Parisi, and Y,-C, Zhang, *Phys. Rev. Lett.*, **56**, 889 (1986).

[27] J. M. Kim and J. M. Kosterlitz, *Phys. Rev. Lett.*, **62**, 2289 (1989).

CHAPTER 18

DYNAMICS NEAR ROBUST HETEROCLINIC CYCLES

Jeff Porter

Department of Applied Mathematics, University of Leeds,
Leeds LS2 9JT, UK

Symmetric dynamical systems can contain robust heteroclinic cycles, a fact that is supported by a range of experiments and direct numerical simulations, as well as theory. These cycles, however, can be easily destroyed by small symmetry-breaking or stochastic terms, or simply through local or global bifurcations, and may then be replaced by new types of heteroclinic or homoclinic behavior, intermittency, periodic or quasi-periodic orbits, chaos, etc. The 1:2 steady state resonance is used as a detailed example to illustrate the types of new heteroclinic and approximate heteroclinic behavior that can result from the destruction of robust cycles. This is done first within an O(2)-symmetric model, and then through the addition of reflection symmetry-breaking terms.

1. Introduction

A heteroclinic cycle is a finite, ordered sequence of (dynamically transitive) invariant sets $\{\xi_1, \xi_2, \cdots, \xi_k\}$ and connecting manifolds $\{\Gamma_1, \Gamma_2, \cdots, \Gamma_k\}$ such that Γ_j is backward asymptotic to ξ_j and forward asymptotic to ξ_{j+1} with $\xi_{k+1} \equiv \xi_1$. The invariant sets (or nodes) ξ_j are typically equilibria but may include higher-dimensional objects such as periodic orbits or chaotic attractors [1, 2]. The connecting manifolds Γ_j typically describe isolated trajectories but could be multi-dimensional surfaces [3] (see Sections 4 and 5). In the special case $k = 1$ the cycle reduces to a *homoclinic orbit* — this same term is often applied when the invariant sets ξ_j lie on the same group orbit, although we favor "heteroclinic" in this article.

Heteroclinic cycles constitute a very special but important class of solutions that has received a great deal of attention in the past thirty years (see the review article by Krupa [4] and references therein). They are associated with intermittent behavior, because long periods near the invariant sets ξ_j are followed by rapid transitions along the connections Γ_j, and with bursting [5], because the excursions along the Γ_j can correspond to dramatic growth in otherwise quiescent modes of the system. Furthermore, heteroclinic or homoclinic cycles often give rise to chaotic dynamics. Despite these interesting features, there are two prominent difficulties with heteroclinic cycles. (1) They are generically of high codimension, i.e., the formation of multiple connections Γ_j requires multiple parameters to be tuned just so. (2) The dynamics of heteroclinic cycles, even when attracting, is ultimately uninteresting. Trajectories that begin nearby simply traverse the cycle spending ever longer periods near the invariant sets ξ_j until the system effectively becomes "stuck" at one of them.

Symmetry is the solution to the first difficulty and the reason that heteroclinic cycles are observed as frequently as they are. When some or all of the connections Γ_j are related by symmetry then the number of conditions required for the cycle is reduced. Additionally, if some of these connections lie in flow-invariant subspaces (forced by symmetry) they may be structurally stable (robust). It often happens, for example, that an invariant subspace S_j contains two nodes ξ_j and ξ_{j+1} that are, respectively, unstable and stable to perturbations within S_j (ξ_{j+1} must be unstable in at least one *transverse* direction). In this case, the intersection of the unstable manifold of ξ_j with the stable manifold of ξ_{j+1}, $\mathrm{W^u}(\xi_j) \cap \mathrm{W^s}(\xi_{j+1}) \equiv \Gamma_j$, is persistent, i.e., such saddle-sink connections are robust with respect to small changes in the system parameters. If all the connections Γ_j are robust then the cycle itself is robust.

We shall deal with the second difficulty by emphasizing dynamics *near* robust heteroclinic cycles, i.e., we take the point of view that imperfect heteroclinic behavior is often more interesting, and arguably more physical, than true infinite-period heteroclinic cycles. This is so because the symmetry on which robust cycles depend is an idealization that seldom applies unreservedly to real physical systems. Neglected modes, imperfections in experimental boundary conditions, and stochastic effects, for example, may destroy the idealized heteroclinic cycle leaving finite-period intermittent behavior in its place. Even without such perturbations, heteroclinic cycles can generate a host of "approximately heteroclinic" dynamics including long-period periodic orbits, homoclinic orbits, and chaotic attractors. We use the specific example of the 1:2 steady state mode interaction in Sections 4 and 5 to illustrate the fascinating dynamics that can emerge when a very simple, experimentally relevant, heteroclinic cycle is perturbed or destroyed. Before analyzing this detailed example, we review in Section 2 some of the important experimental and numerical observations of heteroclinic behavior (see also [4]) and summarize the different ways in which heteroclinic cycles can be perturbed in Section 3.

2. Observations of Heteroclinic Cycles

Heteroclinic cycles have been discovered in a huge variety of symmetric dynamical systems. They occur in steady state bifurcation problems [6–10], Hopf bifurcation problems [11–14], and a range of Hopf-steady state, and Hopf-Hopf mode interactions [15,16]. Applications include convection [17–19], boundary layers of turbulent flows [20–23], magnetoconvection [24,25], Faraday waves [26, 27], and combustion [28–30]. Heteroclinic cycles have also been proposed as an explanation for the reversals of the Earth's dipolar magnetic field [31,32]. Among these examples, two heteroclinic cycles stand out both for historical reasons and for the strength of the experiments in which they are implicated. We briefly discuss each of these.

2.1. *Rotating Rayleigh-Bénard Convection*

In the Rayleigh-Bénard convection experiment, a layer of fluid confined between two plates is heated uniformly from below. As the temperature difference between the top and bottom plates is increased beyond a critical value, the steady conduction state loses stability and convection ensues. If the layer is also rotated about a vertical axis, there is a range of parameters, associated with the Küppers-Lortz [33] instability, where convection is intermittent, dominated by patches of rolls that repeatedly switch their orientation, each time by approximately 60°. Busse and Clever [17] proposed a simple three-mode model

$$\dot{x}_1 = x_1(\mu + a_1x_1^2 + a_2x_2^2 + a_3x_3^2), \tag{1a}$$

$$\dot{x}_2 = x_2(\mu + a_1x_2^2 + a_2x_3^2 + a_3x_1^2), \tag{1b}$$

$$\dot{x}_3 = x_3(\mu + a_1x_3^2 + a_2x_1^2 + a_3x_2^2), \tag{1c}$$

$x_j, a_j, \mu \in \mathbb{R}$, to explain these dynamics, and observed a heteroclinic cycle (see Figure 1). This cycle was placed in a more rigorous mathematical setting by Guckenheimer and Holmes [6]. It exists, for positive μ, when $a_2 < a_1 < 0$, $a_1 < a_3$, and is stable (unstable) when $2a_1 > a_2 + a_3$ ($2a_1 < a_2 + a_3$); see [4] for details. Robustness is a consequence of the three reflection symmetries $\kappa_j : x_j \to -x_j$, $j = 1, 2, 3$, which force the invariance of the three coordinate planes $x_j = 0$ enabling saddle-sink type connections. Equations (1) also possess the permutation symmetry $x_j \to x_{j+1}$, although

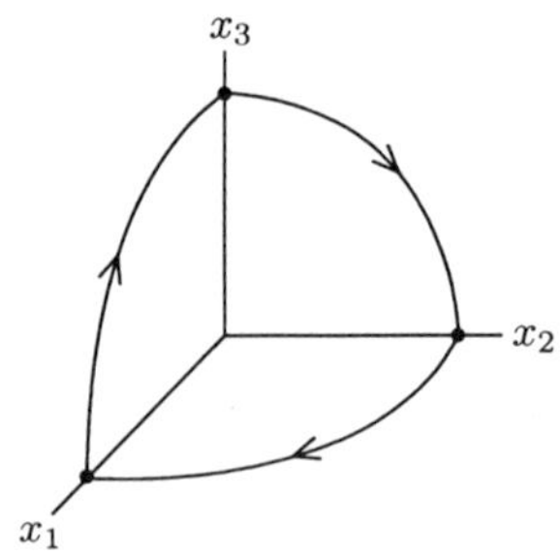

Fig. 1. Sketch of the structurally stable heteroclinic cycle in Eqs. (1).

this is less important than the reflections κ_j. Analogous equations that lack permutation symmetry, and thus have up to nine distinct nonlinear coefficients rather than three, arise in mathematical biology to describe the competition between three species [34, 35] and in the description of triad interactions [36], and exhibit the same basic heteroclinic cycle.

Although Eqs. (1) are too simple to provide a realistic quantitative model for rotating Rayleigh-Bénard convection — they cannot describe the spatial structure of the local domains observed in experiments, or the finite switching time between successive roll states [37, 38] — they capture the gross qualitative behavior surprisingly well. The disagreement with experiment can be reduced by including noise [39] or small symmetry-breaking effects [40, 41] to achieve a finite switching time; see [4] for more discussion.

2.2. *The 1:2 Resonance with O(2) Symmetry*

In O(2)-symmetric systems, the interaction of two modes with wavenumbers in the ratio 1:2 can often be described [7, 42–46] by the pair of complex equations

$$\dot{z}_1 = \mu_1 z_1 + \alpha \bar{z}_1 z_2 + z_1(d_{11}|z_1|^2 + d_{12}|z_2|^2), \tag{2a}$$

$$\dot{z}_2 = \mu_2 z_2 + \beta z_1^2 + z_2(d_{21}|z_1|^2 + d_{22}|z_2|^2). \tag{2b}$$

Solutions of Eqs. (2) will be discussed in detail in Section 4. At this point we need only observe that there is a circle of "pure mode" solutions $(z_1, z_2) = (0, \sqrt{-\mu_2/d_{22}}\, e^{i\varphi_2})$, $\varphi_2 \in [0, 2\pi)$, for $\mu_2 > 0$ when $d_{22} < 0$. Armbruster, Guckenheimer, and Holmes [7] and Jones and Proctor [42, 43] showed that, when $\alpha\beta < 0$, there is an open set of parameters where structurally stable heteroclinic cycles connect diametrically opposite pairs on the circle of pure modes. These robust cycles may also be attracting.

Remarkably, this type of heteroclinic behavior is also observed in higher dimensional systems, and indeed in partial differential equations [19,44,47, 48] where it was in fact discovered [49]. In these spatially extended systems the equilibria $\xi_{1,2}$ correspond to spatially periodic steady states, while the connections $\Gamma_{1,2}$ correspond to translation by half a wavelength. Typically, one finds that the switching time between successive visits to a steady state saturates at a finite value, resulting either in a long-period periodic orbit or sometimes a chaotic orbit [19]. The origin of this behavior in the partial differential equations is not entirely clear but is believed to be due to numerical error, since similar behavior is also found [50] in the normal form (2).

The pure mode heteroclinic cycle of Eqs. (2) was used by Aubry et al. [20] to interpret the observed bursting behavior of turbulent flow along a flat plate [51]. In these experiments, large streak-like eddies form along the boundary in a roughly periodic fashion. After a certain period of time, this configuration undergoes a violent bursting event wherein the eddies are ejected and swept downstream before reforming in a similar configuration, but translated by half a wavelength with respect to the previous pattern. Although the model [20] is quite crude, it does capture the gross features of the turbulent flow in the boundary region. This correspondence can be improved by adding a stochastic perturbation term that takes into account the effect of the pressure field outside the boundary layer, replacing the heteroclinic cycle with intermittent dynamics.

More recently, experiments [30] on premixed flames on a circular porous plug burner have located heteroclinic behavior near the mode interaction point of two Fourier-Bessel modes with azimuthal wavenumbers in the ratio 1:2. In these experiments the O(2) symmetry derives from the geometry of the circular burner. Additional heteroclinic behavior, unrelated to the pure mode cycle of Eqs. (2), is also observed in this system [28,29].

3. Imperfect Heteroclinic Behavior

Heteroclinic cycles can be perturbed or destroyed in a number of ways.

Local bifurcations of ξ_j. A change in stability of one or more of the nodes ξ_j can destroy the heteroclinic cycle. Various scenarios apply, depending on the cycle and the type of bifurcation. For example, new heteroclinic or homoclinic cycles, as well as periodic solutions, may be created [52].

Global bifurcations of Γ_j. The heteroclinic cycle will be destroyed if a connecting manifold Γ_j undergoes a global bifurcation that eliminates the link between ξ_j and ξ_{j+1}. Such global bifurcations may be associated with the emergence of new heteroclinic cycles, as in the example of Section 4.

Forced symmetry breaking. Since robust heteroclinic cycles nearly always rely on symmetry, it is natural to ask what happens when symmetry assumptions are relaxed and appropriately small symmetry-breaking terms are taken into account. For the cycle of Eqs. (1) this results in a long-period periodic orbit [40, 41] or, in some cases, Lorentz-like chaotic dynamics [53]. The effect of symmetry-breaking on Eqs. (2) will be considered in Section 5. In general, some type of approximate heteroclinic behavior is expected. Symmetry-breaking considerations are particularly fitting in the case of normal form symmetries which reflect the structure of the linearized problem and are not "true" symmetries. In the triple Hopf bifurcation, for example, the normal form symmetry (the symmetry of a three-torus) is sufficient to support a robust heteroclinic cycle of the type shown in Figure 1, but with periodic orbits for the nodes ξ_j. When this normal form symmetry is weakly broken the heteroclinic cycle is replaced by intermittent dynamics in a nearby invariant region [16].

Noise. In general, stochastic effects replace heteroclinic cycles with intermittent dynamics [54, 55]. The noise effectively "lifts" trajectories away from the invariant sets ξ_j, thus preventing the system from staying there indefinitely. Noise may also allow trajectories to "jump across" invariant subspaces, giving symmetry-related heteroclinic cycles, which cannot interact in the absence of noise, a chance to "share" trajectories. This stochastic switching from one cycle to another can have unexpected consequences for heteroclinic networks (collections of heteroclinic cycles joined by common nodes). For example, so-called "weak cycles" having a narrow cusp-shaped basin of attraction in the unperturbed case may actually be reinforced, rather than obliterated as one might expect, by the noise [56].

4. The 1:2 Resonance: Novel Heteroclinic Behavior

In this section we consider the heteroclinic behavior of the steady state 1:2 resonance with O(2) symmetry in more detail. In turns out that the pure mode heteroclinic cycles [7, 42, 43] of Section 2.2 are not the only heteroclinic cycles in Eqs. (2). These pure mode cycles exist in a region bounded by a local (symmetry-breaking) bifurcation of the pure mode equilibria and a global bifurcation that generates the necessary global connection (see Section 3). Near the codimension-two point where these bifurcation sets intersect, new types of heteroclinic cycles emerge [45]. This new heteroclinic behavior is organized by a sequence of transitions among (six) distinct heteroclinic cycles, some of which lead to partial cascades of saddle-node and period-doubling bifurcations of Shil'nikov [57] type and the concomitant chaotic dynamics.

4.1. *Normal Form Equations and Basic Solutions*

The normal form equations (2), which describe weakly nonlinear dynamics in a neighborhood of the mode interaction point $(\mu_1, \mu_2) = (0, 0)$, are equivariant under the actions of translation, T_φ, and reflection, $\mathcal{R}$:

$$T_\varphi : (z_1, z_2) \to (e^{i\varphi} z_1, e^{2i\varphi} z_2), \qquad 0 \le \varphi < 2\pi, \tag{3}$$

$$\mathcal{R} : (z_1, z_2) \to (\bar{z}_1, \bar{z}_2), \tag{4}$$

representing the group O(2). The reflection symmetry $\mathcal{R}$ implies that the coefficients in Eqs. (2), which can be computed via a center manifold reduction from the governing equations, are real. Furthermore, as long as α and β are nonzero, the amplitudes z_1 and z_2 can be rescaled (with the cubic coefficients appropriately redefined) to set $\alpha = 1$, $\beta = \pm 1$. The case $\beta = -1$, to which we hereafter restrict attention, is of much greater interest and is the one that arises in hydrodynamics [58].

Equations (2) can be reduced to a three-dimensional system by factoring out the continuous symmetry T_φ. For example, we may write $z_1 = r_1 e^{i\varphi_1}$, $z_2 = r_2 e^{i\varphi_2}$, and $\theta = \varphi_2 - 2\varphi_1$ and work in the (r_1, r_2, θ) phase space. The corresponding "polar" system [7,46], however, is singular when $r_2 = 0$ (a natural consequence of amplitude-phase coordinates) and, because the heteroclinic cycles of interest transect this plane, it is best to use the alternative coordinates [18,45,59] $(x, y) = r_2(\cos\theta, \sin\theta)$ and the equivalent system

$$\dot{r}_1 = \mu_1 r_1 + r_1 x + r_1[d_{11} r_1^2 + d_{12}(x^2 + y^2)], \tag{5a}$$

$$\dot{x} = \mu_2 x + 2y^2 - r_1^2 + x[d_{21} r_1^2 + d_{22}(x^2 + y^2)], \tag{5b}$$

$$\dot{y} = \mu_2 y - 2xy + y[d_{21} r_1^2 + d_{22}(x^2 + y^2)]. \tag{5c}$$

In this representation all standing ($\mathcal{R}$-invariant) solutions are contained within the invariant plane $y = 0$.

Equations (5) contain three types of nontrivial steady states [46].

Pure modes (P) with $r_1 = 0$, $r_2^2 = x^2 + y^2 = -\mu_2/d_{22}$ bifurcate from the trivial state (O) when $\mu_2 = 0$. The eigenvalues of P in Eqs. (2) are 0 (corresponding to translations), $-2\mu_2$ (amplitude perturbations), and

$$\sigma_\pm = \mu_1 - \mu_2 d_{12}/d_{22} \pm \sqrt{-\mu_2/d_{22}}, \tag{6}$$

describing the growth of perturbations away from $z_1 = 0$. In Eqs. (5) there are two fixed points $\mathrm{P}_\pm : (r_1, x, y) = (0, \pm\sqrt{-\mu_2/d_{22}}, 0)$, each representing the entire circle of P states in the full system (2), but with different *orientations* with respect to z_1 perturbations. P_- has the least unstable eigenvalue in the z_1 plane, σ_-, while P_+ has the most unstable eigenvalue σ_+.

Mixed modes (MM) emerge along the line $\mu_1 = 0$ and collide with the pure modes when $\sigma_\pm = 0$. The two branches $\mathrm{MM}_\pm$ satisfy

$$r_1^2 = -(\mu_1 + x + d_{12}x^2)/d_{11}, \qquad y = 0,$$
$$\mu_1 + (1 - d_{21}\mu_1 + d_{11}\mu_2)x + (d_{12} - d_{21})x^2 + (d_{11}d_{22} - d_{12}d_{21})x^3 = 0.$$

Travelling waves (TW) are created in $\mathcal{R}$ symmetry-breaking bifurcations from MM and correspond to steady drift along the group orbit generated by T_φ. In the reduced system (5) they appear as fixed points with

$$r_2^2 = -\frac{2\mu_1 + \mu_2}{d}, \quad r_1^2 = 2r_2^2, \quad \cos\theta = \frac{\mu_2(2d_{11} + d_{12}) - \mu_1(d_{22} + 2d_{21})}{d\sqrt{\beta(2\mu_1 + \mu_2)/d}},$$

where $d \equiv 2(2d_{11} + d_{21}) + (2d_{12} + d_{22})$.

Hopf bifurcations on MM produce standing waves (SW), while Hopf bifurcations on TW lead to modulated travelling waves (MTW).

4.2. *Numerical Investigations*

In this section we illustrate numerically the new type of heteroclinic behavior [45] that emerges, within the O(2)-symmetric model (2), when the pure mode cycles of Section 2.1 are destroyed.

Figure 2 contains selected bifurcation sets in the (μ_1, μ_2) unfolding plane. Each set is labeled by the type of bifurcation: symmetry-breaking (SB), heteroclinic (Het), or Hopf, and by the solution produced. Note that Hopf bifurcations in the three-dimensional system (5) correspond to torus bifurcations in the original system (2). Region 1 (hashmarked) contains the structurally stable $\mathrm{P}_+ \to \mathrm{P}_-$ heteroclinic cycles, while Region 2 (shaded) marks, in an approximate sense, the domain of various heteroclinic cycles [45] involving $\mathrm{P}_\pm$, MM_-, SW, and O. The two regions are separated by the SB bifurcation that produces MM_- from P_-. Representative examples of heteroclinic behavior from each region are shown in Figure 3.

The structurally stable $\mathrm{P}_+ \to \mathrm{P}_-$ cycles in Region 1 are destroyed either by the local bifurcation that produces MM_- and thus destabilizes P_- within the $y = 0$ plane, or by the $\mathrm{P}_+ \to \mathrm{O}$ global bifurcation that occurs when the $\mathrm{P}_+ \to \mathrm{P}_-$ connecting manifold collides with the origin. These

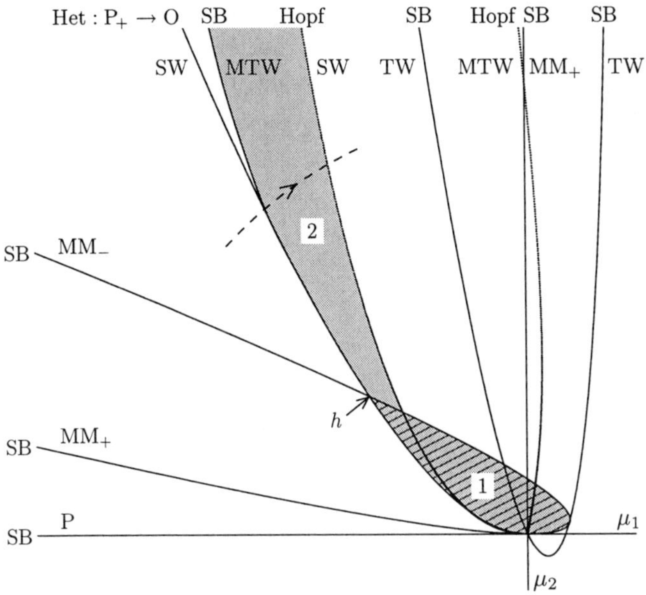

Fig. 2. Bifurcations sets in the (μ_1, μ_2) plane for $d_{11} = -0.4$, $d_{12} = 1.6$, $d_{21} = -6.0$, $d_{22} = -0.5$ (Case I of Ref. [45]) over the range $-1.8 \leq \mu_1 \leq 0.4$, $-0.2 \leq \mu_2 \leq 1.8$. Bifurcations are labelled according to type: symmetry-breaking (SB), heteroclinic (Het), or Hopf, and also by the solution generated. Structurally stable $P_+ \to P_-$ heteroclinic cycles are found in Region 1, while more complicated cycles involving $P_\pm$, MM_-, SW and O appear in Region 2. The dashed path (arc) through Region 2 at $(\mu_1^2 + \mu_2^2)^{1/2} = 1.5$ is used to generate Figure 5.

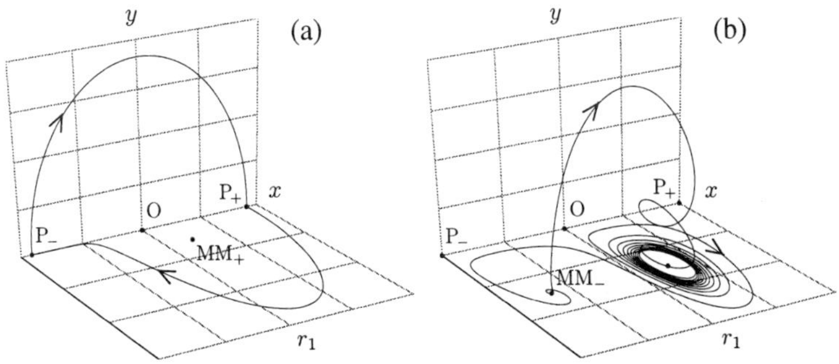

Fig. 3. Examples of the two kinds of heteroclinic dynamics present in the 1:2 resonance: (a) Structurally and asymptotically stable $P_+ \to P_-$ heteroclinic cycle; $\mu_1 \simeq -0.047$, $\mu_2 \simeq 0.167$. (b) Stable periodic orbit associated with one of the heteroclinic cycles in Region 2; $\mu_1 \simeq -0.706$, $\mu_2 \simeq 1.323$.

two boundaries cross in Figure 2 at the point h. Above this point, where the mixed mode MM_- exists, and to the right of the $\mathrm{P}_+ \to \mathrm{O}$ heteroclinic bifurcation where the SW is present, new heteroclinic cycles arise, as suggested in Figure 4, when the two two-dimensional manifolds $\mathrm{W^u(P_-)}$

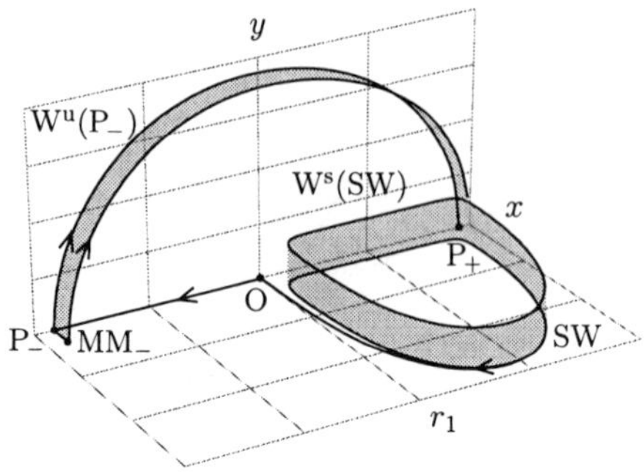

Fig. 4. Sketch of phase space above and to the right of the point h in Figure 2 where newly created MM_- and SW exist. The $\mathrm{P}_+ \to \mathrm{P}_-$ cycle has been destroyed but $\mathrm{W^u(P_-)}$ may intersect $\mathrm{W^s(SW)}$ to create (robust) $\mathrm{O} \to \mathrm{P}_- \to \mathrm{SW} \to \mathrm{O}$ heteroclinic cycles.

and $\mathrm{W^s(SW)}$ intersect; sufficiently close to the $\mathrm{P}_+ \to \mathrm{O}$ global bifurcation, the SW is stable to perturbations transverse to the $y = 0$ plane when $|\mu_1(\sigma_+ - \sigma_-)| > |\mu_2\sigma_+|$, which holds [45] at the point h. This intersection generates a structurally stable cycle of the form $\mathrm{O} \to \mathrm{P}_- \to \mathrm{SW} \to \mathrm{O}$, and leads to additional heteroclinic behavior (discussed below).

For the dotted path in Figure 2, which is well above the codimension-two point h, we locate in the reduced system (5) a series of periodic solutions (quasi-periodic solutions of Eqs. (2)). Figure 5 shows the period as a function of the polar angle α. These solutions form an infinite sequence of isolas that accumulates on the interval $2.087 \lesssim \alpha \lesssim 2.215$. Throughout this interval unstable SW are present. The interval itself is bounded by two global bifurcations (discussed in Section 4.3 The bifurcation at $\alpha \approx 2.215$ corresponds to the accumulation point of the saddle-node (SN) bifurcations in which the individual isolas originate, while the bifurcation at $\alpha \approx 2.087$ (labelled GB) is responsible for the complicated region (hereafter referred to as the twist region) containing multiple SN and period-doubling cascades and exhibiting dramatic variations in period (see Figure 5c). The period-doubling cascades are omitted from the figure for clarity. Corresponding branches on successive isolas are separated (at a fixed value of α) by approximately the period of one SW oscillation, suggesting that the number

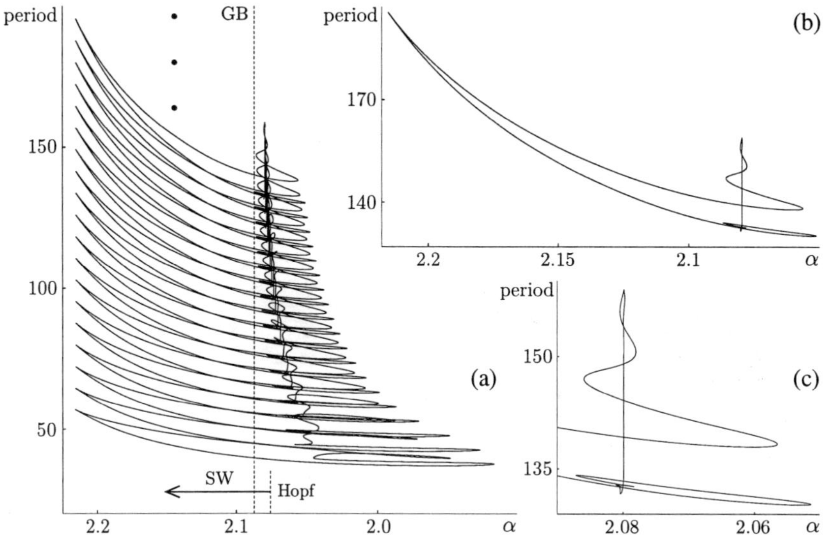

Fig. 5. (a) (Quasi-)periodic solutions forming an infinite stack of isolas. (b) Enlargement of the highest period isola depicted. (c) Further enlargement showing the complicated "twist" region.

of SW oscillations is the basic feature which distinguishes individual isolas. For $\alpha \lesssim 2.087$ (i.e., near the twist regions) typical dynamics include stable chaotic attractors, period one orbits from stable portions of the isolas in Figure 5a, period-doubled orbits, as well as multi-pulse orbits associated with other families of isolas (not depicted); all of these may coexist with one another, as well as with the stable MM_+ and MTW. The competition among these attractors generates a rich variety of hysteretic behavior.

4.3. *Geometric Interpretation*

In Region 2 we have $\mu_1 < 0$, $\mu_2 > 0$, $\sigma_+ > 0$, and $\sigma_- > 0$. SW exist that are unstable within the $y = 0$ plane and stable to MTW ($y \neq 0$) perturbations. For simplicity, we may further suppose that there is only one MM_- state, which is stable within $y = 0$ but unstable to TW perturbations, and one MM_+ state. These conditions correspond to a phase portrait like the one in Figure 6.

Due to the O(2) symmetry of Eqs. (5) there are two invariant subspaces: $y = 0$ representing standing solutions, and $r_1 = 0$ representing pure mode solutions (which are also standing). Within these subspaces the dynamics are two-dimensional and relatively simple. Trajectories with $y = 0$, $r_1 \neq 0$

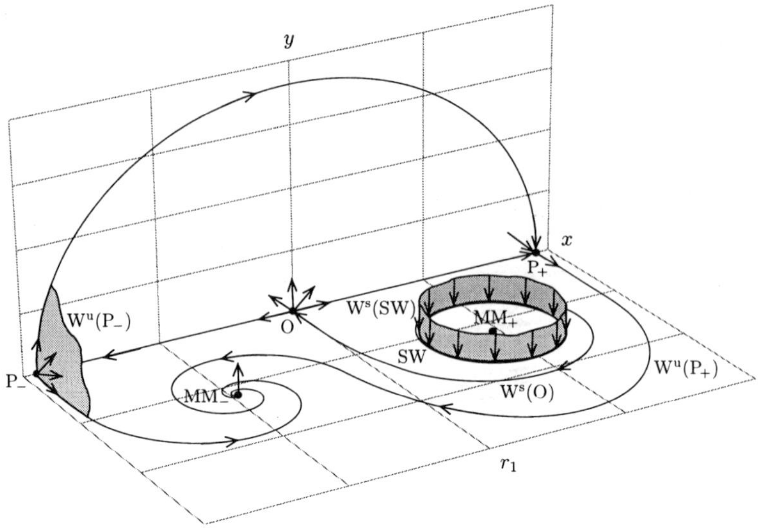

Fig. 6. Sketch of phase space in Region 2.

originating outside the SW limit cycle are attracted to MM_- with the exception of $W^s(O)$. The rest, including $W^u(P_+)$, and the intersection of the $y = 0$ plane with $W^u(P_-)$, are contained in $W^s(MM_-)$. Within the $r_1 = 0$ subspace, trajectories not originating on the half line $y = 0$, $r_1 = 0$, $x \leq 0$ are contained in $W^s(P_+)$. The special trajectory connecting P_- to P_+ along the half circle $x^2 + y^2 = -\mu_2/d_{22}$ corresponds, in the full system (2), to a rotation about a particular pure mode on the circle of pure modes. In Figure 6 we have assumed that MM_- has a complex conjugate pair of eigenvalues, as in Figure 5, although this is not required for the existence of heteroclinic cycles. It is the associated spiraling near MM_-, however, that leads to the interesting behavior found in the twist region.

The two invariant subspaces and their intersection provide for the existence of several persistent heteroclinic connections: SW $\to$ O, SW $\rightrightarrows$ MM_- (double arrows denote a continuum of connections), $P_+ \to MM_-$, $P_- \to MM_-$, O $\rightrightarrows$ P_+, O $\to P_-$, and $P_- \to P_+$. Since these connections are robust, a single additional connection outside of the invariant subspaces can generate heteroclinic cycles. For example, $W^u(P_-)$ may intersect $W^s(SW)$ (see also Figure 4) generating cycles of the form O $\to P_- \to$ SW $\to$ O. These cycles are persistent if the intersection of $W^u(P_-)$ with $W^s(SW)$ is transverse. The first (nontransverse) intersection of $W^u(P_-)$ with $W^s(SW)$ creating heteroclinic cycles of this type corresponds to the accumulation point of the saddle-nodes, at $\alpha \simeq 2.215$, which produce the isolas from the left. The interval $2.215 \gtrsim \alpha \geq \alpha_* \simeq 2.087$ is filled with structurally stable heteroclinic cycles.

A second global bifurcation occurs when $W^u(MM_-)$, which is one-dimensional, falls within $W^s(SW)$. This happens only at an isolated value of the parameter, $\alpha = \alpha_*$, but creates a multitude of (structurally unstable) heteroclinic cycles: $SW \rightrightarrows MM_- \to SW$, $O \rightrightarrows P_+ \to MM_- \to SW \to O$, $O \to P_- \to P_+ \to MM_- \to SW \to O$, and $O \to P_- \to MM_- \to SW \to O$. Periodic orbits approximating each of these cycles coexist within the twist region, indicating that this codimension-one global bifurcation (indicated by the broken vertical line GB in Figure 5 a) is responsible for the complexity found on the rightmost part of the isolas. As one moves up the stack of isolas the corresponding SN bifurcations in this region accumulate at different values of α near α_*, determined by the twisting of $W^u(P_-)$ around MM_-; see [45]. In particular, structurally stable heteroclinic cycles can, and do, exist for $\alpha < \alpha_*$. More complicated possibilities exist also. For example, $W^u(MM_-)$ can cycle several times between neighborhoods of MM_- and SW before reconnecting to create a multi-pulse heteroclinic cycle.

The two global bifurcations described above are intrinsically linked by the fact that $W^u(P_-)$ includes the two connections $P_- \to MM_-$ and $P_- \to P_+$. In other words, if any of the $O \to P_- \to SW \to O$ cycles created in the first global bifurcation happen to remain close, upon leaving the neighborhood of P_-, to either the $y = 0$ or $a_1 = 0$ planes, then these cycles will necessarily come close to the cycles $O \to P_- \to MM_- \to SW \to O$ or $O \to P_- \to P_+ \to MM_- \to SW \to O$, respectively. This requires that one is near to the second global bifurcation at $\alpha = \alpha_*$, and suggests that transitions from one type of heteroclinic orbit to another should occur as α is varied. In fact there are many such transitions on a typical isola, which can be divided into six different regimes, each organized by a particular heteroclinic cycle. In Figure 7is a sketch of an idealized isola with insets illustrating the type of heteroclinic cycle important in each regime.

Much of the detailed structure of the isolas in Figure 5 (see also Figure 7) can be understood by constructing appropriate Poincaré return maps [18, 45]. For example, one can predict the separation in period of successive isolas in regime 1, the presence of a single *stable* periodic orbit in regime 3, and the scaling of the partial cascades of SN and period-doubling bifurcations in regimes 2, 4, and 6. This analysis [45] reveals that, as the number of SW oscillations approaches infinity, regimes 2-6 limit on the second global bifurcation where $W^u(MM_-) \subset W^s(SW)$. Furthermore, in this

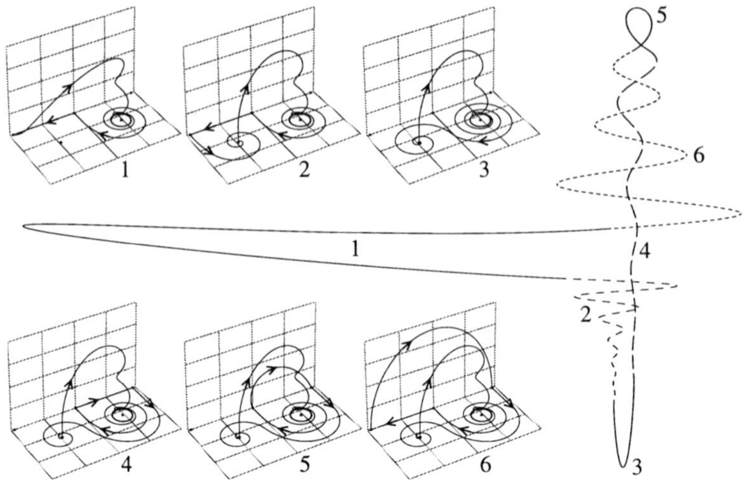

Fig. 7. Sketch of a typical isola showing the different heteroclinic cycles involved in each regime.

limit the separation in period (on a given isola) between regimes 1 and 3, and between regimes 5 and 1, also diverges.

5. 1:2 Resonance with Broken O(2) → SO(2) Symmetry

In this section we consider the effects of small reflection symmetry-breaking perturbations on the dynamics near the heteroclinic cycles. Chossat [60] has shown that breaking the symmetry O(2) down to SO(2) in this way generically destroys the pure mode heteroclinic cycles and replaces them with a quasi-periodic orbit characterized by two small frequencies, one associated with the broken heteroclinic connection and one with a slow drift along the group orbit of translations. Ashwin et al [61] showed that this perturbation must be dispersive: if reflection symmetry is broken by adding a constant throughflow, for example, the cycle will persist. Others [62, 63] have examined the effects of breaking O(2) down to D_4 and O(2) down to D_2. In all cases the cycle is destroyed, although some structurally unstable heteroclinic orbits may be generated at the same time.

Here we return to the problem investigated by Chossat, and consider the generic effects of breaking the reflection symmetry $\mathcal{R}$ of Eq. (4). We confirm Chossat's conclusion, and show that new types of dynamics are introduced into the system above and beyond the quasi-periodic orbit. Specifically we find that secondary Takens-Bogdanov bifurcations, so-called T-points, and degenerate heteroclinic cycles are responsible for the appearance of new types of global bifurcations, and with them of complex dynamics [64].

5.1. *SO(2)-Symmetric Normal Form*

When reflection symmetry is broken the normal form coefficients are no longer forced to be real and can be expected to acquire (small) imaginary parts. In the following we measure the magnitude of the symmetry-breaking by a (small) parameter ε, and replace Eqs. (2) by

$$\dot{z}_1 = (\mu_1 + i\varepsilon\omega_1)z_1 + \tilde{\alpha}\bar{z}_1 z_2 + z_1(\tilde{d}_{11}|z_1|^2 + \tilde{d}_{12}|z_2|^2), \tag{7a}$$

$$\dot{z}_2 = (\mu_2 + i\varepsilon\omega_2)z_2 + \tilde{\beta}z_1^2 + z_2(\tilde{d}_{21}|z_1|^2 + \tilde{d}_{22}|z_2|^2), \tag{7b}$$

where $(\tilde{\alpha}, \tilde{\beta}, \tilde{d}_{jk}) = (1, -1, d_{jk}) + i\varepsilon(\alpha_i, \beta_i, c_{jk})$; $j, k \in \{1, 2\}$, $\alpha_i, \beta_i, c_{jk} \in \mathbb{R}$.

Once again we factor out the continuous symmetry T_φ by working with the real amplitudes r_1 and r_2 and the relative phase $\theta = \varphi_2 - 2\varphi_1$. Writing $\tilde{\alpha} = |\tilde{\alpha}| \exp i\varphi_1$, $\tilde{\beta} = -|\tilde{\beta}| \exp i\varphi_2$, $x = r_2 \cos(\theta + \varphi_1)$, and $y = r_2 \sin(\theta + \varphi_1)$ we obtain, after a simple rescaling, the three-dimensional system (cf. Eqs. (5))

$$\dot{r}_1 = \mu_1 r_1 + r_1 x + r_1[d_{11}r_1^2 + d_{12}(x^2 + y^2)], \tag{8a}$$

$$\begin{aligned} \dot{x} &= \mu_2 x - \nu y + 2y^2 - r_1^2 \cos\varphi \\ &\quad + x[d_{21}r_1^2 + d_{22}(x^2 + y^2)] - y[c_1 r_1^2 + c_2(x^2 + y^2)], \end{aligned} \tag{8b}$$

$$\begin{aligned} \dot{y} &= \mu_2 y + \nu x - 2xy - r_1^2 \sin\varphi \\ &\quad + y[d_{21}r_1^2 + d_{22}(x^2 + y^2)] + x[c_1 r_1^2 + c_2(x^2 + y^2)], \end{aligned} \tag{8c}$$

where $\nu = \varepsilon(\omega_2 - 2\omega_1)$, $c_1 = \varepsilon(c_{21} - 2c_{11})$, $c_2 = \varepsilon(c_{22} - 2c_{12})$, and $\varphi = \varphi_1 + \varphi_2$.

5.2. *General Effects of Symmetry-Breaking*

Breaking the reflection symmetry $\mathcal{R}$ has a number of qualitative effects.

Generic solutions drift. When ε is increased from 0, steady solutions and standing waves begin to drift slowly [65, 66], forming TW and MTW, respectively. The left- and right-travelling waves that are present when $\varepsilon = 0$ persist, but are no longer related by reflection symmetry.

Primary bifurcations are Hopf bifurcations. When $0 < \varepsilon \ll 1$ the primary bifurcations at $\mu_1 = 0$ and $\mu_2 = 0$ become Hopf bifurcations [65, 66], which lead, respectively, to a branch of drifting mixed modes (TW) and a branch of drifting pure modes (P). In the reduced system (8) P also emerges (at $\mu_2 = 0$) as a limit cycle. If $\nu c_2 < 1$ this limit cycle is destroyed, as r_2

increases, in a saddle-node homoclinic bifurcation (also referred to as a "sniper" bifurcation) when $2r_2 = |\nu + c_2 r_2^2|$. A pair of fixed points

$$\mathrm{P}_\pm : (r_1, x, y) = \left(0, \pm\sqrt{r_2^2 - (\nu + c_2 r_2^2)^2/4}\,, (\nu + c_2 r_2^2)/2\right),\ r_2^2 = -\mu_2/d_{22},$$

replaces the limit cycle over the range $2 - \nu c_2 - 2\sqrt{1 - \nu c_2} < c_2^2 r_2^2 < 2 - \nu c_2 + 2\sqrt{1 - \nu c_2}$, provided $c_2 \neq 0$. If $c_2 = 0$ these fixed points exist for all $r_2^2 > \nu^2/4$. In this regime mixed states (slowly drifting TW) are produced by T_π symmetry-breaking bifurcations when one of the eigenvalues (cf. Eq. (6))

$$\tilde{\sigma}_\pm = \mu_1 - \mu_2 d_{12}/d_{22} \pm \sqrt{-\mu_2/d_{22} - (\nu - \mu_2 c_2/d_{22})^2/4} \tag{9}$$

crosses zero.

The relative phase θ can unlock. In the O(2)-symmetric case the planes $\theta = 0$ and $\theta = \pi$ are invariant due to the reflection $\mathcal{R}$. The relative phase θ is therefore bounded and cannot rotate or change sign. With SO(2) symmetry these constraints are lifted. The two-frequency MTW solutions may be "locked" on average so that θ oscillates about a constant value, as they are in the O(2)-symmetric case, or they may be "unlocked", in which case θ evolves in an unbounded fashion.

Waves can reverse their direction of propagation. Along with unlocked MTW, the breaking of reflection symmetry $\mathcal{R}$ allows for more complicated types of travelling solutions in which one (or both) of the constituent modes repeatedly reverses its direction [67,68]. This phenomenon is related to the so-called direction-reversing travelling waves [69, 70].

Structurally stable connections are destroyed, but new possibilities emerge. The structurally stable $\mathrm{P}_+ \to \mathrm{P}_-$ connection in Eqs. (5) relies on the invariance of the $y = 0$ plane; this connection is labelled Γ_1 in Figure 8a. The cycle is completed by "re-orienting" the trajectory (thinking now of the full system (2)) at the pure mode from the stable manifold along which it entered to the unstable manifold so that it can leave again. This rotation can be either up ($y > 0$) or down ($y < 0$), and occurs along (half of) the circle $x^2 + y^2 = -\mu_2/d_{22}$ in Eqs. (5); these two choices are labelled $\Gamma_2^\pm$ in Figure 8a. When the reflection symmetry $\mathcal{R}$ is broken, heteroclinic connections that rely on the invariance of the $y = 0$ plane are no longer robust. The connection $\mathrm{P}_+ \to \mathrm{P}_-$, for example, is now codimension-one (see Figure 8b) since it requires the intersection of the one-dimensional $\mathrm{W^u}(\mathrm{P}_+)$ with the two-dimensional $\mathrm{W^s}(\mathrm{P}_-)$ in three-dimensional space (rather than the two-dimensional plane $y = 0$). Similarly, structurally stable connections

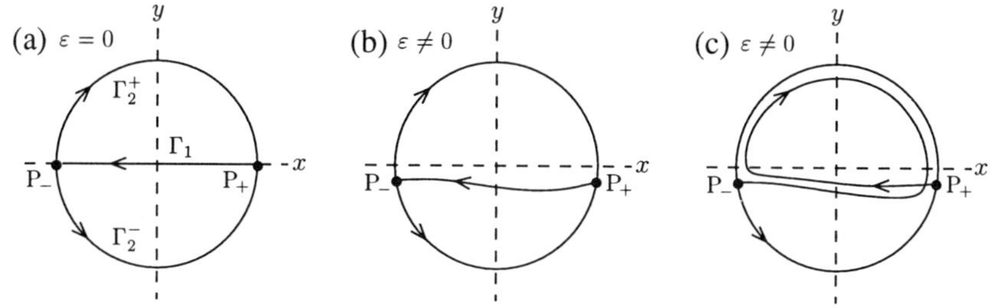

Fig. 8. (a) Schematic representation in the (x, y) plane of a structurally stable $P_+ \to P_-$ heteroclinic cycle with $\varepsilon = 0$. The primary connection Γ_1 is in the $y = 0$ plane. Γ_2 represents a re-orientation. (b) A heteroclinic cycle can still occur with $\varepsilon \neq 0$ but it is no longer structurally stable. (c) Multi-pulse heteroclinic cycles, which are impossible when $\varepsilon = 0$, can now occur as well. These "N-heteroclinics" make N visits to the neighborhood of P_- before connecting up.

in the $y = 0$ plane from P_+ or P_- to MM_-, which play a part in the complicated heteroclinic cycles [45] of Region 2 (see Section 4), are reduced to codimension-one phenomena when $\varepsilon \neq 0$.

Although the structurally stable heteroclinic connections in the $y = 0$ plane are destroyed by the symmetry-breaking, new types of structurally unstable homoclinic and heteroclinic connections are permitted. Besides the direct $P_+ \to P_-$ connection depicted in Figure 8b, more complicated "N-heteroclinics" become possible. Such solutions, like the 2-heteroclinic shown in Figure 8c, occur when $W^u(P_+)$ passes through the neighborhood of P_- several times before connecting up. N-heteroclinics are not possible when $\varepsilon = 0$ because $W^u(P_+)$ and $W^s(P_-)$ are confined to the $y = 0$ plane. In a similar fashion, symmetry-breaking allows for certain homoclinic connections that cannot occur with $\varepsilon = 0$. For example, MM_- may now experience homoclinic connections, whereas such reconnection in the O(2)-symmetric system (5) requires an additional limit set in the $y = 0$ plane.

5.3. *Numerical Example: Broken Symmetry*

We now illustrate the effects of broken reflection symmetry on the numerical example of Section 4.2, retaining $d_{11} = -0.4$, $d_{12} = 1.6$, $d_{21} = -6.0$, and $d_{22} = -0.5$ in Eqs. (8), and setting $\nu = -\varepsilon$, $\varphi = c_1 = c_2 = 0$.

5.3.1. *Weakly Broken Symmetry:* $\varepsilon = 0.01$

Figure 9 shows the changes incurred by Figure 2 when the reflection symmetry $\mathcal{R}$ is broken to a relatively small extent ($\varepsilon = -\nu = 0.01$). There

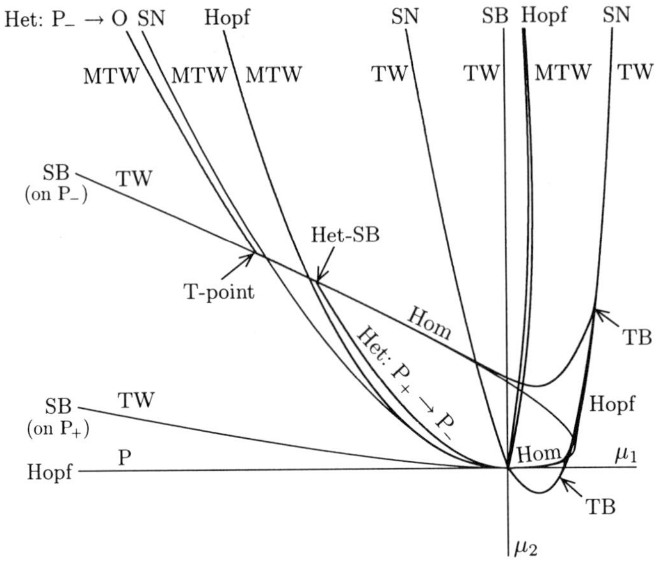

Fig. 9. Bifurcations sets with $\nu = -0.01$, $\varphi = c_1 = c_2 = 0$. The nonlinear coefficients d_{jk} are as in Figure 2. The range displayed is $-1 \geq \mu_1 \geq 0.3$, $-0.2 \geq \mu_2 \geq 1$. The TW homoclinic (Hom) bifurcation and the new codimension-two bifurcations: Takens-Bogdanov (TB), heteroclinic-SB (Het-SB), and the T-point, are discussed in the text.

are significant effects on both global and local bifurcations, as anticipated in Section 5.2. Note the splitting of bifurcation sets (e.g., Hopf to MTW near the positive μ_2 axis) and the introduction of new codimension-two points such as the Takens-Bogdanov (TB) points, the interaction between the $P_+ \to P_-$ heteroclinic cycle and a symmetry-breaking (wavelength-doubling) bifurcation on P_- (Het-SB), and the T-point, all of which organize the behavior of the system in their vicinity. Additional codimension-two points are located too close to the origin to be visible in Figure 9; see [64, 71] for more detail. We focus here on some of the more robust aspects of this "unfolding" that characterize behavior farther from the origin where different types of heteroclinic cycles occur.

Effect on local bifurcations. The primary bifurcation at $\mu_2 = 0$ to P, as mentioned in Section 5.2, is now a Hopf bifurcation. The corresponding limit cycle in Eqs. (8) persists until $\mu_2 = -\varepsilon^2 d_{22}/4$, where it is replaced by the pair of fixed points $P_\pm$ in a saddle-node homoclinic bifurcation that is indistinguishable from $\mu_2 = 0$ in Figure 9 and is not labelled. The $P_\pm$ states change stability at $\tilde{\sigma}_\pm = 0$, producing slowly drifting TW. The parity-breaking bifurcations from MM to TW and SW to MTW of the O(2)-symmetric system are unfolded [72] and replaced in Figure 9 by curves of SN bifurcations.

Effect on global bifurcations. The loss of $\mathcal{R}$ symmetry is manifest most dramatically in the kind of global bifurcations seen in Eqs. (8). In Figure 2, for example, a $P_+ \to O$ heteroclinic bifurcation produces SW and, in Region 1, structurally stable $P_+ \to P_-$ heteroclinic cycles. Since $W^u(P_+)$ and $W^s(O)$ are both one-dimensional their coalescence in a plane is a codimension-one phenomenon but becomes of codimension-two when the $y = 0$ plane ceases to be invariant. Consequently we do not expect to see a $P_+ \to O$ connection once $\varepsilon \neq 0$. This is indeed what happens in Region 1, i.e., below the curve $\tilde{\sigma}_- = 0$. However, when $\tilde{\sigma}_- > 0$, this connection is replaced by a $P_- \to O$ connection. This is possible because in that region $W^u(P_-)$ is two-dimensional and a codimension-one intersection with $W^s(O)$ does not require an invariant plane. In fact, this connection is the remnant of the $P_- \to P_+ \to O$ connection that forms part of the surface of heteroclinic cycles when $\varepsilon = 0$; for small ε this connection still passes very close to P_+.

Chossat [60] shows that the $P_+ \to P_-$ heteroclinic cycles, which exist throughout Region 1 of Figure 2, are in general destroyed and replaced by MTW once $\varepsilon \neq 0$. In fact Figure 9 shows that a single curve of $P_+ \to P_-$ cycles remains. This heteroclinic bifurcation, which is now of codimension one (see Figure 8b), generates two MTW which emerge on opposite sides of the bifurcation to replace the original structurally stable cycle, consistent with the results of [60]. One of these MTW eventually terminates (typically after a SN bifurcation) at one of the two Hopf bifurcations to the right of the μ_2 axis that replace the Hopf bifurcation from TW to MTW when $\varepsilon = 0$. The other is annihilated, typically after one or more SN bifurcations, in either the Hopf bifurcation that replaces the $\varepsilon = 0$ Hopf bifurcation to SW, or in a homoclinic bifurcation involving the TW created when $\tilde{\sigma}_- = 0$ (see below). We do not attempt to describe all possible scenarios for the various MTW branches, nor to calculate all of the SN bifurcation sets.

Homoclinic bifurcations (labeled "Hom" in Figure 9) involving the mixed states are an essential part of the symmetry-breaking scenario, appearing just after the bifurcation from $P_\pm$ to $MM_\pm$ (TW when $\varepsilon \neq 0$). In Figure 2 the SB bifurcation on P_+ is isolated from Region 1 and its $P_+ \to P_-$ cycles and, unlike in the case of P_-, does not give rise to TW homoclinics when $\varepsilon \neq 0$. With other parameters, however, both of the $P_\pm$ bifurcations help to define Region 1 and the TW homoclinics that arise are illustrated in Figure 10. In either case these homoclinic orbits follow the

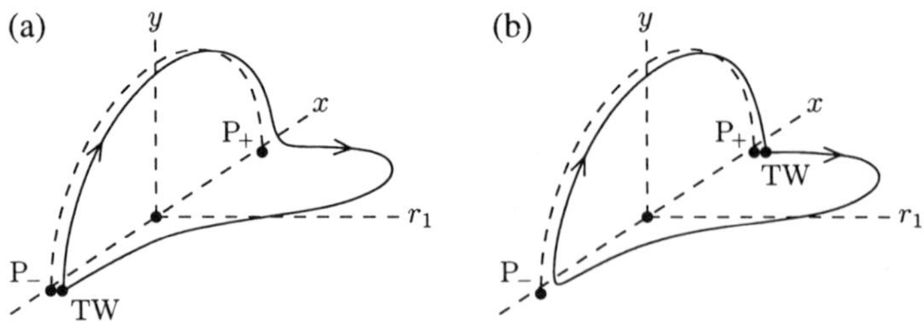

Fig. 10. Sketch of the homoclinic bifurcations that arise near the boundary of Region 1 when $\varepsilon \neq 0$: (a) just after the bifurcation from P_-, as in Figure 9, and (b) after a bifurcation from P_+. We have shown the "up" case, which follows the upper piece of $W^u(P_-)$, but the "down" case also occurs.

path of the $\varepsilon = 0$ heteroclinic cycle, approaching it more and more closely as ε decreases.

The symmetry-breaking has an even more dramatic effect on the various heteroclinic cycles in Region 2, destroying several important heteroclinic connections. In particular, the connections $P_+ \to MM_-$ and $SW \to O$, which were structurally stable with $\varepsilon = 0$, are now of codimension one. The effect of finite ε on each of the heteroclinic cycles of Region 2 (see Section 4.3) is summarized in Table 1. When $\varepsilon \neq 0$ only two of these cycles (the

Table 1. Codimension of various heteroclinic cycles found in Region 2 when $\varepsilon = 0$. A small ε indicates which connections are disrupted by the symmetry-breaking. Recall that MM_- and SW become TW and MTW when $\varepsilon \neq 0$.

Heteroclinic cycle	$\varepsilon = 0$ codim.	$\varepsilon \neq 0$ codim.
$O \to P_- \to SW \xrightarrow{\varepsilon} O$	0	1
$O \to P_- \to P_+ \xrightarrow{\varepsilon} MM_- \to SW \xrightarrow{\varepsilon} O$	1	3
$O \to P_+ \xrightarrow{\varepsilon} MM_- \to SW \xrightarrow{\varepsilon} O$	1	3
$O \to P_- \to MM_- \to SW \xrightarrow{\varepsilon} O$	1	2
$SW \to MM_- \to SW$	1	1

first and last entries) remain of codimension one or less. Consequently, one might expect to encounter only these two cycles when traversing Region 2. However, the situation near the MTW limit cycle is much more subtle. When $\varepsilon = 0$ the connections $P_- \to SW \to O$ and $MM_- \to SW \to MM_-$ include the SW *by necessity*, since a connection to the SW provides the only mechanism for re-entering the invariant $y = 0$ plane. This is no longer the case when the symmetry $\mathcal{R}$ is broken. In addition to the codimension-one cycles of Table 1 (replacing SW with MTW and MM_- with TW) we can expect an infinite number of heteroclinic cycles of the form O $\to$ P_- $\to$ (near MTW) $\to$ O and TW $\to$ (near MTW) $\to$ TW, which merely visit the neighborhood of the MTW, executing a finite number of rotations, before leaving again. The situation is further complicated by the possibility of using either the "up" or the "down" parts of $W^u(P_-)$ or $W^u(TW)$; see also Figure 8. The latter cycles are in fact the most prominent cycles one observes numerically in the parameter regime containing the infinite cascade of isolas (Figure 5) when $\varepsilon = 0$. In Figure 11 we show the period of the various periodic solutions of Eqs. (8) that are related to these isolas.

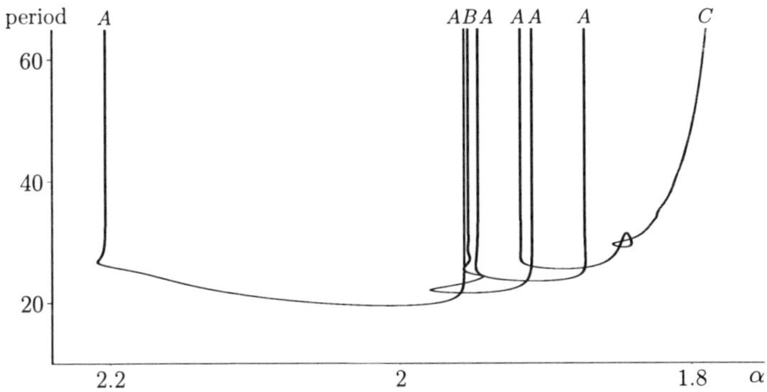

Fig. 11. The period of several periodic solutions of Eqs. (8) with $\varepsilon = 0.01$ as a function of the polar angle α in the parameter regime containing an infinite cascade of isolas (Figure 5) when $\varepsilon = 0$. The global bifurcations that replace the isolas are of three basic types: O $\to$ P_- $\to$ (near MTW) $\to$ O, TW $\to$ (near MTW) $\to$ TW, and TW $\to$ TW, labelled A, B, and C, respectively. The global bifurcation C occurs after a subcritical Hopf bifurcation that destroys the periodic orbit replacing SW, and involves the TW replacing MM_+, in contrast to bifurcation B which involves the TW that replaces MM_-. All the solutions shown are unstable.

New codimension-two points. The new homoclinic and heteroclinic bifurcations discussed above are organized by a series of new codimension-two points. The simplest of these is the TB point, resulting from the intersection of a SN and a Hopf bifurcation with zero frequency [73]. Under appropriate conditions the TB bifurcation generates a curve of homoclinic connections. In Figure 9 the TB points serve as endpoints for the TW homoclinics that emerge near the SB bifurcation from P_- to TW. Note the (new) Hopf bifurcation (which is rather difficult to see in Figure 9) that extends between the two TB points, and the TW homoclinic that appears just below the SB bifurcation on P_- in the lower right part of Figure 9 (also difficult to see).

The point labelled Het-SB in Figure 9 marks the intersection of the $P_+ \to P_-$ heteroclinic cycle with the SB bifurcation on P_- that destroys them. At this point there are two degenerate heteroclinic cycles, one "up" and one "down" (see Figure 12a), involving the nonhyperbolic fixed point P_-. Two corresponding curves of TW homoclinics (up and down) emerge on opposite sides of the Het-SB point, tangent to the SB bifurcation curve.

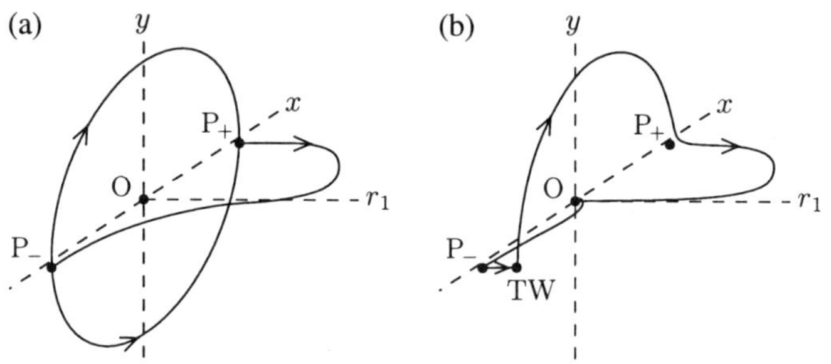

Fig. 12. Sketches of (a) the degenerate $P_+ \to P_-$ cycle(s) at the Het-SB point, and (b) the heteroclinic cycle $TW \to O \to P_- \to TW$ that defines the T-point in Figure 9.

The TW homoclinic bifurcation to the left of the Het-SB point in Figure 9 stays very close to the curve of SB bifurcations that separates Regions 1 and 2 in Figure 2 and then terminates in a T-point [74] at the bottom of the $P_- \to O$ heteroclinic curve. The nature of this T-point is illustrated in Figure 12b. The heteroclinic connection from TW to O is the most important feature; it is a codimension-two event because both $W^u(TW)$ and $W^s(O)$ are one-dimensional. The remaining two segments of the cycle are structurally stable: $O \to P_-$ takes place in the $r_1 = 0$ plane, while the intersection of $W^u(P_-)$ and $W^s(TW)$ is robust because both manifolds are two-dimensional. The unfolding of this T-point contains both the TW homoclinic bifurcation and the $P_- \to O$ heteroclinic cycle [64], as suggested by Figure 9. Because the eigenvalues of O in the $r_1 = 0$ plane are complex (see Eqs. (8)) the curve of TW homoclinics, all of which pass through a neighborhood of O, takes the form of a logarithmic spiral (in a sufficiently small neighborhood of the T-point). However, when $\varepsilon = 0.01$ it is not possible to discern this spiral without substantial magnification of Figure 9.

5.3.2. *Strongly Broken Symmetry:* $\varepsilon = 0.5$

When the much larger value of $\varepsilon = 0.5$ is used, with the remaining coefficients held fixed, one obtains Figure 13. The T-point, identified in Figure 9,

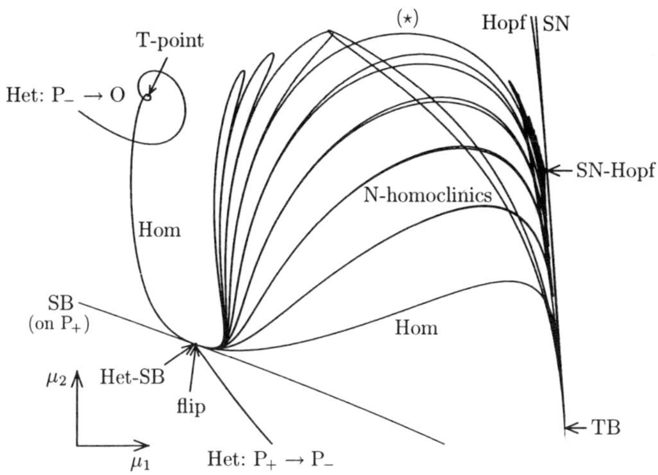

Fig. 13. Selected bifurcation sets related to the T-point, SN-Hopf, Het-SB, and flip bifurcations. The range shown is $-0.7 \leq \mu_1 \leq 0.05$, $0.25 \leq \mu_2 \leq 1.1$. The unpaired curve ($\star$) is special because it is the only N-homoclinic that actually connects to both the flip and SN-Hopf bifurcations.

has shifted away from the curve of SB bifurcations on P_- where it originated, and now exhibits clear spiraling on both the TW homoclinic curve and the $P_- \to O$ heteroclinic curve that emerge from it. The (logarithmic) spiraling of the set of $P_- \to O$ connections indicates that the TW has acquired complex eigenvalues, in contrast to the case depicted in Figure 12b.

Another new feature ofFigure 13is the formation near the upper Takens-Bogdanov (TB) point in Figure 9 of a true SN-Hopf interaction, with a nonzero Hopf frequency at the interaction. This codimension-two singularity has a three-dimensional normal form [73] containing a set of heteroclinic connections (between the two TW) in its unfolding. This connection, however, relies on a normal form symmetry which is not a feature of the full problem. Indeed, as shown by Kirk [75, 76], when appropriate symmetry-breaking corrections are taken into account, one finds that the curve of heteroclinic cycles breaks apart, leaving two curves of *homoclinic* connections, one for each of the TW. These two curves oscillate with exponentially decreasing amplitude as they approach the SN-Hopf point, intertwining in a characteristic "out-of-phase" fashion [75–77]. In Figure 13 one of these wiggly curves of TW homoclinic connections eventually terminates at the TB point. The second curve of TW homoclinics ($\star$) veers away from the region of the SN-Hopf toward the Het-SB point. It cannot terminate here, however, but does so close by, at a homoclinic flip bifurcation of the original "down" (single-loop) TW homoclinic (see Figure 12a) that emerges toward the right from the Het-SB point. The unfolding of this homoclinic flip bifurcation contains very complicated dynamics, including "N-homoclinics" for any integer N and regions of chaotic shift-dynamics [78]. A series of these N-homoclinics, which make N visits to the neighborhood of the TW before reconnecting, is shown in Figure 13. It appears likely that the N-homoclinics created in the homoclinic flip bifurcation connect to the various homoclinics involved in closing the resonance tongues present near the SN-Hopf bifurcation [76]. In fact all but one of the N-homoclinics begin and end at the flip bifurcation and only the curve ($\star$) terminates in the SN-Hopf point. Near the SN-Hopf point we find a variety of interesting dynamics including stable tori and periodic orbits typical of the perturbed normal form [79].

6. Conclusion

In this article we have explored a range of dynamics associated with robust heteroclinic cycles. Such cycles are most easily found in symmetric systems where the structural stability derives from the presence of invariant subspaces supporting saddle-sink type connections. This is the case for the two prominent heteroclinic cycles discussed in Section 2: the three-state cycle [6,17,34] in Eqs. (1), associated with the Küppers-Lortz [33] instability in rotating Rayleigh-Bénard convection, and the pure mode cycle [7,42,43] in the 1:2 steady state resonance with O(2) symmetry, of relevance for turbulent boundary layer breakdown [20,80], cellular flame patterns on a circular porous plug burner [30], Rayleigh-Bénard convection in a plane layer [19,43], and von Kármán flow between two exactly counter-rotating disks [44].

Heteroclinic cycles, robust or not, are delicate objects. They depend on a collection of nodes ξ_j, which can undergo local bifurcations, and connecting manifolds Γ_j, which can experience global bifurcations. Most crucially, they rely on assumptions of symmetry which neglect experimental imperfections, stochastic effects, and other symmetry-breaking processes. If these effects are taken into account the heteroclinic cycle is typically replaced by some type of finite-period intermittent dynamics. Dramatic new behavior such as bursting [81] or random switching between different cycles [55] may also occur.

In Section 4 we looked in detail at new heteroclinic behavior related to the destruction of the structurally stable pure mode cycles familiar from previous studies [7,42,43] of the 1:2 steady state resonance with O(2) symmetry. The emergence of this new behavior was traced to the interaction of a local symmetry-breaking (wavelength-doubling) bifurcation producing a mixed mode MM_- and a global bifurcation generating the necessary $P_+ \to P_-$ connection joining pure modes and their pi-translates. Near this interaction point, where the pure mode heteroclinic cycles are interrupted both by the presence of MM_- and the destruction of the $P_+ \to P_-$ connection, heteroclinic cycles arise involving standing waves, pure modes, mixed modes, and the trivial state. As Region 2 is traversed beyond this point, one observes in Eqs. (5) a series of periodic solutions distinguished by the number of standing wave oscillations and organized into a countably infinite family of isolas (Figure 5) by a sequence of transitions among distinct heteroclinic cycles. These isolas assume a very complicated shape with several regions exhibiting (partial) Shil'nikov-like cascades of saddle-node and period-doubling bifurcations. A variety of complex heteroclinic behavior can be found in this region including chaotic attractors.

The new type of heteroclinic dynamics described in Section 4 relies heavily on the invariance of the $y = 0$ plane, and hence is very sensitive to symmetry-breaking terms, as a comparison of Figures 5 and 11 demonstrates. The robust pure mode heteroclinic cycles in Eqs. (2) are likewise destroyed by the addition of small reflection symmetry-breaking terms and are replaced by modulated travelling waves, except at isolated parameter values where structurally unstable $P_+ \to P_-$ connections remain. While this result is in accord with earlier analyses [60], we have seen that much else happens besides. In particular, the loss of reflection symmetry is responsible for the introduction of homoclinic connections and interesting codimension-two global bifurcations into the dynamics. The behavior at two of these global bifurcations, the Het-SB bifurcation and the T-point, is summarized in Figure 12. Both of these bifurcations are associated with the disappearance of the heteroclinic cycle in the O(2)-symmetric system when the pure modes lose stability to mixed modes. The third is a (pair of) Takens-Bogdanov bifurcations on the branch of drifting mixed modes. When the forced symmetry-breaking is weak, these two sets of phenomena are unrelated, but as the symmetry-breaking increases in strength they become interconnected in a remarkable way. In particular we have identified near the Het-SB bifurcation a homoclinic flip bifurcation; this bifurcation is responsible for the generation of much complex dynamics, although we have focused here only on the so-called N-homoclinics.

In this article we have tried to give a sense of the types of new heteroclinic or approximately heteroclinic behavior that can arise when robust heteroclinic cycles are destroyed through bifurcation or via the addition of symmetry-breaking terms. The results presented in Sections 4 and 5, which focus on the 1:2 steady state mode interaction, clearly cannot address this problem in a complete or rigorous fashion. Furthermore, we considered in Section 5 only generic low order terms that break the reflection symmetry while preserving the continuous symmetry. As in other problems of this type, reliable results are expected when the strength of these terms is small ($\varepsilon \ll 1$). No guarantee can be given, however, that higher order terms will not produce new dynamics that is absent from our formulation. Likewise, we can make no guarantee that the behavior found at larger ε will be seen generically and that nothing else can occur.

Acknowledgments

Preparation of this article was supported by EPSRC Grant GR/R52879/01. The author is indebted to E. Knobloch for collaboration in the research reported here and to A. M. Rucklidge and M. Higuera for valuable comments.

References

[1] M. Dellnitz, M. J. Field, M. Golubitsky, A. Hohmann and J. Ma, *Int. J. Bifur. Chaos* **5**, 1243 (1995).
[2] P. Ashwin and A. M. Rucklidge, *Physica **D*** **122**, 134 (1998).
[3] P. Ashwin and P. Chossat, *J. Nonlinear Sci.* **8**, 103 (1998).
[4] M. Krupa, *J. Nonlinear Sci.* **7**, 129 (1997).
[5] E. Knobloch and J. Moehlis, in *Nonlinear Instability, Chaos and Turbulence*, Vol. 2, Eds. L. Debnath and D. N. Riahi (WIT Press, Southhampton, 2000), p. 237.
[6] J. Guckenheimer and P. Holmes, *Math. Proc. Cambridge* **103**, 189 (1988).
[7] D. Armbruster, J. Guckenheimer and P. Holmes, *Physica **D*** **29**, 257 (1988).
[8] D. Armbruster and P. Chossat, *Physica **D*** **50**, 155 (1991).
[9] M. J. Field and J. W. Swift, *Nonlinearity* **4**, 1001 (1991).
[10] J. Guckenheimer and P. Worfolk, *Nonlinearity* **5**, 1211 (1992).
[11] J. W. Swift and E. Barany, *Eur. J. Mech. **B**-Fluid* **10**, 99 (1991).
[12] M. Silber and E. Knobloch, *Nonlinearity* **4**, 1063 (1991).
[13] M. Silber, H. Riecke and L. Kramer, *Physica **D*** **61**, 260 (1992).
[14] E. Knobloch and M. Silber, *Physica **D*** **94**, 213 (1993).

[15] I. Melbourne, P. Chossat and M. Golubitsky, *Proc. Roy. Soc. Edinb.* ***A*** **113**, 315 (1989).
[16] I. Melbourne, *J. Dyn. Diff. Eqn.* **1**, 347 (1989).
[17] F. M. Busse and R. M. Clever, in *Recent Developments in Theoretical and Experimental Fluid Mechanics*, Eds. U. Müller, I. G. Roessner and B. Schmidt (Springer-Verlag, New York, 1979), p. 376.
[18] J. Porter and E. Knobloch, *Physica* ***D*** **143**, 138 (2000).
[19] I. Mercader, J. Prat and E. Knobloch, *Int. J. Bif. Chaos* **12**, 2501 (2002).
[20] N. Aubry, P. Holmes, J. L. Lumley, E. Stone, *J. Fluid Mech.* **192**, 115(1988).
[21] G. Berkooz, P. Holmes and J. L. Lumley, *J. Fluid Mech.* **230**, 75 (1991).
[22] G. Berkooz, P. Holmes, J. L. Lumley, *Th. Comp. Fl. Dyn.* **4**,255(1993).
[23] S. Sanghi and N. Aubry, *J. Fluid Mech.* **247**, 455 (1993).
[24] A. M. Rucklidge and P. Matthews, *Nonlinearity* **9**, 311 (1996).
[25] T. Clune and E. Knobloch, *Physica* ***D*** **74**, 151 (1994).
[26] M. Higuera, J. Porter and E. Knobloch, *Physica* ***D*** **162**, 155 (2002).
[27] J. Porter and M. Silber, *Physica* ***D*** **190**, 93 (2004).
[28] M. Gorman, M. el-Hamdi, K. A. Robbins, *Comb. Sci. Tech.* **98**, 79(1994).
[29] E. Stone, M. Gorman, M. el-Hamdi and K. A. Robbins, *Phys. Rev. Lett.* **76**, 2061 (1996).
[30] A. Palacios, G. H. Gunaratne and M. Gorman, *Chaos* **7**, 463 (1997).
[31] I. Melbourne, M. R. E. Proctor and A. M. Rucklidge, in *Dynamo and Dynamics, a Mathematical Challenge*, Eds. P. Chossat, D. Armbruster and I. Oprea (Kluwer, Dordrecht, 2001) p. 363.
[32] P. Chossat and D. Armbruster, *Proc. R. Soc. Lond.* ***A*** **459**, 577 (2003).
[33] G. Küppers and D. Lortz, *J. Fluid Mech.* **35**, 609 (1969).
[34] R. M. May and W. Leonard, *SIAM J. Appl. Math.* **29**, 243 (1975).
[35] J. Hofbauer and K. Sigmund, *The Theory of Evolution and Dynamical Systems* (Camb. Univ. Press, Cambridge, 1988).
[36] J. Guckenheimer and A. Mahalov, *Physica* ***D*** **54**, 267 (1992).
[37] Y. Tu and M. C. Cross, *Phys. Rev. Lett.* **69**, 2515 (1992).
[38] Y. Hu, R. E. Ecke and G. Ahlers, *Phys. Rev. Lett.* **74**, 5040 (1995).
[39] F. M. Busse, in *Turbulence and Chaotic Phenomena in Fluids*, Ed. T. Tatsumi (Elsevier, Amsterdam, 1984), p. 197.
[40] J. W. Swift, *Contemporary Mathematics (AMS)* **28**, 435 (1984).
[41] A. M. Soward, *Physica D* **14**, 227 (1985).
[42] C. A. Jones and M. R. E. Proctor, *Phys. Lett.* ***A*** **121**, 224 (1987).
[43] M. R. E. Proctor and C. A. Jones, *J. Fluid Mech.* **188**, 301 (1988).
[44] C. Nore, L. S. Tuckerman, O. Daube, S. Xin, *J. Fluid Mech.* **477**, 51(2003).
[45] J. Porter and E. Knobloch, *Physica* ***D*** **159**, 125 (2001).
[46] G. Dangelmayr, *Dynam. Stabil. Syst.* **1**, 159 (1986).
[47] S. M. Cox, *Physica* ***D*** **95**, 50 (1996).
[48] J.H.P. Dawes, C.M. Postlethwaite, M.R.E. Proctor, *Physica* ***D191***,1(2004).
[49] I. G. Kevrekidis, in *Supercomputer Research in Chemistry and Chemical Engineering*, Eds. K. F. Jensen and D. G. Truhlar (American Chemical Society, Minneapolis, 1987), p. 284.
[50] G. Berkooz, P. Holmes, J. L. Lumley, N. Aubry and E. Stone, *Phys. Fluids*

6, 1574 (1994).
[51] S. J. Kline, W. C. Reynolds, F. A. Schraub and P. W. Rundstadler, *J. Fluid Mech.* **30**, 741 (1967).
[52] P. Chossat, M. Krupa, I. Melbourne, A. Scheel, *Physica D* **100**,85(1997).
[53] B. Sandstede and A. Scheel, *Nonlinearity* **8**, 333 (1995).
[54] E. Stone and P. Holmes, *Physica D* **37**, 20 (1989).
[55] E. Stone and D. Armbruster, *Chaos* **9**, 499 (1999).
[56] D. Armbruster, E. Stone and V. Kirk, *Chaos* **13**, 71 (2003).
[57] L. P. Shil'nikov, *Sov. Math. Dokl.* **6**, 163 (1965).
[58] P. Chossat, *Dynam. Cont. Dis. Ser. A* **8**, 575 (2001).
[59] S. Y. Vyshkind and M. I. Rabinovich, *Eksp. Teor. Fiz.* **71**, 557 (1971).
[60] P. Chossat,*Nonlinearity* **6**, 723 (1993).
[61] P. Ashwin, K. Böhmer and Z Mei, *J. Comput. Appl. Math.* **70**, 297 (1996).
[62] S. A. Campbell and P. Holmes, *Physica D* **59**, 52 (1992).
[63] G. Dangelmayr and E. Knobloch, *Nonlinearity* **4**, 399 (1991).
[64] J. Porter and E. Knobloch, preprint (2004).
[65] R. E. Ecke, F. Zhong and E. Knobloch, *Europhys. Lett.* **19**, 177 (1992).
[66] E. Knobloch, in *Lectures on Solar and Planetary Dynamos*, Eds. M. R. E. Proctor and A. D. Gilbert (Camb. Univ. Press, Cambridge, 1994) p. 331.
[67] J. D. Rodriguez and M. Schell, *Phys. Lett. A* **146**, 25 (1990).
[68] S. Rüdiger and E. Knobloch, *Fluid Dyn. Res.* **33**, 477 (2003).
[69] A. S. Landsberg and E. Knobloch, *Phys. Lett. A* **159**, 17 (1991).
[70] E. Knobloch, A.S. Landsberg, J. Moehlis, *Phys. Lett. A* **255**,287(1999).
[71] E. Knobloch and M. R. E. Proctor, *Proc. R. Soc. Lond. A* **415**, 61 (1986).
[72] M. Golubitsky and D. G. Shaeffer, *Singularities and Groups in Bifurcation Theory, Volume 1* (Springer-Verlag, New York, 1985).
[73] J. Guckenheimer and P. Holmes, *Nonlinear Oscillations, Dynamical Systems, and Bifurcations of Vector Fields* (Springer-Verlag, New York, 1986).
[74] P. Glendinning and C. Sparrow, *J. Stat. Phys.* **43**, 479 (1986).
[75] V. Kirk, *Phys. Lett. A* **154**, 243 (1991).
[76] V. Kirk, *Physica D* **66**, 267 (1993).
[77] A. R. Champneys and V. Kirk, *Physica D* **195**, 77 (2004).
[78] A. J. Homberg and B. Krauskopf, *J. Dynam. Diff. Eqs.* **12**, 807(2000).
[79] W. F. Langford, in *Proc. Int. Symp. on Applied Mathematics and Information Science* (Kyoto University, Kyoto, 1982).
[80] P. Holmes, J. L. Lumley and G. Berkooz, *Turbulence, Coherent Structures, Dynamical Systems and Symmetry* (Camb. Univ. Press, Cambridge, 1996).
[81] J. Moehlis and E. Knobloch, *Physica D*, **135**, 263 (2000).

CHAPTER 19

INTERNAL DYNAMICS OF INTERMITTENCY

Rob Sturman* and Peter Ashwin†

* *Department of Mathematics, University of Bristol, Bristol BS8 1TW, UK*

† *Department of Mathematical Sciences, Laver Building, University of Exeter, Exeter EX4 4QE, UK*

By examining the behavior of intermittent trajectories near invariant subspaces for the dynamics of flows or maps we introduce and discuss the concept of *internal dynamics* of an intermittent attractor to an invariant subspace. For a smooth planar mapping, we give examples where the internal dynamics is minimal (as in on–off intermittency), has a single attracting and a single repelling invariant set (as in in–out intermittency) and more complicated but nonetheless typical examples with several isolated invariant sets in the internal dynamics. In particular we discuss and analyze an example of generalized in–out intermittency in which there is one form of 'in'-dynamics and two of 'out'-dynamics. We contrast this behavior with another structurally stable example of intermittency, that exhibits non-ergodic intermittency.

1. Introduction

There are many examples of dynamical systems that possess constraints in the form of invariant subspaces forced often, but not always, by symmetries. Consider the dynamics generated by a mapping $f : M \to M$ with an invariant subspace $N \subset M$. In such systems one can find chaotic attractors that show dynamical intermittency to states that have more symmetry but are not attractors in themselves; for instance on–off intermittency in which a chaotic saddle lies within N that is an attractor within N but is transversely repelling on average. This type of intermittency has been observed in many systems (see for example [10, 13, 16, 21]) since it was identified [22]. It is associated with a loss of transverse stability of a riddled basin attractor at blowout bifurcation [2,20]. It was later noted that an intermittent attractor may contain much larger invariant sets within N that are not attractors within N; this case was referred to as 'in–out' intermittency [3] and has been seen in a number of systems.

The intermittent dynamics we discuss here arises robustly in some sense, due to the presence of N. Intermittent dynamics may occur without the presence of some N but typically this is codimension one. Examples of the latter type of intermittency have been discussed by many authors [23]; we do not discuss it here except to mention that this is caused by an attractor interacting with an invariant set with minimal dynamics.

In a previous paper [3] it was noted that one can identify different dynamics associated with approach to N (the 'in phase') and departure from N (the 'out phase'). This paper reviews and examines this effect. We propose that the 'internal dynamics' of an intermittent attractor may be much more complicated than this even for robust attractors. There may be a variety of different 'in' and 'out' dynamics, as well as saddles within N. We give examples of on–off, in–out, and what we call generalized in–out intermittency. These are characterized by comparing their time series, phases portraits and structure. For the generalized in–out case we construct a Markov chain model that gives the scaling of the invariant density of the transverse variable for this intermittency. Finally we contrast these 'chaotic' intermittencies with 'regular' but non-ergodic intermittency ('cycling chaos' [12]) found robustly in systems with several invariant subspaces forced by symmetry.

2. Internal Dynamics of Intermittency

Suppose $f : M \to M$, $M = \mathbb{R}^m$ is a smooth map. We consider A a *minimal Milnor attractor* for f. Recall that A is a minimal Milnor attractor if it is an invariant set such that (i) the basin of attraction of A has positive measure and (ii) all proper invariant subsets have zero measure basin. We say A is a *weak attractor* [9] if the basin of A has positive measure, i.e. (i) is satisfied but not necessarily (ii).

If the map f leaves a linear subspace N of M (forward) invariant (i.e., $f(N) = N$) then one can define $A_0 = A \bigcap N$. Note that this must be an invariant set. If $A \neq A_0$ and $A_0 \neq \emptyset$ then we say A is *intermittent* to N. We refer to the dynamics of f restricted to A_0 as the *internal dynamics* of A on N. Since A is an attractor, typically A_0 will be a weak attractor for $f|_N$. However A_0 needs not be a minimal attractor. The form of the internal dynamics governs the form of the intermittency in the full system in the following way:

Definition 1: If A_0 is a minimal attractor for $f|_N$ then the intermittency is called *on–off*.

Definition 2: If A_0 is not a minimal attractor for $f|_N$ then the intermittency is called *(generalized) in–out.*

We say the in–out intermittency is *simple* if A_0 contains a single minimal chaotic attractor which is transversely repelling, and a single repellor that is transversely attracting [3]. However the dynamics on N need not be restricted to a single attractor but may decompose into two or more isolated invariant subsets. For generalized in–out intermittency it is possible to decompose A_0 into attractors $P_1, ..., P_k$ and repellors $Q_1, ..., Q_l$ within A_0. If A is assumed to be a minimal attractor this means that the P_i must be transversely repelling and Q_i transversely attracting. In such a case we refer to the intermittency as *l-in, k-out* for obvious reasons. Figure 1 shows a schematic diagram of an example of generalized in–out intermittency, in this case 3-in, 3-out.

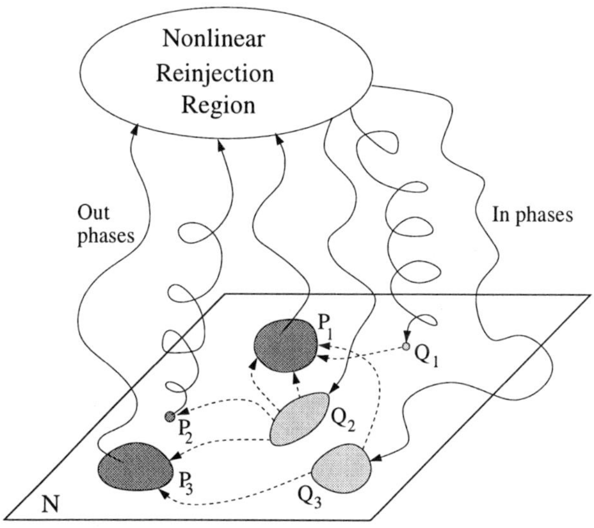

Fig. 1. Schematic diagram showing elements of a generalized in–out intermittent attractor. For $f|_N$, A_0 decomposes into attractors and repellors. In this diagram P_1, P_3 are attracting chaotic sets, P_2 is an attracting periodic orbit, while Q_1, Q_2, Q_3 are repelling sets. The dotted lines represent sample trajectories within N. The P_i are transversely unstable, while the Q_i are transversely stable. The transverse directions are represented by solid lines. A nonlinear reinjection region completes the intermittent attractor, which in our terminology is 3-in, 3-out.

3. Internal Dynamics of Intermittency for a 2-d Mapping

We consider three examples of attractors for a smooth mapping that illustrate the behavior of three types of intermittency; on–off, simple in–out, and generalized in–out dynamics.

3.1. *Example Mappings*

The smooth two-dimensional mapping $f : \mathbb{R}^2 \to \mathbb{R}^2$ we consider is given by:

$$\begin{aligned} x &\to g_1(x, y^2) \\ y &\to y g_2(x, y^2) \end{aligned} \tag{1}$$

where x, y are real variables and g_1, g_2 are functions we will specify. The mapping has a symmetry $(x, y) \mapsto (x, -y)$ which implies there is an invariant subspace $\mathcal{X} = \{x, 0\}$ on which the system reduces to the one-dimensional map which we fix as being

$$x \to g_1(x, 0) = h(x) = x(c_0 + c_1 x)e^{-c_2 x^2} \tag{2}$$

where we use a number of different values of the parameters c_0, c_1 and c_2. If $c_2 = 0$ and $c_0 = -c_1$, then $h(x)$ is the well-known logistic map with parameter c_0. We will examine three different examples of intermittent dynamics to $\mathcal{X}$. In order to do this we consider the dynamics of h in relation to the dynamics of (1).

3.2. *Intermittent Dynamics*

1. On–off intermittency If we consider the dynamics of (1) with

$$g_1(x, y^2) = 3.8285x(1 - x) - 0.3xy^2, \quad g_2(x, y^2) = 1.82e^{-x} - y^2, \tag{3}$$

this has an attractor that exhibits on–off intermittency [3]. Figure 2 shows a time series of 10^4 iterates for x (top panel) and $\log |y|$ (bottom panel). The behavior of $\log |y|$ shows the trajectory with long 'laminar phases' near the invariant subspace and intermittent excursions to order 1. Plotting the logarithm of y resolves the tiny values of y when close to N. The dynamics within the invariant subspace is the minimal chaotic attractor on $[0, 1]$, as shown in the time series for x. Note that the fluctuations in y are not driven by x visiting isolated invariant sets within N, but rather by x nearing invariant sets *embedded* in the minimal A_0. Figure 3 shows a phase portrait in xy-space for the on–off intermittent attractor, for 10^5 iterates for the time series in Figure 2. The attractor for $f|_N$ is an interval corresponding to the intersection of the full attractor with N.

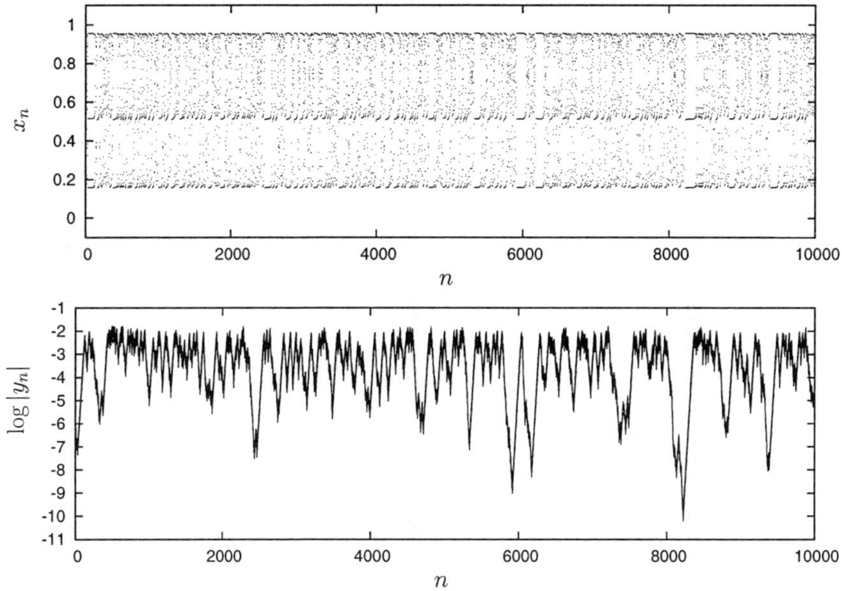

Fig. 2. Time series showing on–off intermittency for the mapping (1,3). The top panel shows the chaotic behavior of the x variable, while the bottom panel shows the y variable approaching and leaving N on a logarithmic scale. An initial condition in $y = 0$ typically shows the same behavior as x_n shown here.

2. In–out intermittency By contrast, if we consider (1) with

$$g_1(x, y^2) = 3.88615x(1 - x) - 0.3xy^2, \quad g_2(x, y^2) = 1.82e^{-x} - y^2 \tag{4}$$

we find an example of in–out dynamics [3]. Figure 4 is the corresponding time series to Figure 2. In the top figure we can see the two qualitatively differing dynamics in the invariant subspace. A periodic orbit is attracting within N and transversely repelling, and so we see a linear growth during the 'out' phases in the $\log|y|$ plot modulated with the periodic of the periodic orbit. A chaotic repellor within N is transversely attracting, giving the chaotic 'in' phases. The phase portrait in Figure 5 shows a high density of points near the 'out' dynamics; a periodic orbit within N.

3. Generalized in–out intermittency (1-in, 2-out) We consider the map (1) with

$$\begin{aligned} g_1(x, y^2) &= x(4 + bx)e^{-x^2-20y^2} + (cx - 16x^2)(1 - e^{-20y^2})e^{-x^2} \\ g_2(x, y^2) &= (1.88 - 1.5e^{-x^2}) - y^2 \end{aligned} \tag{5}$$

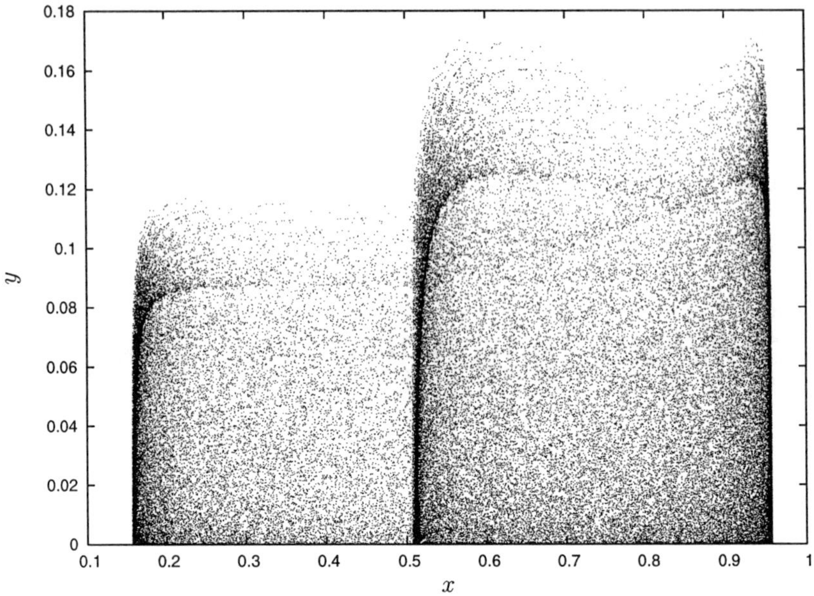

Fig. 3. Phase portrait of the on–off intermittency displayed by mapping (1,3). The attractor $A_0 \subset N$ is identical to the intersection of the intermittent attractor A with N. The high density near a period 3 orbit simply reflects high density on the attractor for the map restricted to $y = 0$.

where we vary the parameters b, c. For

$$b = 1.6, \quad c = -8.0$$

this gives an example of generalized in–out intermittency. Within $\mathcal{X}$ there are two attracting invariant sets; a chaotic attractor $\mathcal{C}$ in $x > 0$ and a period-2 orbit $\mathcal{P}$ in $x < 0$. These invariant sets are shown in Figure 6, showing 100 iterates of (2) (with $c_0 = 4.0$, $c_1 = 1.6$, $c_2 = 1.0$) for two different initial conditions; either $x_0 = -0.1$ or $x_0 = 0.1$.

The invariant sets $\mathcal{P}$ and $\mathcal{C}$ are both transversely unstable to perturbations in the y-direction, whereas the fixed point $\mathcal{O}$ is transversely stable. The intermittent trajectory approaches $\mathcal{O}$, and once close to $\mathcal{X}$, leads to either $\mathcal{P}$ or $\mathcal{C}$ depending on whether $x < 0$ or $x > 0$. The 'out' dynamics of each of $\mathcal{P}$ and $\mathcal{C}$ (the growth of y away from $\mathcal{X}$) behave in different ways. These can be seen in Figure 7. Whilst close to $\mathcal{P}$ the y-dynamics grow in a linear manner (modulated with the periodicity of the x-dynamics) with the rate of linear growth governed by the transverse Lyapunov exponent at $\mathcal{P}$. In contrast when $x > 0$ and the trajectory is close to $\mathcal{C}$ there is a different

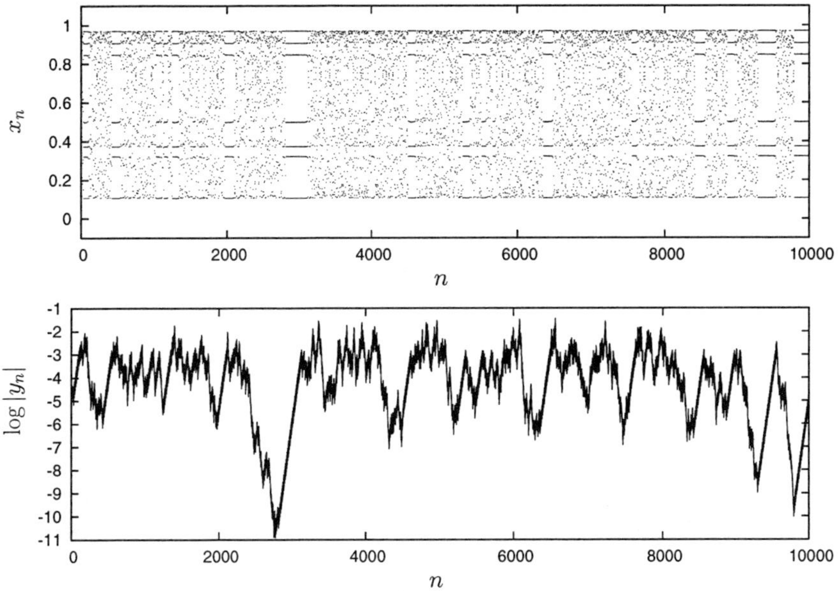

Fig. 4. Time series showing in–out intermittency for the mapping (1,4). The top panel shows x alternating between chaotic behavior and a period 14 orbit. The bottom panel, as for the on–off trajectory in Figure 2, shows the y variable approaching and leaving N. In this case the dynamics leaving N (the 'out' dynamics) is different from the dynamics approaching N (the 'in' dynamics). Here the 'in' dynamics is chaotic while the 'out' dynamics is periodic.

rate of average growth governed by the transverse Lyapunov exponent at $\mathcal{C}$. At order one distance from $y = 0$ the nonlinear map diverts both 'out' dynamics towards the 'in dynamics' after some time.

Figure 8 shows the intermittent attractor in (x, y)-space. This shows the periodic 'out' dynamics for $x < 0$, the chaotic 'out' dynamics for $x > 0$ and the 'in' dynamics along $x = 0$. The nonlinear reinjection region far from N leads both 'out' trajectories back towards the 'in' phase.

3.3. *Transverse Dynamics*

Lyapunov exponents for (1) are given, for each $(x, y) \in \mathbb{R}^2$ and non-zero $(u, v) \in T_{(x,y)}\mathbb{R}^2$ by

$$\lambda_{(x,y)}(u, v) = \lim_{n\to\infty} \frac{1}{n} \log \|Df^n_{(x,y)}(u, v)\|$$

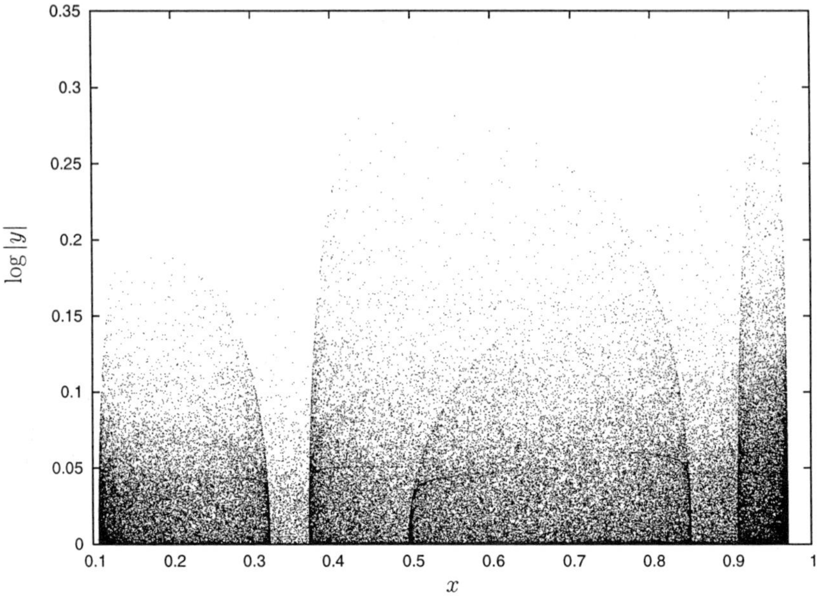

Fig. 5. Phase portrait of the in–out intermittency displayed by mapping (1,4). In contrast to the phase portrait in Figure 3, the intersection of the intermittent attractor A with N is not equal to the attractor for $f|_N$. Note the high density of points around the period 14 transversely unstable set in N.

For an ergodic attractor this converges for almost all (x, y) in its basin and all (u, v) [19]. If $y \neq 0$ this gives a pair of Lyapunov exponents λ_1, λ_2 for the intermittent attractor. For $(x, 0)$ the internal Lyapunov exponent is given by $\lambda_i = \lambda_{(x,0)}(u, 0)$, and the transverse Lyapunov exponent [19] by $\lambda_T = \lambda_{(x,0)}(u, v)$ $(v \neq 0)$. Note that λ_i is the Lyapunov exponent for the one-dimensional map (2) restricted to the invariant subspace.

Figure 9 shows Lyapunov exponents for the generalized (1-in, 2-out) intermittency on varying the parameter b. The top panel shows λ_1 and λ_2, the pair of exponents for the full system. The middle panel shows the transverse exponent (dotted line) and internal exponent (solid line) for a trajectory initially at $(x > 0, y = 0)$. The internal exponent λ_i is positive, indicating the chaotic attractor $\mathcal{A}$, and λ_T is also positive, indicating transverse instability. In the bottom panel we have λ_T and λ_i for $(x < 0, y = 0)$. Here $\lambda_T > 0$ again, as the periodic orbit $\mathcal{P}$ is transversely unstable, and $\lambda_i < 0$ confirms that it is not chaotic. Observe that there is a range of values of b that result in similar Lyapunov exponents and thus the intermittency can

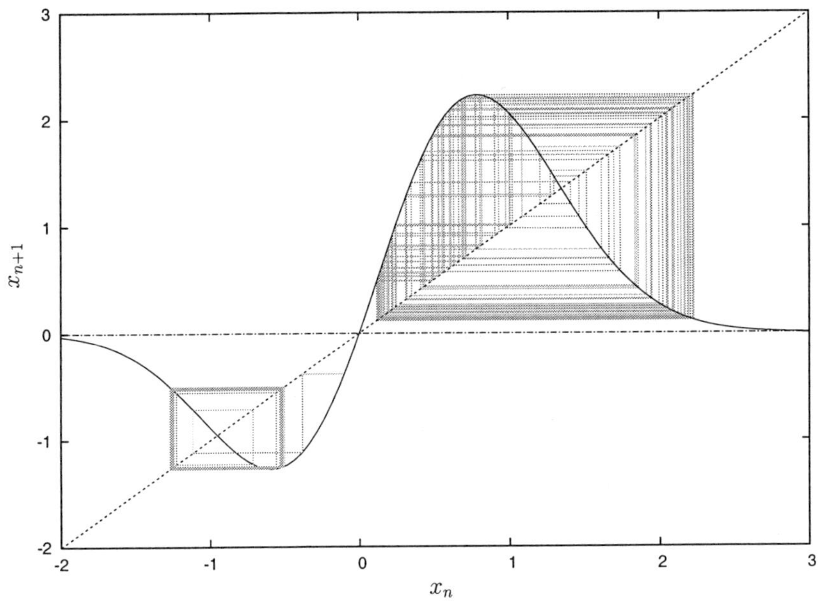

Fig. 6. The 1-dimensional mapping (2), with 100 iterates starting from $x_0 = -0.1$, leading to an attracting period-2 orbit $\mathcal{P}$, and from $x_0 = 0.1$, leading to a chaotic attractor $\mathcal{C}$.

be robust.

4. Markov Model

We construct a Markov chain model [3, 8] for generalized in–out intermittency, as shown in Figure 10. The 'in' chain is represented by a semi-infinite chain of states p_i, and the periodic (resp. chaotic) 'out' chain by states q_i (resp. r_i). The ith state of each chain represents a transverse distance ρ^i from the invariant subspace $\mathcal{X}$, for some $0 < \rho < 1$. We assume a leakage ε (resp. δ) from the 'in' chain onto the periodic (resp. chaotic) 'out' chain. Referring to the time series in Figure 7, we see that during the 'in' and periodic 'out' phases, the behavior of $\log|y|$ is roughly linear, whereas during the chaotic 'out' phase $\log|y|$ can fluctuate. For this reason we model the q_i and p_i chains as allowing one-way transitions only, while transitions in the r_i chain can go in either direction. Note that this model does not take into account the length of time spent at each state, only the permitted transitions between states (there is work of the authors in progress in improving this model).

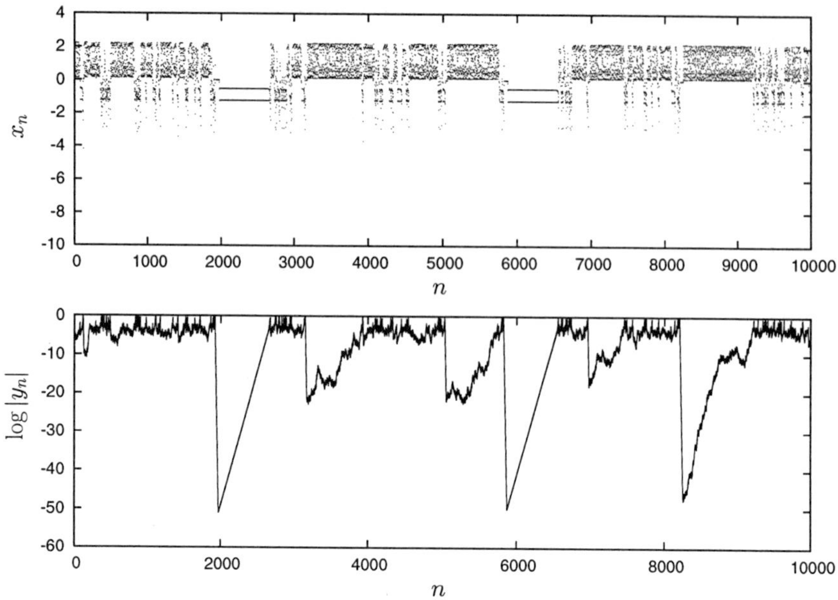

Fig. 7. Time series showing generalized in–out intermittency for the mapping (1,5). Here the y dynamics are characterized by two different types of 'out' dynamics. When x is attracted to the transversely unstable period 2 orbit for $x < 0$, we see a linear growth of y. When x is close to the chaotic attractor in N for $x > 0$ we see chaotic 'out' dynamics. The 'in' dynamics here consist of very sudden approaches to the invariant subspaces.

The equilibrium probability distribution $\mathbb{P}$ can be calculated by solving

$$\begin{aligned}
\mathbb{P}(q_i) &= \mathbb{P}(q_{i+1}) + \varepsilon\mathbb{P}(p_i) \\
\mathbb{P}(p_i) &= (1 - \varepsilon - \delta)\mathbb{P}(p_{i-1}) \\
\mathbb{P}(r_i) &= \beta_D\mathbb{P}(r_{i-1}) + \beta_U\mathbb{P}(r_{i+1}) + \delta\mathbb{P}(p_i) \\
\mathbb{P}(p_1) &= \mathbb{P}(q_1) + \beta_U\mathbb{P}(r_1) \\
\mathbb{P}(r_1) &= \beta_U\mathbb{P}(r_2),
\end{aligned}$$

giving

$$(\mathbb{P}(p_i), \mathbb{P}(q_i), \mathbb{P}(r_i)) = (A\mu^i, \frac{\varepsilon A}{\varepsilon_\delta}\mu^i, B\mu^i + C\nu^i),$$

where $\mu = 1 - \varepsilon - \delta$, $\nu = \frac{1-\sqrt{1-4\beta_U\beta_D}}{2\beta_U}$, and A, B and C are constants that can be found via the equations for $\mathbb{P}(p_1)$, $\mathbb{P}(r_1)$, and the normalisation requirement

$$\sum_i (\mathbb{P}(p_i) + \mathbb{P}(q_i) + \mathbb{P}(r_i)) = 1.$$

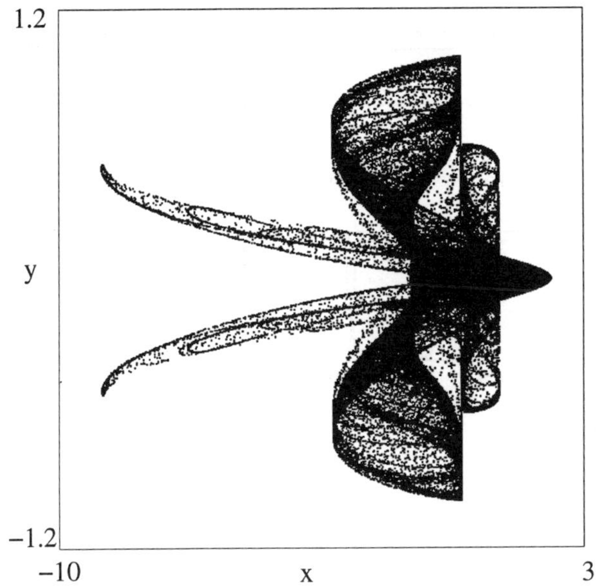

Fig. 8. Phase portrait for the generalized intermittency time series in Figure 7, showing 10^7 iterates. This shows the complexity of the nonlinear reinjection region for y away from zero. The map restricted to the invariant subspace $y = 0$ contains two attractors separated by a repellor, though this is not visible in this attractor.

Note that the scaling behavior near $y = 0$ corresponding to $i \to \infty$ is determined by constants $0 < \mu < 1$ and $0 < \nu < 1$. In the case that $\mu < \nu$ we note that the scaling behavior is different in the two 'out' chains, while for $\mu \geq \nu$ it is the same. This implies that the scaling of the invariant density of the attractor may be different transverse to the two different 'out' dynamics.

4.1. *Numerical Investigation of Invariant Densities*

To estimate the invariant density of the intermittent trajectory, we computed 10^6 iterates and constructed a pair of histograms, y^+ and y^-, of values of $\log|y| \in [0, -50]$; y^+ if $x > 0$ and y^- if $x < 0$. These estimate the proportion of time spent in each of the 'out' phases, and the density of transverse distances from N. The 'in' phases are much faster (see Figure 7) so we ignore contributions from this. Figure 11 shows a histogram of these frequencies. The boxed plusses (y^+) and the empty boxes (y^-) are for the parameters (5) used above. The differing gradients of the laminar regions

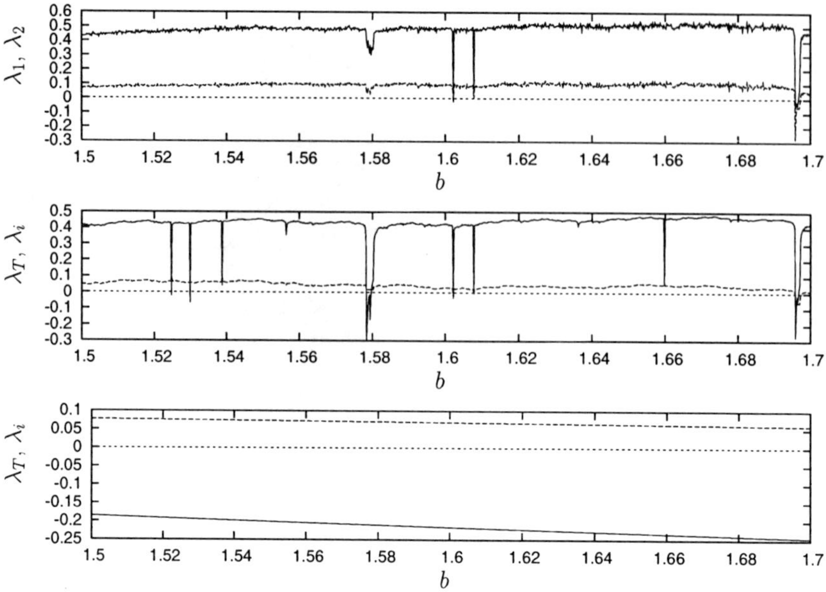

Fig. 9. Lyapunov exponents for a typical generalised intermittent trajectory of mapping (1) with parameters (5), as functions of the parameter b. The top panel shows λ_1, λ_2, the pair of Lyapunov exponents for the full system. The middle panel is for a trajectory beginning at $(x, y) = (0.1, 0)$. This shows the internal Lyapunov exponent λ_i as a solid line (as $\mathcal{C}$ is a chaotic attractor this is positive with negative portions at periodic windows). Also in the middle panel is the transverse Lyapunov exponent λ_T plotted as a dotted line, also positive confirming that $\mathcal{C}$ is transversely unstable. The bottom panel also shows λ_i (solid) and λ_T (dotted), this time for a trajectory initially at $(x, y) = (-0.1, 0)$. Since $x < 0$ leads to a transversely unstable periodic orbit, we see $\lambda_T > 0$ and $\lambda_i < 0$.

$(-45 < \log|y| < -10)$ suggest that the chaotic 'out' dynamics is more common than the periodic 'out' dynamics. This can be verified by examining long time series of a typical trajectory. By analogy with the Markov model, this behavior suggests that $\nu > \mu$, and so the equilibrium probability distribution gives more weight to the r_i chain than the q_i chain. In contrast, in Figure 11, the plusses (y^+) and the dashes (y^-) are for a time series for the same mapping and identical parameters, except that $c = -10.0$. Here the laminar regions have roughly the same gradients, analogously to the Markov model with $\mu > \nu$, where $\mathbb{P}(r_i)$ and $\mathbb{P}(q_i)$ have the same dominant scaling.

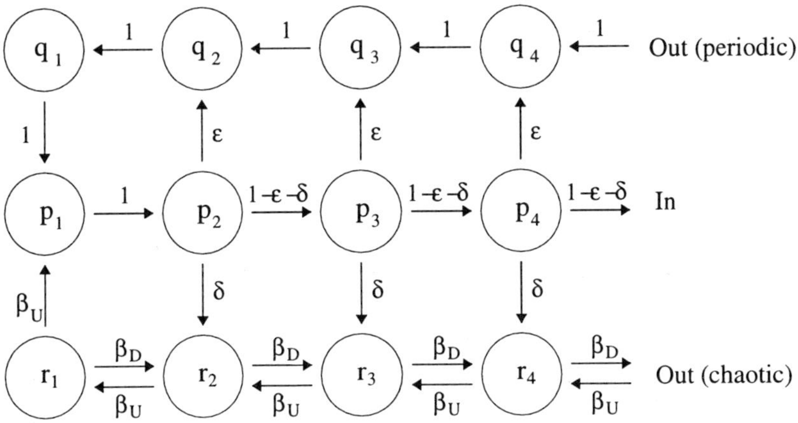

Fig. 10. A Markov chain model for the one in, two out intermittency example. The states p_i represent transverse distances ρ^i (for some $0 < \rho < 1$) from N near the 'in' dynamics, while q_i (resp. r_i) represents distances ρ^i from N near the periodic (resp. chaotic) 'out' dynamics. The leakage onto the periodic (resp. chaotic) 'out' chain is assumed to occur at a uniform rate ε (resp. δ). The chaotic 'out' chain permits transitions in both directions ($\beta_U + \beta_D = 1$) while the periodic 'out' and the 'in' chains allow one-way transitions only.

5. Non-Ergodic Intermittency

Another type of intermittent behavior may occur and can be robust in systems which possess invariant subspaces forced by symmetry. Structurally stable heteroclinic cycles between equilibria are well documented in symmetric systems, and recently examples of heteroclinic cycles between chaotic saddles have been observed (for example in coupled oscillators [12] and planar magnetoconvection [5]) and studied in detail (for example [6, 7]). This type of intermittency differs fundamentally from those discussed above. In on–off and in–out intermittency, chaotic dynamics within the invariant subspace make approaches to and departures from the subspace, and occur at unpredictable times. Here, invariant subspaces are approached at regular, geometrically increasing, intervals.

As an example, consider the mapping of $\mathbb{R}^3$ given by

$$(x, y, z) \to (F(x)e^{-\gamma z}, F(y)e^{-\gamma x}, F(z)e^{-\gamma y}) \tag{6}$$

where $F(x) = rx(1-x)$ is the logistic map and γ is a parameter determining the strength of the coupling. Each of the coordinate axes $x = 0$, $y = 0$, $z = 0$ and the coordinate planes $xy = 0$, $xz = 0$, $yz = 0$ are invariant subspaces. Moreover each of the axes contains a chaotic attractor $\mathcal{A}_x$, $\mathcal{A}_y$, $\mathcal{A}_z$ (for

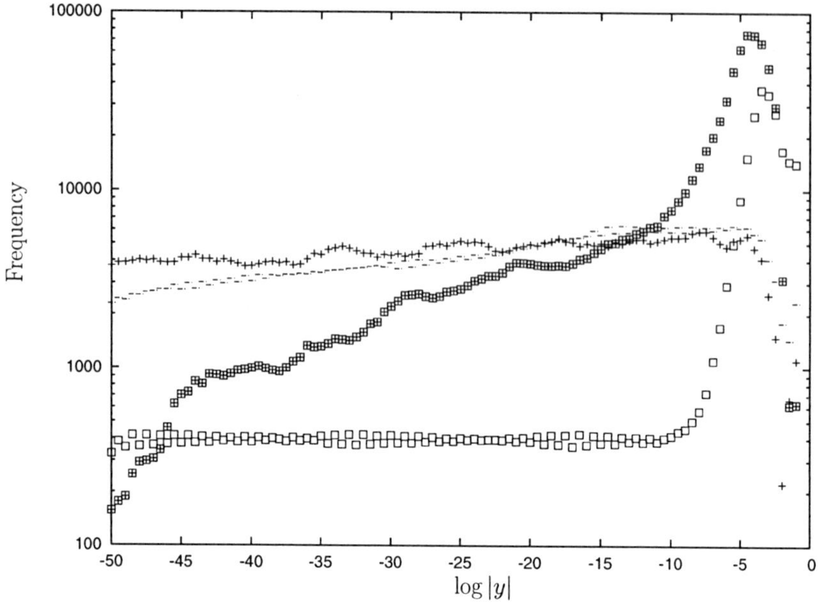

Fig. 11. Histogram for a trajectory of mapping (1) with parameters (5). A trajectory of 10^6 iterates was computed and box-counted into a pair of grids y^+ (if $x > 0$) and y^- (if $x < 0$) consisting of 100 strips in $\log|y| \in [-50, 0]$. The boxed plusses show y^+ and the empty boxes show y^- for the parameters (5). The differing gradients of the laminar regions indicate a preference for the chaotic 'out' over the periodic 'out'. The plusses show y^+ and the dashes show y^- for same the parameters, except with $c = -10.0$.

an appropriate choice of r), and each of the planes contains a connection between these attractors (for an appropriate choice of γ). The attractor $\mathcal{A}$ for the full system consists of a union of each of the attractors within the axes, and the connections between them. Since $\mathcal{A} \neq \mathcal{A}_{x,y,z}$ it is clear that $\mathcal{A}$ is intermittent to each of the invariant coordinate axes. However the cyclic nature of the coupling produces heteroclinic-type behavior as each variable cycles through 'out', 'active' (performing the logistic map dynamics when the coupling term is negligible), and 'in' phases. The time series in Figure 12 shows such a cycle (we actually iterate the logarithms of the variables in order to resolve a neighbourhood of the invariant subspaces clearly), with the length of subsequent phases increasing approximately geometrically as the invariant subspaces are approached. The lack of ergodicity in the system is evident from Figure 12. Note that long-term averages of typical observables of the system will fail to converge as the cycling phases get systematically longer.

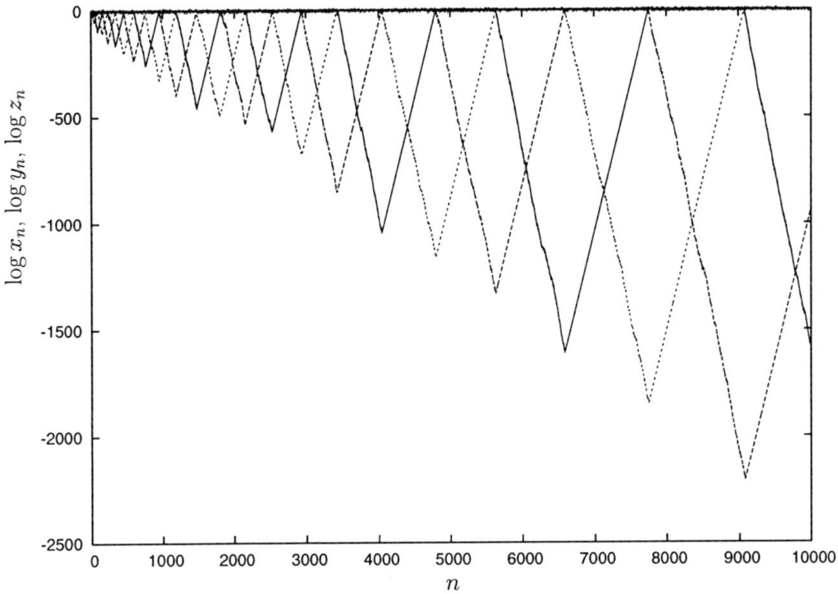

Fig. 12. Time series for attracting cycling chaos for (6), with parameters $r = 4.0$, $\gamma = 6.0$. The logarithms of the variables x, y, z are plotted as solid, dashed and dotted lines respectively against n. The trajectory cycles through 'out', 'active' and 'in' phases for each variable. Observe that the length of phase increases approximately geometrically.

6. Discussion and Conclusions

Various forms of intermittency have been found in applications including electronic systems [15], solar cycles [18], surface waves [17] and semiconductor lasers [24] to mention a few; these are discussed elsewhere [4, 8, 11]. We have demonstrated that there remain a range of new and relevant phenomena to be investigated. As previously considered in similar contexts, Markov-type models can be very useful for modelling the scaling behavior of trajectories near an invariant subspace [8].

A particular new case examined here is an example of a generalized in–out intermittency where A_0, the intersection of the attractor with the invariant subspace contains one in and two out phases. For other cases it is possible that A_0 contains not only attractors and repellers but also saddles, and these attractors may even display some form of intermittency in the internal dynamics. This could occur in cases where there are several different invariant subspaces to which an attractor is intermittent. It is even possible that the internal dynamics for an intermittent but ergodic attractor may include an attractor that is non-ergodic. In most cases one should still be able to classify and predict scaling behavior close to A_0 using transverse Lyapunov exponents. Similar structures to those discussed here for attractors should also be visible in chaotic transients [14] for trajectories close to a chaotic saddle that is only weakly unstable.

Acknowledgements

PA was partially supported by a Leverhulme Research Fellowship. The authors thank Eurico Covas, Alastair Rucklidge and Reza Tavakol for interesting conversations related to this work.

References

[1] P. Ashwin and M. Breakspear, *Phys Lett.* ***A*** **280**, 139 (2001).
[2] P. Ashwin, J. Buescu, and I. Stewart, *Nonlinearity* **9**, 703 (1996).
[3] P. Ashwin, E. Covas and R. Tavakol, *Nonlinearity* **12**, 563 (1999).
[4] P. Ashwin, E. Covas and R. Tavakol, *Physical Review* ***E*** **64**, 066204 (2001).
[5] P. Ashwin and A.M. Rucklidge, *Physica* ***D*** **122**, 134 (1998).
[6] P. Ashwin, A.M. Rucklidge and R. Sturman, *Phys. Rev.* ***E*** **66**, 035201(R) (2002).
[7] P. Ashwin, M. Field, A. M. Rucklidge and R. Sturman, *Chaos* **13**, 973 (2003).
[8] P. Ashwin, A.M. Rucklidge and R. Sturman, *Physica* ***D*** **194**, 30 (2004)
[9] P. Ashwin and J. Terry, *Physica* ***D*** **142**, 87 (2000).
[10] A. Çenys, A. N. Anagnostopoulos, and G. L. Bleris, *Phys. Lett.* ***A*** **224**, 346 (1997).
[11] E. Covas, R. Tavakol, P. Ashwin, A.Tworkowski and J.M. Brooke, *Chaos* **11**, 404 (2001).
[12] M. Dellnitz, M. Field, M. Golubitsky, A. Hohmann and J. Ma, *Intl. J. Bifn. and Chaos* **5** 1243–1247 (1995).
[13] M. Ding, and W. Yang,*Phys. Rev.* ***E*** **56**, 4009 (1997).
[14] V. Dronov and E. Ott, *Chaos* **10**, 291 (2000).
[15] A. Hasegawa, M. Komuro, and T. Endo, in Proceedings of ECCTD'97, Budapest, Sept. 1997 (sponsored by the European Circuit Society).
[16] Y.-C. Lai, *Phys. Rev.* ***E*** **54**, 321 (1996).
[17] C. Martel, E. Knobloch and J.M. Vega, *Physica* ***D*** **137**, 94 (2000).
[18] D. Moss and J. Brooke, *Mon. Not. R. Astron. Soc.* **315**, 521 (2000).
[19] E. Ott. *Chaos in Dynamical Systems, 2nd edition.* (Cambridge University Press, 2002).
[20] E. Ott, and J. Sommerer, *Phys. Lett.* ***A*** **188**, 39 (1994).
[21] N. Platt, S. M. Hammel, and J. F. Heagy, *Phys. Rev. Lett.* **72**, 3498 (1994).
[22] N. Platt, E. A. Spiegel, and C. Tresser, *Phys. Rev. Lett.* **70**, 279 (1993).
[23] Y. Pomeau and P. Manneville, *Comm. Math. Phys.* **74**, 189 (1980).
[24] A. Prasad, Y.C. Lai, A.Gavrielides and V. Kovanis. Complicated basins in external-cavity semiconductor lasers. *Phys Lett.* ***A*** **314**, 44 (2003).

CHAPTER 20

EXPERIMENTS WITH DICTYOSTELIUM DISCOIDEUM AMOEBAE IN DIFFERENT GEOMETRIES

Camilla Völtz and Eberhard Bodenschatz

Laboratory of Atomic and Solid State Physics, Cornell University, Ithaca, NY 14853, USA

Experiments with *Dictyostelium discoideum*amoebae in different geometries were performed. The formation of *Dictyostelium discoideum* aggregation mounds upon starvation was investigated when the amoebae were confined to layers of different heights, starting from a monolayer of cells. The average mound area was observed to grow approximately linearly with the height. For small heights, the distribution of the mound area was found to be narrow, while for larger heights it was found to be wide with secondary peaks at large mound areas. The results show that the number of cells in a mound is not constant but depends on the geometrical restrictions.

1. Introduction

Several factors affect the evolution of animal size, such as the strength of bones, the ability of the heart to pump blood up to the brain, and the surface-to-volume ratio, which affects the ability of animals to retain heat [1]. But on the other side, little is known of how multicellular organisms regulate the size and morphology of their tissue [2–5]. An example of size regulation occurs in the simple eukaryote *Dictyostelium discoideum.* Here, as a paradigm, we investigate the aggregation dynamics of this social amoeba. We report on experiments which investigate the development of *Dictyostelium discoideum* mounds in confined geometries. It could be shown that there is a relation between the confinement in the third dimension and the areas covered by the mounds. Hence the number of cells in a mound depends on the geometrical restrictions.

1.1. *Dictyostelium Discoideum Life Cycle*

Dictyostelium discoideum is the most studied species of the social amoebae, which are also known as cellular slime molds. In nature, they live in the soil and feed on bacteria. While food, *i.e.* bacteria, are abundant *Dictyostelium discoideum* amoebae live a solitary life as single cells. Once the amoebae have consumed all of their prey, the onset of starvation forces a major revision in the life cycle. The amoebae become chemotactically active. The signals that *Dictyostelium discoideum* cells use are relayed, which allows the organism to collect cells from a wide area. A single cell in a field of starving cells releases a pulse of cAMP (cyclic adenosine 3'-5' monophosphate), and other cells detect the cAMP as it binds to their cell surface receptors. The cells respond in two ways – by releasing another pulse of cAMP, and by moving up the gradient of cAMP released by the first cell. This is how the amoebae collect into aggregates, forming a mound consisting of approximately 20,000 cells. After differentiation into prespore and prestalk cells, a slug is formed, as shown in Figure 1. This slug then develops into a fruiting body by a process called culmination. The fruiting body consists of spores suspended on a stalk. The spores are then released, and the life cycle is ready to start all over again [6–9].

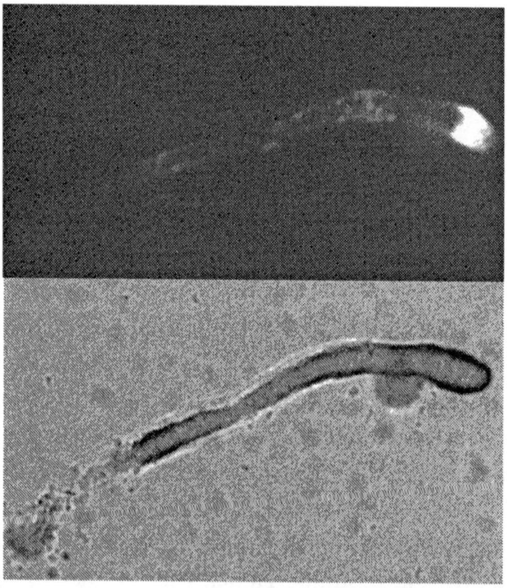

Fig. 1. Dictyostelium discoideum slug. The first image is taken in fluorescent microscopy, the second in bright field microscopy. Using a Dictyostelium strain where the prestalk cells are fluorescently labeled, the fluorescent microscopy image clearly shows the anterior part of the slug. The width of the figures is 900 μm.

1.2. *Experiments with Dictyostelium Discoideum in Confined Geometries*

The development of *Dictyostelium discoideum* cells has been intensely studied. But the amoebae have rarely been put into confined geometries so far. Previous experiments that studied Dictyostelium cells in a confined geometry observed cells in glass capillaries with diameters between 20 and 170 μm [10]. The main result of that study was that the process of differentiation was accelerated. In a different experiment [11] flat slugs were produced in a glass-mineral oil interface. The goal of the experiments was to follow individual cells in the migrating slugs. However, sometimes the slugs were able to move up into the third dimension.

Here a different method of confining the Dictyostelium cells is reported, which does not allow the cells to extend into the third dimension. Instead of confining the slugs, we applied this method to the starving cells before they start to aggregate [12]. The cells were confined to layers of different heights with the smallest height confining the cells to a monolayer. Here we report these experiments in much more detail than in [12].

The paper is organized as follows: In the following section the experimental setup is described. In Section 3 the experimental results are discussed. Sections 4 and 5 contain a discussion of the results and a conclusion.

2. Experimental Setup

To study the mound formation chambers of different heights were built in a sandwich structure, as shown in Figure 2. They consisted of a microscope cover glass as the bottom of the chamber, Mylar Polyester film sheets of different thicknesses (15 μm, 20 μm, 75 μm, 82 μm, 128 μm) as spacers, and as a top a membrane from an OptiCell®Culture Chamber (model 1100). The Opticell®membrane allows for gas exchange which is crucial for the development of the cells [13]. The mylar is attached to the cover glass and to the OptiCell®membrane using a silicone lubricant (Dow Corning High Vacuum Grease). First pieces of mylar sheet of 1.5 cm $\cdot$ 1.5 cm size were cut and a circular hole of 7 mm diameter was punched out in the middle.

The cells (cotB/GFP strain in AX4 with prespore labelling (TL155)) were washed in a centrifuge with a Developmental Buffer (see appendix) in order to remove nutrients and hence to initiate starvation. Therefore the cells diluted in Developmental Buffer and centrifuged for 60 seconds. This was repeated three times.

After having glued the mylar sheet to the cover glass, 70 μl of the washed cell suspension was pipetted into the hole. The density of the cell suspension was $2.5 \cdot 10^6$ cells/ml. After a waiting time of 30 min – in order to give the cells sufficient time to attach to the bottom of the glass – 35 μl of the suspension was pipetted away. Then the chambers were closed by attaching the Opticell®membrane to the mylar sheet, thereby pressing down the membrane to the mylar, so that the additional spacing due to the vacuum grease was minimized.

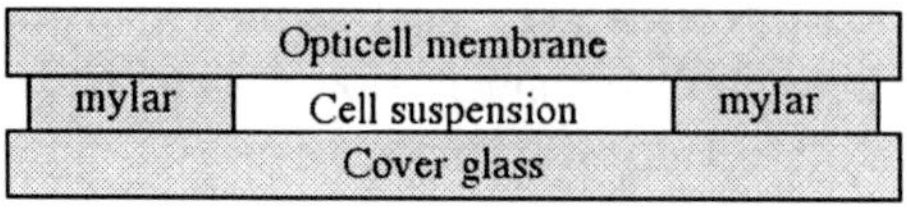

Fig. 2. Sandwich structure of the experimental chambers. By varying the thickness of the mylar sheets, the height of the experimental chamber is varied.

The heights of the layers were measured with a micrometer screw. This gives the actual heights of the cavities which total 15 μm, 21 μm, 77 μm, 84 μm and 129 μm, with a standard deviation of 0.4 μm. In order to prevent the cells from dehydration during the long lasting experimental runs, the assembled chambers were put into petri dishes, and filter paper, which was soaked with Developmental Buffer, was added into the petri dishes. The petri dishes were closed and sealed with Parafilm M®Laboratory film.

For the experiments, the chambers were placed on an inverted microscope (Nikon Diaphot 200) and images were taken with a 10 x objective by a CCD-camera connected to a frame-grabber in the computer.

3. Experimental Results

Images of the development of a slug in an experiment without a lid on top of the chambers are taken. Figure 3 shows this process. The first image shows the cell distribution 20 minutes after the initiation of starvation. The amoebae develop normally and form a slug. Later they also developed a fruiting body.

Then the amoebae were confined at different heights. Sections of images of the cells during the process of forming a mound for different heights at $t = 21.5$ hours after starvation are shown in Figure 4. For the control experiment without a lid, where the mound is free to extend into three dimensions, a fruiting body has formed and extends into the third dimen-

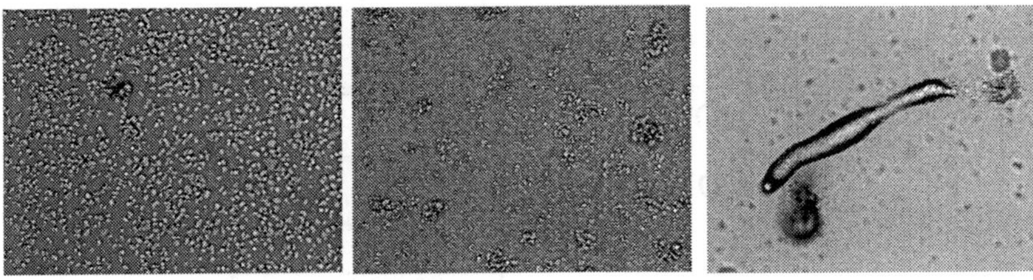

Fig. 3. Images of development for amoebae which are not confined at 20 min, 2 h 45 min and 19 h 15 min after the initiation of starvation. The width of the figures is 900 μm.

sion. In contrast, in the confined geometries the cells have formed mounds, whose area depends on the height of the chamber.

The areas in the images covered by the mounds were measured and the mound area distribution was calculated. These mound area distributions for all heights are shown in Figure 5a)-e) for t=21.5 hours up to an area of 21000 μm^2. (For the cell height of 84 μm there existed one mound with a larger area than shown.) The data show an obvious trend: For higher chambers the highest peak in the distribution becomes smaller and the area distribution becomes wider. In addition, at large heights the distribution has secondary peaks at large areas. Also, the more restricted the geometry (i.e. the shallower the chamber), the more distinct is the distribution. All this means that the number of cells in the mound is not constant but depends on the geometrical restrictions. It is evident that the number of mounds at the maximum peak decreases monotonically with the height. The distribution becomes less distinctive with increasing height.

The center of gravity of the area distribution in Figure 5 is calculated and shown in Figure 6. It shows that the center of gravity of the area distribution moves to smaller numbers of mounds and larger areas.

Figure 7 shows the dependence of the mean mound area on the chamber height for two different times after starvation. A straight line is fitted into the data. It shows a clear tendency that the mean mound area increases with the chamber height.

From the area distribution the volume distribution is calculated and is shown in Figure 8. The difference in the distribution for the different heights is even more pronounced for the volume distribution than for the area distribution. The mound with the smallest volume contains about 50 cells and the mound with the largest volume contains about 2600 cells. Checking the samples 44.5 hours after starvation showed that the mound

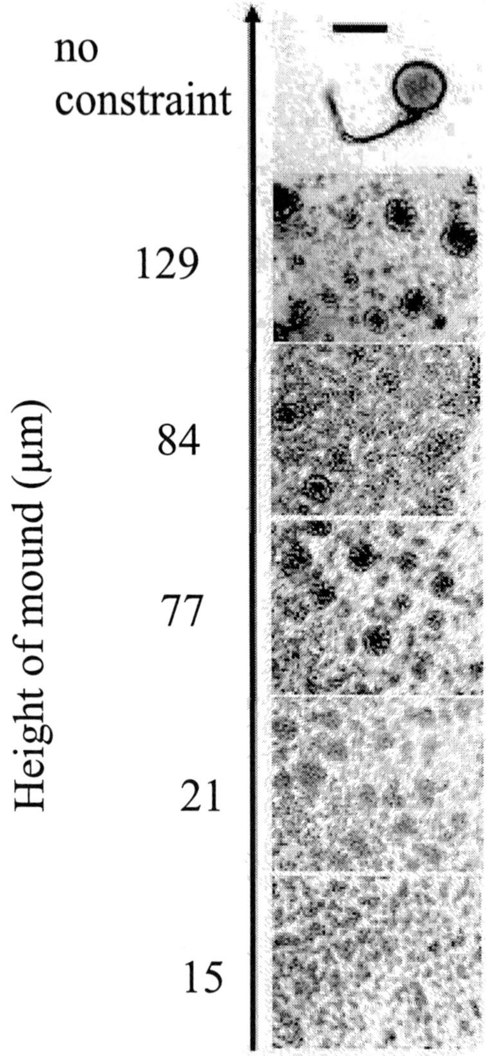

Fig. 4. Images of the mounds 21.5 hours after starvation for different chamber heights. From bottom to top: 15 μm, 21 μm, 77 μm, 84 μm, 129 μm and open top (fruiting body). The dark spots represent the mounds, whose area grows with the chamber height. The black bar indicates 200 μm.

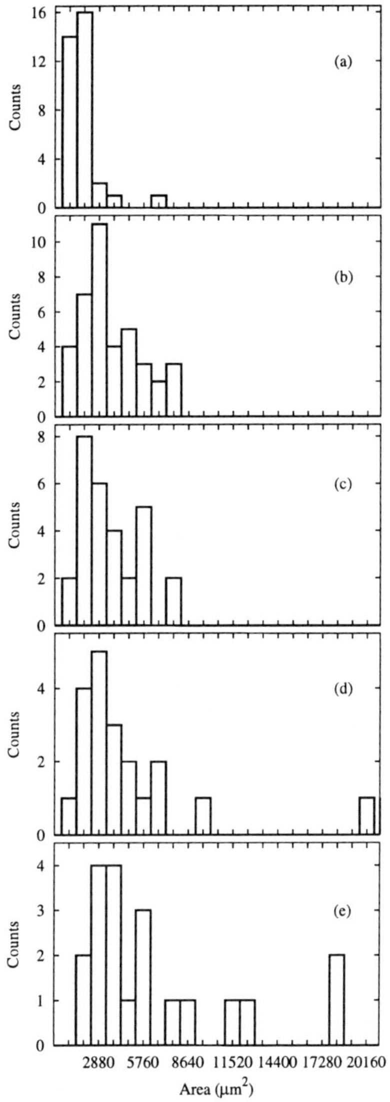

Fig. 5. Mound area distribution at t =21.5 hours after starvation for five different heights: (a) 15 μm, (b) 21 μm, (c) 77 μm, (d) 84 μm, and (e) 129 μm.

areas had not increased any more after the first 24.5 hours. This suggests that the confinement did not just slow down the formation of slugs and

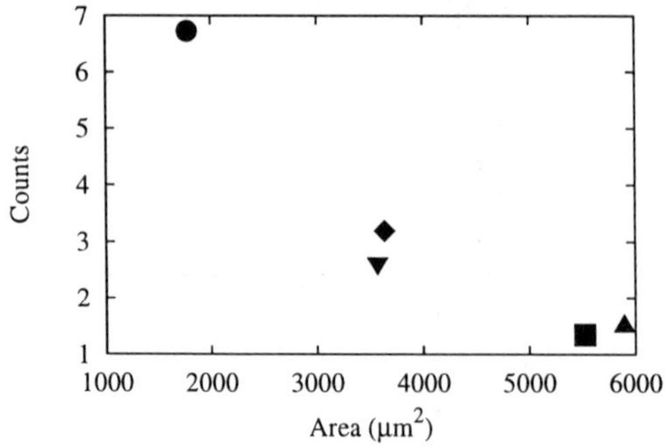

Fig. 6. Center of gravity of the size distribution for five heights: 15 μm (circle), 21 μm (rhombus), 77 μm (triangle upside down), 84 μm (triangle), and 129 μm (square).

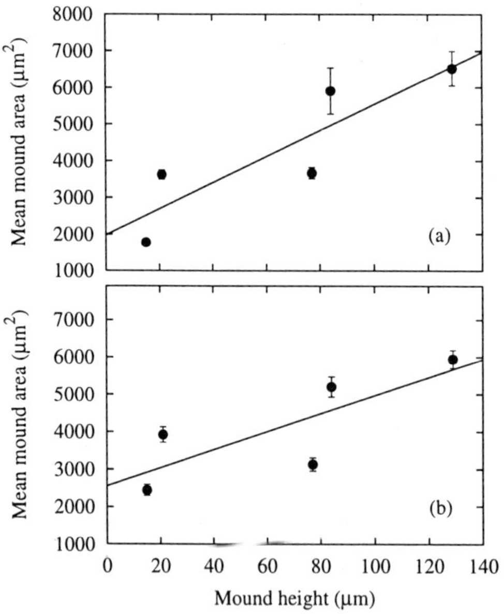

Fig. 7. Dependence of the mean mound size on the chamber height, shown for two times: (a) 21.5 h after starvation, (b) 24.5 h after starvation. The error bars give the standard deviation of the mean value.

fruiting bodies, but rather inhibited it.

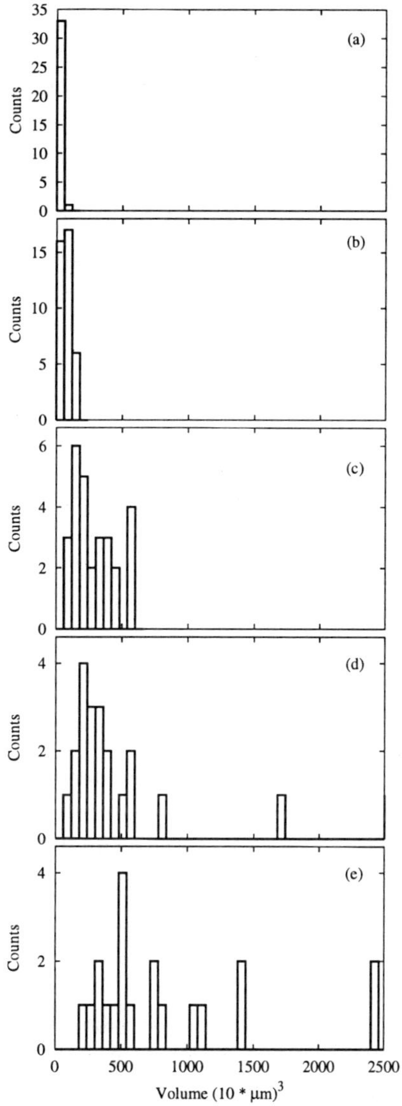

Fig. 8. Mound volume distribution at t =21.5 hours after starvation for five different heights: (a) 15 μm, (b) 21 μm, (c) 77 μm, (d) 84 μm, and (e) 129 μm.

4. Discussion

The mechanism for the regulation of the mound size is not yet known. In the literature [14] it is assumed that the mound size of *Dictyostelium discoideum* cells might be regulated by a secreted 450-kDa protein complex called counting factor which adjusts the mound size by regulating cell motility and cell-cell adhesion. In a different work [15, 16] it was postulated that randomly located cells secrete small pulses of PDI (the inhibitor of phosphodiesterase). The secretion of PDI has the effect of reducing the degradation of cAMP by inhibiting PDE (phosphodiesterase), which leads to a local increases in cAMP. This would increase the local excitability of the cells and might influence the evolution of territory size [15,17]. It might be speculated that the geometrical confinement applied in our experiments might influence the PDI secretion and hence have an impact on the mound size.

A different mechanism which could explain the scaling of the area covered by the mounds with the mound height would assume that the chemotactic signal becomes weaker for smaller mounds.

An alternative mechanism would suggest that the cells in three dimensions want to maintain a self-similar shape of the mounds. This would mean that the mounds become smaller when the constraint in three dimensions becomes larger. Additional experiments are clearly necessary in order to detect the underlying mechanism of the scaling of the mounds.

Theoretical models and simulations like reaction-diffusion models [18] or automata models [19] give a different approach to the understanding of the mound formation of *Dictyostelium discoideum*. These approaches do not use a parameter for the maximum height of the mounds. It might be interesting to include this parameter into the models and to see whether the models give the same results as found in the experiments presented here.

After having investigated the development of *Dictyostelium discoideum* at different heights, the next consequent idea would be to investigate the development of *Dictyostelium discoideum* cells in different geometrics. One example might be a circular track. First experiments were already performed, where starving cells in Developmental Buffer were put into a circular track which had a lid on top. For the fabrication of the annulus the Cornell Nanofabrication facility was used. The tracks were cast from a silicone elastomer, polydimethylsiloxane (PDMS). As lid for the channels Opticell membrane was used. Figure 9 shows two different geometries of the tracks. First preliminary results show that the amoebae, which were homogeneously distributed inside the tracks at the beginning of the experiment, finally aggregate at one side. It might be interesting to continue with these experiments and to see whether different patterns at the end of the experiment are reserved. After all, the amoebae act as an excitable medium and might form different patterns than just one single aggregation center in the end.

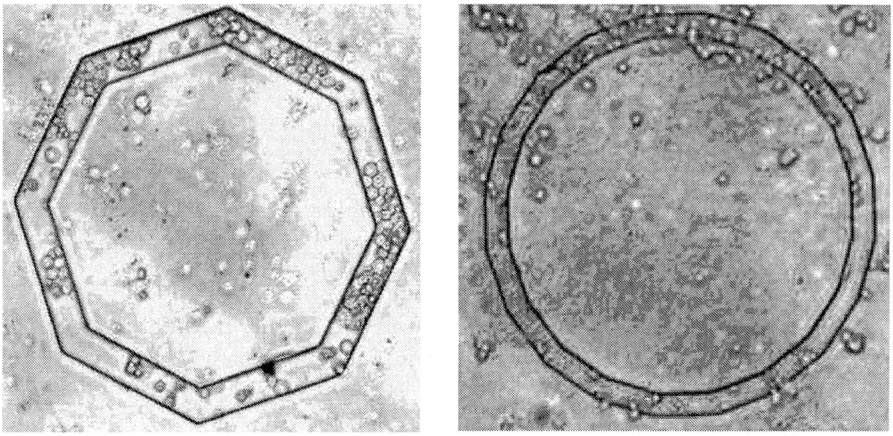

Fig. 9. Tracks for Dictyostelium development with different annular geometries.

5. Conclusions

The development of *Dictyostelium discoideum* mounds was investigated when the development of the amoebae was confined to layers of different heights, starting at a monolayer of cells. It was found that the size distribution depends on the height of the mounds. The mean mound size increases with the height. The data also show that the more restricted the cells are, the more uniform is the size of the mounds, with a narrow size distribution and high peaks at small mound areas. At large heights the mound size distribution has secondary peaks at large mound areas. These results show that the number of cells in a mound is not constant. It rather depends on the geometrical restrictions and hence the environment. Preliminary experiments confining the amoebae to circular tracks were also performed.

Acknowledgements

It is a pleasure to thank Herbert Levine (UCSD) and W. Loomis (UCSD) for helpful discussions. The work was supported by the NSF through the Biocomplexity Program.

Appendix

Developmental Buffer is made from:
0.710 g Na_2HPO_4, 0.690 g $NaH_2PO_4 \cdot H_2O$, 0.241 g $MgSO_4$, 0.029 g $CaCl_2 \cdot 6\, H_2O$. The pH is adjusted to 6.2.

References

[1] J.B.S.Haldane, in *Possible Worlds and other Papers*, 20-28, Harper & Brothers, New York (1928).
[2] R.H.Gomer, *Nat. Rev.* **2**, 48-54 (2001).
[3] I.Conlon and M.Raff, *Cell* **96**, 235-244 (1999).
[4] L.Tang, T.Gao, C.McCollum, W.Jang, M.G.Vicker, R.R.Ammann, and R.Gomer, *PNAS* **99**, 1371-1376 (2002).
[5] T.Gao, K.Ehrenmann, L.Tang, M.Leippe, D.A.Brock, and R.H.Gomer, *Journ. of Biolog. Chemistry* **277**, 32596 (2002).
[6] R. H. Kessin, *Dictyostelium- Evolution, Cell Biology, and the Development of Multicellularity*, Cambridge University Press, ISBN 0-521-58364-0 (2001).
[7] W. Loomis, *The Development of Dictyostelium Discoideum* (Academic, New York, 1982).
[8] J. T. Bonner, *The Cellular Slime Molds*, Princeton University Press, Princeton, NJ, 1967.
[9] H. Levine, *Physica* ***A*** **249**, 53-63 (1998).
[10] J.T. Bonner, K. B. Compton, E.C. Cox, P. Fey, and K. Y. Gregg, *Proc. Natl. Acad. Sci. USA* **92**, 8249 - 8253 (1995).
[11] J.T. Bonner, *Proc. Natl. Acad. Sci. USA* **95**, 9355 - 9359 (1998).
[12] C.Völtz and E. Bodenschatz, submitted to *Phys. Rev.* **E** (Rapid Communications)(2004).
[13] Y. Sawada, Y. Maeda, I. Takeuchi, J. Williams, and Y. Maeda, *Develop. Growth Differ.* **40**, 113 - 120 (1998).
[14] R.H. Gomer, *Nature Rev. Mol. Cell Biol.* **2**, 48 - 54 (2001); D.A.Brock, R.D. Hatton, D. Giurgiutu, B.Scott, R.Ammann and R.H. Gomer, *Development* **129**, 3657 - 3668 (2002); L.Tang, R.Ammann, T.Gao, and R.H. Gomer, *J. Biol. Chem.* bf 276, 27663 - 27669 (2001); L. Tang, T. Gao, c. McCollum, W. Jang, M.G. Vicker, R.R.Ammann, and R.H. Gomer, *Proc. Natl. Acad. Sci. USA* **99**, 1371-1376 (2002).
[15] E.Palsson and E.C.Cox, *Proc. Natl. Acad. Sci. USA* **93**, 1151 - 1155 (1996).
[16] E.Palsson and E.C.Cox in *Dictyostelium - A model system for cell and*

developmental biology, ed. Y.Maeda, K. Inouye, and I. Takeuchi, Universal Academy, Tokyo, pp. 411 - 423 (1997).

[17] E.Palsson, K. Lee, R.E. Goldstein, J.Franke, R.H. Kessin, and E.C.Cox, *Proc. Natl. Acad. Sci. USA* **94**, 13719 - 13723 (1997).

[18] B.N.Vasiev, P.Hogeweg, and A.V.Panfilov, *Phys. Rev. Lett.* **73** 23, 3173 - 3176 (1994).

[19] H.Levine, L. Tsimring, and D.Kessler, *Physica* **D** 106, 375 - 388 (1997).

INDEX